Isaac Newton's Natural Philosophy

DIBNER INSTITUTE FOR THE HISTORY OF SCIENCE AND TECHNOLOGY

Dibner Institute Studies in the History of Science and Technology
Jed Z. Buchwald, general editor, Evelyn Simha, governor

Jed Z. Buchwald and I. Bernard Cohen, editors, *Isaac Newton's Natural Philosophy*

Anthony Grafton and Nancy Siraisi, editors, *Natural Particulars: Nature and the Disciplines in Renaissance Europe*

Frederic L. Holmes and Trevor H. Levere, editors, *Instruments and Experimentation in the History of Chemistry*

Agatha C. Hughes and Thomas P. Hughes, editors, *Systems, Experts, and Computers: The Systems Approach in Management and Engineering, World War II and After*

N. M. Swerdlow, editor, *Ancient Astronomy and Celestial Divination*

Isaac Newton's Natural Philosophy

edited by Jed Z. Buchwald and I. Bernard Cohen

The MIT Press
Cambridge, Massachusetts
London, England

This book was set in Bembo on '3B2' by Asco Typesetters, Hong Kong and printed in the United States of America

Library of Congress Cataloging-in-Publication Data
Isaac Newton's natural philosophy / edited by Jed Z. Buchwald and I. Bernard Cohen.
p. cm. — (Dibner Institute studies in the history of science and technology)
Includes bibliographical references and index.
ISBN 0-262-02477-2 (hc: alk. paper)
1. Newton, Isaac, Sir, 1642–1727. 2. Science—England—History—17th century. I. Buchwald, Jed Z. II. Cohen, I. Bernard, 1914– III. Series.
Q143.N495.I73 2000
509.42′09′032—dc21 99-042985

Contents

Introduction

In the last century, and especially in the last half century, our views concerning Isaac Newton have undergone radical changes. Today, we have a deeper understanding of Newton's science and mathematics, and we have become aware of his full creative stature and the many dimensions of his complex personality. The present volume displays some of these recent changes in our understanding of Newton's scientific thought, the results of new analyses of manuscripts and printed documents.

A convenient place to begin examining changes in Newton scholarship is the two-volume biography by David Brewster, published in 1855. At that time, the fashion was to write of historic personages in an adulatory mode. Brewster thus calls Newton the "High Priest of Science."[1] Brewster's biography is notorious as well for the treatment of Newton's alchemy and his religious beliefs. Brewster summed up his discussion of Newton's alchemy by saying that he simply could not "understand how a mind of such power, and so nobly occupied with the abstractions of geometry, and the study of the material world, could stoop to be even the copyist of the most contemptible alchemical poetry." Although he published some extracts from Newton's unfinished theological writings, he either did not see or purposely ignored Newton's statements of his unitarian beliefs.

In retrospect, what is most disappointing about Brewster's biography, however, is not the attitude toward alchemy and religion, but rather the failure to illuminate our understanding of Newton's actual science and mathematics. The two volumes are notable for the absence of mathematical equations or illuminating discussion of Newton's physical science. Thus in today's world, this work is not very useful to scholars and pales by comparison with the meticulously edited and copiously annotated edition of the Newton-Cotes correspondence, produced by J. Edleston, as useful and important a tool for Newton scholars today as when first published in 1850.

One of the positive features of Brewster's two volumes is that they did make available some selections from the manuscripts belonging to the family of the Earl of Portsmouth, Newton's collateral descendants.[2] Until the latter part of the nineteenth century, the greatest collection of manuscripts by or relating to Newton was still in their possession. These consisted of Newton's own records of correspondence (letters both to and from Newton), drafts of major and minor works, and various kinds of essays on many different subjects, together with the documents assembled by John Conduitt for a planned biography of Newton (Conduitt was the husband of Newton's niece). These papers were inherited by the Conduitts' only child, who married Viscount Lymington, whose son was the second Earl of Portsmouth. This collection, generally known as the "Portsmouth Papers," was kept in Hurstbourne Castle until the Earl and his family decided that Newton's scientific manuscripts would be better preserved in some public repository. Accordingly, in the 1870s, what was called the "scientific portion" of the manuscripts was deposited in the Cambridge University Library.[3]

Over the next decades, this extraordinary hoard of Newtoniana attracted little scholarly attention and was hardly used. One of the few who even deigned to look at any of these manuscripts was W. W. Rouse Ball, known today chiefly for his popular book on "mathematical recreations."[4] His books, however, did not make the world cognizant of the great treasures awaiting study in the library.

Between the time of the gift of the Portsmouth Papers and the 1930s, a few others did make some use of this vast collection, although the work of these scholars did not declare to the world at large the importance of studying Newton in the original sources. In trying to understand why this was so, we must remember that in those decades there was as yet no real discipline of the history of science and of mathematics. The number of individuals producing lasting historical contributions in the history of science and mathematics was small, including such heroic figures as J. L. Heiberg, G. Eneström, Thomas Little Heath, and Paul Tannery.

In 1934, Louis Trenchard More, Dean of the Graduate School of the University of Cincinnati, published a 675-page biography of Newton. Even though More did make some use of manuscript sources, quoting or citing many hitherto unnoticed documents from the Portsmouth Papers (both those in Cambridge and those still remaining in the possession of the family), his book did not spark a new interest in Newton—either in the personality of this extraordinary man or in his scientific work. One reason for this was More's failure to produce a deep analysis of Newton's mathematics and physics, as demonstrated by the absence of equations and diagrams in this massive biography.[5]

As all Newton scholars are aware, the sale at public auction at Sotheby's in 1936 of the vast horde of Newton papers still belonging to the family of the Earl of Portsmouth (provoked by the need to satisfy the payment of death duties) changed the availability of Newtonian sources almost overnight. The sale dispersed Newton's papers to the far corners of the earth. A notable scholarly catalogue was produced for it, containing many generous extracts from documents hitherto unknown or inaccessible to scholars. The descriptions of the various manuscripts and other items put up for auction, together with copious extracts, revealed more about Isaac Newton as a person and about various aspects of the development of his scientific thought than More's biography of a few years earlier. Prior to the late 1940s, however, the only publication in which these newly available manuscripts were used was a curious paper by John Maynard Keynes. Keynes had assembled a considerable mass of the Newton papers disseminated in the Sotheby sale; these are now in King's College Library. In particular, Keynes owned a number of manuscripts dealing with alchemy and with theological subjects.

On the basis of an examination of these manuscripts, Keynes produced a paper, first read to a group at Trinity College and later to the Royal Society Club. Keynes was no longer living when in 1946 the Royal Society somewhat belatedly celebrated the 300th anniversary of the birth of Isaac Newton. Keynes's brother Geoffrey—surgeon, bibliophile, and book collector—read this paper at the celebratory meetings, and it was published in the Royal Society's volume. Entitled "Newton the Man," this essay has become famous for its radical portrayal of Newton.

Keynes insisted that his reading of Newton's manuscripts revealed a Newton who was not "the first and greatest of the modern age of scientists, a rationalist, one who taught us to think on the lines of cold and untinctured reason." Rather than being "the first of the age of reason," Keynes presented Newton as "the last of the magicians," the "last wonder-child to whom the Magi could do sincere and appropriate homage." In short, Keynes denigrated Newton's mathematics and physics and his founding of celestial dynamics, denying that he had been "the first and greatest of the modern age of scientists." The new Newton was to be considered "a magician" who "looked on the whole universe and all that is in it *as a riddle*, as a secret which could be read by applying pure thought to . . . certain mystic clues which God had laid about the world to allow a sort of philosophers' treasure hunt to the esoteric brotherhood."[6]

The next works to be based on the Newton manuscripts were a pair of scholarly articles published by Rupert Hall in 1948 and 1957. The first of these dealt with one of Newton's early notebooks; the second, "Newton on Central Forces," used manuscript sources to explore the genesis and development of Newton's concepts in dynamics. Despite the fact that Hall had shown how studying the manuscripts in the Cambridge University Library could yield insights into the development of Newton's scientific thought, others did not follow his example at once. Indeed, not until 1962, when Rupert and Marie Hall published *Unpublished Scientific Papers of Isaac Newton*, did the scholarly world at large, the world of historians of science and of scientists and mathematicians interested in historical questions, become aware of some of the extraordinary insights to be found by examining the unpublished Newton materials.

The seminal importance of the Halls' work can best be illustrated by a few examples of the ways in which it changed our general thinking about Newton's scientific ideas. Of course, the primary significance of this work is that Rupert and Marie Hall actually found, identified, and interpreted a large number of documents, the very existence of which was then not generally known, and explained through interpretive essays and commentaries the significance of the new documents they had found and were presenting.

One of the most notable of these documents is Newton's essay beginning "De gravitatione et aequipondio fluidorum." The Halls recognized its significance as Newton's response to a first contact with Descartes's *Principia*. Here we find Newton formulating major concepts concerning space, time, and motion, and also force and inertia, in a Cartesian framework, just as he did for his mathematics. The publication of this essay, together with the Halls' introductory commentary, documented Newton's first full encounter with the philosophy of Descartes and wholly changed our idea concerning Descartes's influence on Newton.[7]

A second important contribution of *Unpublished Scientific Papers* was the publication of a set of documents relating to Newton's early thoughts about motion. These include the first English translation made of the tract *De Motu*. Although this tract had been published at least twice before, in the original Latin, the Halls not only gave the first English translation but also listed the various extant versions and included extracts indicating the differences among them. Thus for the first time, this earnest of the great *Principia* to come was made available in a form for scholars' general use.

A third revelation of *Unpublished Scientific Papers*—in some sense the most important of all—was the existence of preliminary versions of an introduction and conclusion planned for the first edition of the *Principia*, including some thoughts that finally appeared in the second edition of the

Principia in the concluding General Scholium and in later Queries of the *Opticks*. The Halls also found documentary fragments relating to these rejected texts. Thus they solved a long-standing scholarly puzzle: why did the first *Principia* end abruptly in a discussion of comets? How could Newton have written so magisterial a work on natural philosophy without a proper conclusion? The Halls found the answer. He had planned a general conclusion in which he would indicate, inter alia, how his findings might be extended into other domains of natural philosophy, suggesting how his work in rational mechanics and celestial dynamics might be carried into studies of the constitution of matter and of the action of short-range forces between its constituent elements. In retrospect, we can understand why Newton decided not to burden his *Principia* with debatable and intimate speculations on such points. There were enough topics in that work that were bound to arouse hostility, such as the introduction of a gravitating force "acting at a distance," a concept abhorrent to all scientists who were followers of the reigning "mechanical philosophy."

The Halls solved another long-standing puzzle about the *Principia*. In the eventual General Scholium, which appeared for the first time in the second edition of the *Principia* (1713), a final paragraph discusses what Newton calls a "spiritus," or spirit. What did he mean? No one was quite sure. It had even been proposed that Newton may have had in mind the "spirit of God." The Halls discovered some preliminary drafts, which they published for the first time, that suggested that the "spirit" that Newton had in mind was an aspect of the new science of electricity then being developed by Francis Hauksbee, whose relations with Newton were later the subject of several important studies by the late Henry Guerlac.[8]

An important use of Newton's manuscripts and other writings was made by Alexandre Koyré, whose seminal *Newtonian Studies* was published in 1965. At about the same time, J. E. McGuire used manuscript sources to reveal a different Newton from the one we were accustomed to think of.

By 1962, several other scholars had begun projects based on the use of Newton manuscripts. The Royal Society had undertaken the edition of the *Correspondence* of Isaac Newton, the first volume of which appeared in 1959, three years before *Unpublished Scientific Papers*. I. B. Cohen had begun, in close collaboration with Alexandre Koyré, to study Newton's manuscripts and annotated books in order to learn about the genesis and development of the concepts and methods of his great *Principia*, eventually leading to their edition of the *Principia*, which included variant readings. Others, notably John Herivel (in 1959), had begun to publish the results of explorations in the Newton manuscripts.[9] Even more important was the enterprise of D. T. Whiteside, eventually resulting in the eight magnificent volumes of the *Mathematical Papers of Isaac Newton* (the first of which appeared in 1967).

It is difficult to think of any work of scholarship produced in our time of comparable magnitude to Whiteside's *Mathematical Papers*. Not only do we find here in full display the documents that mark the development of Newton's thinking in mathematics, but there are also extensive annotations and commentaries that provide more information concerning the development of the exact sciences in the seventeenth century than is available in most treatises.

The eight volumes of *Mathematical Papers* are remarkable in many different ways, not least because they resulted from the scholarly activity of a single individual. Whiteside not only discovered or identified a vast quantity of documents, but his presentation of them (in the original Latin, often accompanied by English versions) is graced with an illuminating commentary. Whiteside provides not only interpretive glosses on the texts he has edited but also a running historical commentary that is without any doubt the most extensive and important historical presentation of seventeenth-century mathematics produced in our times. Whiteside's extraordinary project involved identification of documents, the recognition that certain fragments in different parts of the manuscript collection belonged together, and the ability to date Newton's manuscripts on the basis of changes in handwriting.

Among Whiteside's most dramatic findings is the importance of Descartes in the early formulation of Newton's mathematics. Before the edition of *Mathematical Papers*, our knowledge of this subject was based on what seemed to be irrefutable evidence of Newton's disdain for Descartes. There was the statement recorded by Pemberton that Newton regretted having begun his study of mathematics by reading the moderns rather than the ancients. This was interpreted as indicating his regret at wasting time with Descartes's *Géométrie* rather than studying Euclid and Apollonius. The second was the statement by Brewster that Newton's copy of Descartes's *Géométrie* was marked throughout "Error. Error. Non Geom." For some time it was thought that Brewster might have been exaggerating, since there seemed to be no such copy in existence. But finally—indeed, in part as a result of Whiteside's astuteness—this puzzle was resolved. Newton did make such marks in a copy of Descartes. When such evidence concerning Newton's attitude toward Descartes was coupled with the conclusion of Book II of the *Principia*, the implication seemed certain: Descartes was not one of those who had exerted a formative influence on Newton.

Whiteside changed that view entirely insofar as mathematics is concerned. He showed how Newton's early views of the calculus were forged while making a close study of Descartes's *Géométrie*—not the edition in French in which he gleefully noted the errors, but the edition in Latin with the commentaries of Frans van Schooten and others. This introduced Newton not only to the mathematical concepts and methods of Descartes himself, but also to the important innovations of the Dutch school, who were van Schooten's pupils, notably Hudde and Huygens.

The revelation of this seminal role of Descartes, enriched by van Schooten, was paralleled by Alexandre Koyré's recognition, at more or less the same time, of the ways in which Newton's reading of Descartes conditioned some of his concepts concerning motion. Koyré showed us how Newton took from Descartes the concept of "state" of motion or of rest and developed Descartes's ideas in his own formulation of Definition 3 and Law 1. Today, we are aware that the "axiomata sive leges motus" of Newton's *Principia* were a kind of transformation of what Descartes called "regulae quaedam sive leges naturae" in his *Principia*.

Whiteside made a second revelation in his gloss on Proposition 41 of Book I of the *Principia*. Until recently, most scholars had limited their study of the *Principia* to the definitions and laws and then the first three sections of Book I. They skipped all the rest of Book I and also the whole of Book II, finally studying the first part of Book III and the concluding General Scholium. This was in fact Newton's own suggestion to readers in the beginning of Book III.

Today, however, it is becoming generally recognized that we don't begin to see Newton as the master of mathematical physical science until farther along in Book I. Here we needed someone like Whiteside to guide us, to make clear, at the start, the exact nature of Newton's dependence on what he called "the quadrature of certain curves," or the ability to perform the integration of certain functions. Even more important, Whiteside's guidance made it evident that in Proposition 41, as elsewhere, Newton was in fact using the calculus: that his very language, when read carefully, permits no other reading. For example, when Newton writes about "the line element IK" that is "described in a given minimally small time," he clearly has in mind what we would call a distance *ds* described in a time *dt*. In short, the cluster of propositions around Proposition 41 are written in the language of the differential and integral calculus; their analytic character is masked only partially by their synthetic form of expression.

Among more recent scholarship, special notice should be taken of the magnificent biography of Newton by the late R. S. Westfall (1980).

This work, *Never at Rest,* based on an extensive study of Newton's manuscripts, does not merely chronicle the events in Newton's life but illuminates almost every aspect of Newton's life and thought, providing a rich and valuable commentary on Newton's scientific achievement. Another project certain to assume an important place in the treasury of scholarly editions based on manuscript sources is the edition of Newton's papers on optics, edited by Alan Shapiro, an earnest of which appeared as volume 1 in 1984. In the past decade, monographs by Michel Blay, Bruce Brackenridge, S. Chandrasekhar, François de Gandt, Dana Densmore, Betty Jo Teeter Dobbs, Herman Ehrlichson, Niccolò Guicciardini, Michael Nauenberg, George Smith, and Curtis Wilson—to name but a few—have enriched the scholarly study of Newton's science.

Recent Newton scholarship has illuminated not only his work in mathematics and the exact and experimental physical sciences but the full scope of his work, his public and private life, and his personality. This scholarship has cast light, for example, on his philosophical and religious beliefs, notably his ideas concerning theology, prophecy, biblical history and the interpretation of Scripture, and a tradition of ancient knowledge. There have also been important studies of Newton's alchemy and his general philosophy of nature, including his views on "matter" and "spirit" and their relation to the operations of nature. Newton's ideas concerning the books of Daniel and Revelation have been studied, as well as his concern for the issue of prophecy. Researchers are now examining the links between what we today consider to be Newton's scientific work and the general religious and philosophical background of the times in which he lived.[10]

The present volume singles out two strands in contemporary Newtonian studies for special consideration. One of these concentrates on the intellectual background to Newton's scientific thought; the other concerns both specific and general aspects of Newton's technical science. The contributions to both strands offer new, and even startling, claims concerning Newton's mathematical methods, experimental investigations, and motivations, as well as the effect that his long presence had on the pursuit of science in England.

Each of the papers in part I offers, among other riches, a new claim concerning Newton's motivations or the sources of his method. Maurizio Mamiani traces the immensely influential *regulae philosophandi,* which achieved final form in the beginning of Book III of the *Principia,* to an entirely novel source, namely, a 1618 treatise on logic and rhetoric by Robert Sanderson that Newton studied while an undergraduate at Cambridge. I. B. Cohen uses evidence not previously offered concerning Newton's reluctance to put his name on his *Opticks* (1704) to find a connection between the *Opticks* and Huygens's *Traité de la Lumière*. He offers other examples in which Newton on more than one occasion did not put his name on the title page of a work he wrote. Cohen relates this practice to an ambivalence on Newton's part to publish the *Opticks*. New evidence on this ambivalence is presented by Alan Shapiro, who argues, in his discussion of Newton's troubled work on diffraction, that it was more Newton's failure to arrive at a satisfying account of this phenomenon than his desire to avoid controversy with Robert Hooke that led him to put off until 1704 publication of the *Opticks*. Mordechai Feingold asserts, among other things, that antagonistic reactions to Newton's first publications in optics derived as much from the stance he took concerning the kind of knowledge that ought to count as proper (specifically, mathematical knowledge) as they did from the specific content of Newton's claims. Feingold locates Newton's long withdrawal from public participation in scientific argument precisely here, arguing that the partisans of mathematics at the Royal Society, in particular the emerging body of Newtonians, constituted a sort of rapid deployment force–well organized, eager for battle, and anxious for hegemonic control–whose legacy included persistent and demoralizing battles over the kinds of natural knowledge worth pursuing.

Part II consists of five essays that explore Newton's mathematical philosophy and his development of rational mechanics and celestial dynamics. These include detailed examples of Newton's mechanics in two apparently different areas: first, his mathematics for orbital mechanics and lunar motion, and then his investigations in Book II of the *Principia* of motion in resisting media. Bruce Brackenridge argues, in respect to the first area, that Newton may have used curvature methods quite early in finding the force that acts on a body given its orbit (this being the so-called direct problem), which underscores a similar claim made by Michael Nauenberg on the basis of a 1679 letter sent by Newton to Hooke. The method deployed in the *Principia* finds the resultant motion by compounding a continuing inertial velocity with an impulsively added one directed to the center of force, and passing then to the orbital path as a limit of infinitely many impulses. The curvature method instead decomposes the acceleration at a point into components normal and parallel to the velocity there and then applies Huygens's measure for acceleration in a circular path. This earliest method does not provide a simple route to Kepler's area law, whereas the *Principia*'s method does so.

Michael Nauenberg's and Curtis Wilson's contributions both concern Newton's treatment of lunar motion. Each discusses a manuscript in the Portsmouth collection that describes a method for analyzing the motion of the lunar apse, one that, according to Nauenberg, "corresponds" to the method of variation of orbital parameters developed many years later by Euler and then completed by Lagrange. Although Wilson does not agree with Nauenberg's assertion on this point, both emphasize the differences between the Portsmouth and *Principia* approaches to lunar motions (Newton being concerned in the *Principia*, however, with mean rather than apsidal motion). For his part, Nauenberg examines a text he calls the "Portsmouth manuscript" for differential equations that are equivalent to ones obtained many years later by Clairaut and d'Alembert. He finds that Newton was dissatisfied with the method in the Portsmouth manuscript and compares it to the one he did use in the *Principia*. Wilson, on the other hand, argues that the problems Newton faced were connected to the kind of mathematics that he used, because Newton worked directly with the geometric properties of orbits instead of with successive approximations to the governing differential equations. According to Wilson, the major changes in theories of orbital motion that developed during the eighteenth century were due precisely to the replacement of geometric by analytic methods. In this regard, Michel Blay's contribution provides a specific example of the manner in which analytic methods were brought to bear early in the eighteenth century on problems in mechanics. Blay discusses, in particular, Varignon's deployment of the methods of Leibnizian calculus, with its specific notion of continuity, in addressing the problem of central forces, which Newton had treated using a very different, impulsive model of action.

Wilson's and Nauenberg's contributions provide striking examples of Newton at work. In them, we see Newton trying to choose appropriate procedures for attacking a problem that required adroit approximations adjusted to the demands of astronomical data. George Smith's article concerns a subject that at first appears to be quite different from this one, namely, Newton's account in Book II of the *Principia* of motion in resisting media. Yet there are strong methodological links between Smith's subject and lunar motion. In both cases Newton worked through a sequence of approximations, with the evidence from the empirical world driving the sequence from one idealized situation to the next (although much depended upon whether the approximations involved geometrical considerations or ones linked to the governing differential equations proper). Moreover, we find that the Newton of the *Principia* is entirely similar to the Newton of the *Opticks*, for as Shapiro remarks in his contribution, Newton in his analysis of diffraction "plays off his measurements against his mathematical descriptions, which allows him both to change his measurements and to revise his laws."

Finally, Richard S. Westfall explores an aspect of the age of Newton by examining some of the different ways in which mathematics came to be used in pursuits and domains other than theoretical or rational mechanics. Alas, Westfall died in 1996 and so was unable to make any final revisions to his chapter. In particular, he had planned to add a note on the use of mathematics in the analysis of the bills of mortality by John Graunt and on the work of Sir William Petty on "political arithmetic," together with some account of Edmond Halley's tables of mortality—all indications that in the age of Newton numerical considerations were being introduced into new areas of human activity.

All of the chapters in this volume originated in papers that were presented and discussed at a series of meetings held at the Dibner Institute for the History of Science and Technology in Cambridge, Massachusetts. The occasion was a symposium to honor the Grace K. Babson Collection of Newtoniana, formerly housed in the Horne Library of Babson College and now on permanent deposit in the Burndy Library, which is located with the Dibner Institute on the campus of the Massachusetts Institute of Technology. The extraordinary resources of the Babson Collection, in conjunction with the Newton and Newton-related volumes and manuscripts in the Burndy Library (originally assembled by Bern Dibner), constitute one of the most important scholarly resources available for the study of Isaac Newton's career and contributions to science as well as the science that Newton's work engendered.

NOTES

1. Brewster lifted this phrase, "High Priest of Science," from the eighteenth-century accounts of Newton written by William Stukeley, an antiquary who had actually conducted a sort of oral-history interview with Newton.

2. It was long thought that Brewster had based his biography on a complete examination of these manuscripts, and that he himself had made the selection of those he published or mentioned. The researches of D. T. Whiteside, however, revealed that Brewster had not actually had free access to all the Newton papers, but had only certain selections made for him by a younger son.

3. This enormous collection, rich in all kinds of materials relating to Newton's life, his thought, his work in mathematics and physics and astronomy, and much else, was described in a published catalogue (1888) in which the descriptions are so brief and

laconic that they hardly reveal either the quality or the extent of the items in question. A positive feature of this catalogue was an introductory essay, partly written by astronomer John Couch Adams, that revealed some of the astronomical and mathematical texts, including materials on the lunar theory and the solid of least resistance.

4. Rouse Ball made some use of Newton's manuscripts in his *Essay on Newton's Principia*, publishing a transcript of a version of Newton's essay *De Motu*, previously published by Rigaud from the copy in the Royal Society, and also some samples of correspondence, notably Newton's correspondence with Halley concerning the *Principia*. In another volume, called *Cambridge Studies*, he included an essay by Newton on a plan for education at the university.

5. More's biography displays an extreme arch-conservatism in science that defeats any attempt to understand Newton's science and mathematics on their own terms. For example, after lauding Newton's *Principia* as a work of equal greatness with Aristotle's *Organon*, he deplored the fact that "these two works, probably the two most stupendous creations of the scientific brain," were both "now under attack." The enemies were, respectively, "the relativists in physics," led by "Professor Einstein," and what he called "modern symbolists in logic"—a curious name for those developing mathematical logic or, as it was then often called, symbolic logic. More insisted that "Aristotle and Newton will be honoured and *used* when the modernists are long forgotten."

6. This paper is a curious production. For example, there is no specific reference to a single manuscript or text, or indeed to any particular document. Furthermore, it includes not a single quotation from Newton. In considering this paper, we should keep in mind that this was not written as a serious scholarly contribution intended for publication. Keynes wrote it for a Cambridge audience who would welcome a presentation that was brilliant, challenging, daring, and unorthodox.

7. The late Betty Jo Dobbs published a study of this manuscript and concluded that it dates from a time just before the *Principia* rather than earlier. Corroborative evidence in support of this conclusion may be found in I. B. Cohen's *Guide to the "Principia,"* accompanying the new translation of the *Principia* made by I. B. Cohen and Anne Whitman (University of California Press, 1999).

8. Anyone engaged in Newtonian research and, indeed, in research on any aspect of the science of the seventeenth century depends heavily on the extraordinary edition of the correspondence of Henry Oldenburg, first secretary of the Royal Society, produced by Rupert and Marie Boas Hall.

9. In the preface to *Unpublished Scientific Papers*, Rupert and Marie Hall called attention to a bare handful of scholars who had made use of the Newton manuscripts. The list is short, but it does include a beginning made by Alexandre Koyré and the first essays of John Herivel, whose later volume of texts is regularly cited by Newton scholars.

10. Frank E. Manuel has made important studies of Newton as historian and has also analyzed Newton's texts on prophecy. The researches of Betty Jo Dobbs and Karin Figala have illuminated Newton's alchemical explorations. J. E. McGuire has studied Newton's concern for ancient knowledge.

Contributors

J. Bruce Brackenridge is Alice G. Chapman Professor of Physics Emeritus at Lawrence University. He is the author of *The Key to Newton's Dynamics: The Kepler Problem and the Principia* (1996).

Michel Blay is Directeur de recherche au CNRS and has been Editor-in-Chief since 1984 of the journal *Revue d'Histoire des Sciences.* He is the author of *La naissance de la mécanique analytique. La science du mouvement au tournant des XVIIe et XVIIIe siècles* (1992), *Les raisons de l'infini. Du monde clos à l'univers mathématique* (1993), and *Les 'Principia' de Newton* (1995).

Jed Buchwald is Director of the Dibner Institute for the History of Science and Technology and Bern Dibner Professor of the History of Science at MIT. His books include *The Creation of Scientific Effects: Heinrich Hertz and Electric Waves* (1994); *The Rise of the Wave Theory of Light* (1989); and *From Maxwell to Microphysics* (1985).

I. Bernard Cohen is Victor S. Thomas Professor (emeritus) of the History of Science, Harvard University. His most recent books are *Howard Aiken: Portrait of a Computer Pioneer* (1999) and the first completely new translation of Newton's *Principia* in the last 270 years (1999).

Mordechai Feingold is Professor of Science Studies at Virginia Tech. His publications include *The Mathematicians' Apprenticeship: Science, Universities and Society in England, 1560–1640* (1984) and (as editor) *Before Newton: The Life and Times of Isaac Barrow* (1990).

Maurizio Mamiani is Professor of History of Science and Technology at the University of Udine (Italy). Among his books and papers about Isaac Newton's natural philosophy are *I. Newton filosofo della natura* (1976); *Il prisma di Newton* (1986), an essay about the scientific invention; *Introduzione a Newton* (1990); and *Newton* (1995). He has also transcribed and edited Newton's *Treatise on Apocalypse.*

Michael Nauenberg is Professor of Physics (emeritus) at the University of California at Santa Cruz where he has been teaching since 1966. He has written on Newton, Hooke, and Huygens; his article "Curvature in Newton's Dynamics" (with J. B. Brackenridge) is included in the *Cambridge Companion to Newton* (2000).

Alan E. Shapiro is Professor of the History of Science and Technology at the University of Minnesota and has written widely on Newton and optics in the seventeenth and eighteenth centuries. He is the editor of *The Optical Papers of Isaac Newton* and the author of *Fits Passions and Paroxysms: Physics, Method and Chemistry and Newton's Theories of Colored Bodies and Fits of Easy Reflection* (1993).

George E. Smith is chair of the Philosophy Department of Tufts University. He has published papers on Newton's methodology in the *Principia,* as well as on Book 2, and he is coeditor of the *Cambridge Companion to Newton* (2000). He is also a practicing engineer, specializing in aerodynamically induced vibration and resulting metal fatigue failures in jet engines and other turbomachinery.

Richard S. Westfall (1924–1996) was Professor of History and of History and Philosophy of Science at Indiana University. His publications include *Science and Religion in Seventeenth-Century England* (1958, 1973), *The Construction of Modern Science: Mechanism and Mechanics* (1971), *Force in Newton's Physics: the Science of Dynamics in the Seventeenth Century* (1971), *Never at Rest: A Biography of Isaac Newton* (1980), *Essays on the Trial of Galileo* (1989), and *The Life of Isaac Newton* (1993). In 1983 the History of Science Society (which he served as president in 1977 and 1978), honored him with its Pfizer Award for *Never at Rest.*

Curtis Wilson is a Tutor at St John's College, Annapolis. He is the author of *Astronomy from Kepler to Newton: Historical Studies* (1989) and coeditor (with René Taton and Michael Hoskin) of the four-volume *Planetary Astronomy from the Renaissance to the Rise of Astrophysics.*

I

Motivations and Methods

1

To Twist the Meaning: Newton's *Regulae Philosophandi* Revisited

Maurizio Mamiani

The theme of this investigation is a principle developed in various publications by I. Bernard Cohen: that "a dynamic rather than a static point of view" should guide analyses of the development of scientific ideas.[1] One consequence of this principle is that many important innovations in science prove, on close examination, to consist, to some degree, of radical "transformations" of existing ideas, concepts, and methods. In many cases, however, finding a specific or causal link among scientific ideas involves considerable difficulty. For example, it is historically certain that Newton's "New Theory of Light and Colors" of 1672 is in part a transformation of Robert Boyle's ideas about colors of 1664. There is, however, no direct inductive or deductive link between these two sets of ideas. Clearly, Boyle himself did not see any such inductive or deductive transformation; this was Newton's great move forward.

There is a paradox here. If scientific innovation tends to be a transformation, what is the process (neither induction nor deduction) that leads from the old ideas to the new ones? There is no simple and straightforward path that leads from the old to the new ideas. The actual stages of the transformation and the cause are usually not obvious. In a sense, the history of the scientific ideas hides the actual transformation. But if such history depends on transformations, there seems to be a charge of circularity.

To escape the vicious circle, it is necessary to look for the actual cause or occasion of the growth of the scientific concepts. Where do we look? Often beyond science, namely, beyond the scientific tradition of the age we are studying. For instance, Boyle produced his ideas about colors in 1664 in a philosophical (i.e., physical) context. In the *Lectiones Opticae*, Newton modified the boundaries between physics and mathematics, which enabled him to twist the meaning of the same kind of the experiments on colors that he and Boyle had performed.

In what follows, my goal is to focus attention on a particular transformation that marked the migration of categories and methods from one discipline to another. I do not intend to discuss some kind of vague influential metaphysics, something that is "in the air" like some kind of elusive ghost. Rather, I want to trace the transformation of a specific set of concepts and their integration into a wholly different kind of system of thought, thus disclosing a link between very different traditions of thought. The case study I wish to explore is the set of famous *regulae philosophandi* that appear at the head of Book III in the later editions of Newton's *Principia*.

There can be little doubt that these *regulae* are a transformed version of a set of "rules" that Newton composed somewhat earlier. These rules, sixteen in number, appear in Newton's *Treatise on the Apocalypse*.[2] Between the time of composition of the study of the Apocalypse and the writing of the *Principia*, Newton reduced the number of rules. Only two of the final set of rules from the *Treatise on the Apocalypse* appear in the first edition of the *Principia* (1687), where they are part of the introductory "Hypotheses." In the second edition (1713), they are joined by a third *regula*, and in the third and final edition there is an additional fourth *regula*.[3]

A simple comparison of the wording of the two sets of rules reveals the direct lineage between them; thus there is no difficulty in seeing how Newton transformed the rather diffuse rules of the *Treatise on the Apocalypse* into the more concise *regulae* of the *Principia*. This transformation is in many ways remarkable because the backgrounds of these two sets of rules involve different concepts deriving from logic, rhetoric, mathematics, theology, and the philosophy of nature. But no one, so far as I am aware, has sought to find the common source of the longer and earlier set. In what follows, I shall show how the sixteen rules for interpreting the Apocalypse were in turn a transformation of some rules and principles that Newton studied while an undergraduate at Cambridge.

As far as logic and rhetoric are concerned, Robert Sanderson's *Logicae Artis Compendium*[4] is the primary source of Newton's rules. In this work Sanderson followed the Ramists rather than the scholastics, stressing the theory of method.[5] Newton owned a copy of this work, in the flyleaf of which he inscribed his name and the date "1661." The "Trinity Note-Book" shows indisputably that in 1664 Newton carefully read Sanderson's *Logic* along with various works of Descartes. Although both Sanderson and Descartes deal with method, they do so in a very different manner. Newton's study of their works would have given rise to an intellectual difficulty in arriving at any conceptual harmony between them.

In the *Treatise on the Apocalypse*, Newton subdivided his rules into three sections: "Rules for Interpreting ye Words and Language in Scripture," "Rules for Methodizing/Construing the Apocalypse," and "Rules for Interpreting the Apocalypse." However, the numbering of the rules is continuous. The order of succession is from the most to the least general, according to Sanderson's suggestion. Yet their literary style is similar to the four precepts we find in Descartes's *Discours de la méthode*. Like Descartes, Newton introduces his rules as precepts: "To observe diligently," "To assign but one meaning," "To choose those constructions w^{ch} . . . reduce things to the greatest simplicity," and so on.

The twelfth rule is clearly borrowed from Descartes's *Discours*. "Every truth I found," Descartes stated, "is the rule that I need afterwards to find other truths." For Newton, this rule took the form: "The construction of ye Apocalypse after it is once determined must be made the rule of interpretations."[6] Despite certain such links with Descartes's *Discours*, however, the sixteen rules of the *Treatise on the Apocalypse* are more closely tied to Sanderson's *Logic* than to Descartes's *Discours* and, on close examination, prove to be an expansion of the methodological laws Sanderson listed, a remarkable instance of conceptual transformation.

Sanderson and Descartes agree about the general meaning of method. Method is synonymous with order. Sanderson makes a clear distinction between the method of discovering knowledge and the method of presenting or teaching it. The first of these is called the method of invention and the second the method of doctrine, which is twofold: the method of composition [methodus compositiva] and the method of resolution [methodus resolutiva].[7] The two varieties of the method of doctrine, according to Sanderson, are applied differently, one (composition) to the theoretical sciences and the other (resolution) to the practical ones.

Sanderson lists five laws as common to both resolution and composition. On the contrary, the method of invention has no law, but four means or steps: sense, observation or history, experience, and induction [sensus, observatio sive historia, experientia, inductio]. The method of invention, according to Sanderson, has nothing in common with the method of resolution or analysis.

In the *Treatise on the Apocalypse*, Newton twists the meaning of Sanderson's distinctions. The method Newton follows is subdivided into three parts, which he calls "Rules," "Definitions," and "Propositions." This is apparently analogous to the geometrical method or *mos geometricus* that Newton described to Oldenburg in a letter of 21 September 1672:

> To comply wth your intimation . . . I drew up a series of such Expts on designe to reduce ye Theory of colours to Propositions & prove each

> Proposition from one or more of those Expts by the assistance of common notions set down in the form of Definitions & Axioms in imitation of the Method by wch Mathematitians are wont to prove their doctrines.[8]

However, when writing of the order of the propositions in the *Treatise*, Newton uses the same terms by which Sanderson defined the method of resolution. According to Sanderson, the method of resolution begins with the notion of an end [finis] and searches for the substance [subiectum] and the means [media].[9] Newton interprets Sanderson in this manner: "I compare ye pts of the Apocalyps one wth another & digest them into order by those internal characters [the media] wch ye Holy-ghost hath for this end [the finis] impresт upon them. And this I do by drawing up the substance [the subiectum] of ye Prophecy into Propositions."[10]

The link among the rules of the *Treatise* and the laws of method of Sanderson's *Logic* is even closer. The first law of method, according to Sanderson, is the law of brevity [lex brevitatis]: "Nothing should be left out or be superfluous in a discipline" [Nihil in disciplina desit, aut redundet].[11] Newton transforms this law in the second and third rule of the *Treatise*: "To assigne but one meaning to one place of scripture" and "To keep as close as may be to ye same sense of words."[12] The second law is the law of harmony [lex harmoniae]: "The individual parts of each doctrine should agree among themselves" [Doctrinae singulae partes inter se consentiant].[13] Newton expresses this law in many rules: the first, "To observe diligently the consent of Scriptures";[14] the eighth, "To choose those constructions w^{ch} ... reduce contemporary visions to ye greatest harmony of their parts,"[15] from which the ninth and the fourteenth rules also depend.

Simplicity is a consequence of the law of harmony, as Newton makes clear in the ninth rule:

> To choose those constructions wch without straining reduce things to the greatest simplicity. The reason of this is manifest by the precedent Rule. Truth is ever to be found in simplicity, & not in ye multiplicity & confusion of things. As ye world, wch to ye naked eye exhibits the greatest variety of objects, appears very simple in its internall constitution when surveyed by a philosophic understanding, & so much ye simpler by how much the better it is understood, so it is in these visions.[16]

A number of rules in Newton's *Treatise* correspond to the third law of Sanderson's *Logic*, the law of unity or homogeneity [lex unitatis, sive homogeniae]: "No doctrine should be taught that is not homogeneous with subject or end" [Nihil in doctrina praecipiatur, quod non sit subiecto aut fini homogeneum].[17] It will be sufficient to consider the fifteenth rule, which claims that the prophecies must be interpreted according to the end for which they are designed. Finally, consider the eleventh rule of the *Treatise*: "To acquiesce in that construction of ye Apocalyps as ye true one wch results most naturally & freely from ye characters imprinted by the holy ghost on the severall parts thereof for insinuating their connexion."[18] There may be no doubt that this rule is a direct translation of the fifth law of Sanderson's *Logic*, the law of connection [Lex connexionis]: "The individual parts of a doctrine ought to be connected by opportune transitions" [Singulae partes doctrinae aptis transitionibus connectantur].[19]

What is new in Newton's rules? Neither the content nor the expression. It is true that Sanderson's laws are very concise whereas Newton's rules are verbose and redundant. We must wait for the rules of the *Principia* in order to find a conciseness equivalent to the laws of Sanderson's *Logic*. There is, however, a great difference among Newton's rules for interpreting the Apocalypse and Sanderson's laws. Sanderson is repeating the precepts of a dead tradition for presenting or teaching acquired knowledge. Newton is proposing rules to be used in discovering new knowledge. Here is a real transformation, in which Newton is transposing the old concepts beyond logical and rhetorical limits.

We may see a clear example of this transposition in Newton's use of the term "construction." In his *Treatise on the Apocalypse*, Newton titled the second section of rules premised to the definitions, "Rules for Methodizing the Apocalypse."[20] Afterwards he corrected the title by adding "construing" above "methodizing." In all the rules there are fourteen occurrences of the terms "construing" or "construction." What does this tell us? Newton himself declares the origin of his concept: grammatical analysis: "a man acquiesces in ye meaning of an Author how intricate so ever when he sees ye words construed or set in order according to ye laws of Grammar, notwithstanding y^{t} there may be a possibility of forceing ye words to some other harsher construction."[21] However, Newton does not limit himself to this notation. The order of the laws of grammar agrees with the mechanical order: "For . . . of an Engin made by an excellent Artificer a man readily beleives y^{t} ye parts are right set together when he sees them joyn truly with one another notwithstanding that they may be strained into another posture."[22] The conclusion is the same for both the comparisons: "a man ought wth equal reason to acquiesce in the construction of these Prophesies when he sees their parts set in order according to their suitableness & the characters imprinted in them for that purpose."[23]

It is evident that Newton has put together different conceptual entities. And these are the means (the *media* of Sanderson's method of resolution) by which "the Language of ye Prophets will become certain & ye liberty of wresting it to private imaginations be cut of. The heads to wch I reduce these words I call Definitions."[24] In this phrase we see a consequence of the above mentioned transposition of concepts. For it is surely very inappropriate to affirm that a language will become certain, since certainty pertains to knowledge, not to speech.

The same ambiguity is in the term itself of "definition." The definitions listed by Newton in his *Treatise on the Apocalypse* appear to be linguistic definitions. But Newton also considers them mathematical definitions. Therefore these propositions, in which he draws up the substance of the prophecy according to Sanderson's method of resolution, are proved by subjoining the reason for their truth, as if they were mathematical propositions. This goes beyond grammar and logical order. Newton is mobilizing concepts from the logical and rhetorical tradition as if they belonged to the mathematical tradition.

However, if we take into consideration the propositions of the *Treatise on the Apocalypse*, we may notice that Newton does not prove them solely by means of the definitions and rules, but by adding lists of particulars. For example, he proves the eighth proposition ("The Dragon & Beast are ye Kingdome whose symptomes are declared in ye Seales & Trumpits, whereof ye Dragon begins wth ye Seales & ye Beast wth ye Trumpets")[25] by eight particulars, and the meaning of each particular refers to the definitions. Newton twists the meaning both of the traditional *mos geometricus* and the logical and rhetorical methods. In *De Gravitatione*, as well as in the above quoted letter to Oldenburg concerning optics, we find the same philosophical discussions as in the *Treatise on the Apocalypse*; the difference between them is one of detail and reference, since the letter and *De Gravitatione* refer to natural experiments and not to parts of the Scripture.

In the *Treatise on the Apocalypse*, Newton fuses and integrates many different methodological procedures: grammatical analysis, rules of construction and of interpretation, definitions, double demonstration of propositions by means both of particulars and common notions.

In actual fact, in his *Treatise on the Apocalypse*, Newton cancels out the traditional distinction, present in Sanderson's *Logic*, between the methods of invention and of doctrine. Indeed, according to Sanderson, the particulars pertain to the method of invention and not to the method of doctrine, to which the analysis or method of resolution pertains. Consequently, Newton is no longer able to distinguish between analysis and method of invention.

In the seventeenth century the term "analysis" became more and more ambiguous. Sanderson lists many meanings of this term according to logical, rhetorical, and grammatical traditions. Analysis is the method (the *methodus resolutiva*) of practical sciences, but it is also a logical operation [operatio logica], analogous to the procedures with the same name in grammar and rhetoric. Logical analysis, which may be applied both to the theoretical and practical sciences, is twofold: *simplex* or *methodica*. Methodical analysis may be, in turn, *thematica*, *problemathica*, or *methodica stricte*.[26] Only problematical analysis is concerned with demonstration, the kind of demonstration Newton is concerned with in the *Treatise on the Apocalypse*. However, Sanderson does not mention the analysis of mathematics, which was also a demonstrative procedure. This analysis is very different from the method of resolution, because the former is a method of demonstration, the latter of explanation. As we will see, Newton made a fusion, or confusion, of nearly all the meanings of the term "analysis."

Newton was not the only one to have done so. We may find a sibylline account of analysis and synthesis in Descartes's reply to the second objections to the *Meditations*. The Latin text of this work, the one that Newton read,[27] is very different from Clerselier's French translation. According to Descartes the analysis is a method of invention: "The analysis shows the true way by means of which a thing is found methodically and as it was a priori."[28] The ancient authors made no public use of it: "Not because they did not simply know it, but, as I think, because they judged it so important to reserve it to themselves as a secret."[29] Descartes does not mention resolution. However, an echo of Descartes's statements may be found in a handwritten passage of Newton's, often quoted. Newton likely wrote it for the second edition of the *Principia*, but it also is very similar to the statement Newton made in the concluding portion of his (anonymous) review of the *Commercium Epistolicum*, published in the *Philosophical Transactions*:

> The Ancients had two Methods in Mathematicks wch they called Synthesis & Analysis, or Composition & Resolution. By the method of Analysis they found their inventions & by the method of Synthesis they [published them] composed them for the publick. The Mathematicians of the last age have very much improved [Analysis &] Analysis [& laid aside the Method of synthesis] but stop there [in so much as] & think

> they have solved a problem when they have only resolved it, & by this means the method of Synthesis is almost laid aside. The Propositions in the following book were invented by Analysis. But considering that [they were] the Ancients (so far as I can find) admitted nothing into Geometry [but wha] before it was demonstrated by Composition I composed what I invented by Analysis to make it [more] Geometrically authentic & fit for the publick.[30]

Like Descartes, Newton considers analysis and synthesis two mathematical methods. However, he adds their names (resolution and composition) according to the logical tradition. It is odd that Clerselier added the same names to the French translation of Descartes's replies to the second objections. Evidently there is a contamination of the linguistic sources that prefigure the conceptual transformations. In Newton's language, analysis and synthesis, or resolution and composition, become, respectively, the method of invention and the method of doctrine. Sanderson made this distinction too, with the difference that he included resolution in the method of doctrine.

Finally, we must consider briefly the conceptual sources of the *regulae philosophandi* of the second and third edition of the *Principia* and their place in Newton's method. Neither the *regulae philosophandi* nor the rules of the *Treatise on the Apocalypse* agree with the scheme we find in Newton's paragraph, quoted above, concerning the mathematical methods of the ancients. In the *Principia* the rules have the function of linking together the particular phenomena in the same manner that the rules of the *Treatise* assemble the meanings of scriptural language. Both are linked to the laws and to the method of invention as described by Sanderson, rather than to mathematical analysis or synthesis. For this reason it is possible to find a correspondence between the *regulae philosophandi* and the rules of the *Treatise*, notwithstanding the great temporal gap between them.

Table 1.1 describes the conceptual links and analogies among the laws of Sanderson's logic, the rules of the *Treatise on the Apocalypse*, and the *regulae philosophandi*.

In this table the interconnection and fusion of Sanderson's methods of invention and of resolution are clear. There is no doubt that the transformations Newton gradually introduced are very great and significant. In the eighteenth century, the *regulae philosophandi* were correctly considered the canon of experimental science, the result of a revolution that had given an established configuration to science. Nevertheless, their conceptual source is Sanderson's *Compendium*, a manual that combines the scholastic matter of the logical and rhetorical traditions with the liberal arts

Table 1.1

Sanderson's *Compendium*	Rules of the *Treatise on the Apocalypse* (ca. 1672)	*Regulae philosophandi*
Law of brevity (*lex brevitatis*): "Nothing should be left out or be superfluous in a discipline (Nihil in disciplina desit, aut redundet)."	"2. To assigne but one meaning to one place of scripture." "3. To keep as close as may be to the same sense of words."	Regula I (1687) "Causas rerum naturalium non plures admitti debere, quam quae et verae sint & earum phaenomenis explicandis sufficiant."
Law of harmony (*lex harmoniae*): "The individual parts of each doctrine should agree among themselves (Doctrinae singulae partes inter se consentiant)."	"1. To observe diligently the consent of Scripture." "8. To choose those constructions w^{ch} . . . reduce contemporary visions to ye greatest harmony of their parts." "9. To choose those constructions w^{ch} . . . reduce things to the greatest *simplicity*."	Comment to Regula I "Natura enim *simplex* est & rerum causis superfluis non luxuriat."
Law of unity or homogeneity (*lex unitatis, sive homogeniae*): "No doctrine should be taught that is not homogeneous with subject or end (Nihil in doctrina praecipiatur, quod non sit subiecto aut fini homogeneum)."	Rules 4, 6, 7, 10, 12, 14, 15	Regula II (1687) "Ideoque effectuum naturalium eiusdem generis eaedem assignandae sunt causae, quatenus fieri potest."
Law of connection (*lex connexionis*): "The individual parts of a doctrine ought to be connected by opportune transitions (Singulae partes doctrinae aptis transitionibus connectantur)."	"5. To acquiesce in that sense of any portion of Scripture as the true one wch results most freely & naturally from ye use & propriety of ye Language & tenor of the context in that & all other places of Scripture to that sense." "11. To acquiesce in that construction of ye Apocalyps as ye true one wch results most naturally & freely from ye characters imprinted . . . for insinuating their connexion."	Regula III (1713) "Qualitates corporum quae intendi & remitti nequeunt, quaeque corporibus omnibus competunt in quibus experimenta instituere licet, pro qualitatibus corporum universorum habendae sunt."

Table 1.1 (continued)

Sanderson's *Compendium*	Rules of the *Treatise on the Apocalypse* (ca. 1672)	*Regulae philosophandi*
"Induction, by which we make up a universal conclusion summoning many experiences (*Inductio*, qua collectas plures Experientias ad *universalem* conclusionem adhibemus)."	"[2]. If two meanings seem equally probable he is obliged to beleive no more then in general ye one of them is genuine untill he meet wth some motive to prefer one side."	Regula IV (1726) "In philosophia experimentalis, propositiones ex phaenomenis per *inductionem* collectae, non obstantibus contrariis hypothesibus, pro veris aut accurate aut quamproxime haberi debent, donec alia occurrerint phaenomena, per quae aut accuratiores reddantur aut exceptionibus obnoxiae."

of the *trivium*. Thus, the transformation of concepts is the key to understanding the innovative procedures of the "new science." We may see in this example how, historically, the genesis of experimental method involves a complexity of interactions that, from a static point of view, are hidden and thus irremediably lost.

I wish to conclude with an observation of Koyré's, in his *Newtonian Studies*. After having studied the *De Gravitatione*, Koyré concluded that the ways followed by human thought in the search for truth are indeed very odd. However, it is not very odd that a new way is often a rectification of an old one.

Notes

1. I. Bernard Cohen, *Introduction to Newton's Principia* (Cambridge: Harvard University Press, 1971), 21.

2. The work of Newton's that I have called his *Treatise on the Apocalypse* is an untitled work (part of the Yahuda MSS in the University of Jerusalem). A transcription of this work, together with an Italian translation and commentary, may be found in my edition, I. Newton: *Trattato sull'Apocalisse* (Torino, Italy: Bollati Boringhieri, 1994). Frank E. Manuel edited another partial version in Appendix A of his *The Religion of Isaac Newton: The Freemantle Lectures* (Oxford: Clarendon Press, 1974, 107–125); here the abbreviations have been spelled out and the details of Newton's alterations are not given. The parts of this "treatise" containing the rules have subtitles: "Rules for Inter-

preting the Words and Language in Scripture," "Rules for Methodizing/Construing the Apocalypse" and "Rules for Interpreting the Apocalypse."

3. On this topic see I. B. Cohen, "Hypotheses in Newton's Philosophy," *Physis* 8 (1966): 63–184.

4. R. Sanderson, *Logicae Artis Compendium* (Oxoniae, 1618); now in anastatic reprint, edited by E. J. Ashworth (Bologna, Italy: Clueb, 1985). Newton's library included a copy of the third edition of Sanderson's *Compendium* (Oxoniae, 1631) with Newton's signature and date ("Isaac Newton Trin Ccli Cant 1661") and a few signs of dog-earing. Cf. John Harrison, *The Library of Isaac Newton* (Cambridge: Cambridge University Press, 1978), 231.

5. W. S. Howell, *Logic and Rhetoric in England, 1500–1700* (New York: Russell & Russell, 1961), 302.

6. Newton, *Trattato sull'Apocalisse*, 30.

7. Sanderson, *Logicae artis compendium*, 226–7.

8. H. W. Turnbull, ed., *The Correspondence of Isaac Newton*, (Cambridge: Cambridge University Press, 1959), 1:237. Cf. M. Mamiani, *Isaac Newton filosofo della natura* (Firenze: La Nuova Italia, 1976), 184–212.

9. Sanderson, *Compendium*, 227, 231.

10. Newton, *Trattato sull'Apocalisse*, 18.

11. Sanderson, *Compendium*, 227.

12. Newton, *Trattato sull'Apocalisse*, 20–2.

13. Sanderson, *Compendium*, 227–8.

14. Newton, *Trattato sull'Apocalisse*, 20.

15. Ibid., pp. 26–8.

16. Ibid., p. 28.

17. Sanderson, *Compendium*, 228.

18. Newton, *Trattato sull'Apocalisse*, 28.

19. Sanderson, *Compendium*, 228.

20. Newton, *Trattato sull'Apocalisse*, 26.

21. Ibid., p. 30.

22. Ibid.

23. Ibid.

24. Ibid., p. 18.

25. Ibid., p. 128.

26. Sanderson, *Compendium*, "Appendix prima, De Analysi Logica," 75–84.

27. Cf. Harrison, *The Library of Isaac Newton*, 132, and *Certain Philosophical Questions*, ed. J. E. McGuire and M. Tamny (Cambridge: Cambridge University Press, 1983), 23.

28. "Analysis veram viam ostendit per quam res methodice, et tanquam a priori inventa est." R. Descartes, *Meditationes de Prima Philosophia* (Amstedolami, 1642), 172.

29. "Non quod . . . plane ignorarent, sed, quantum judico, quia ipsam tanti faciebant ut sibi solis tanquam arcanum quid reservarent." Descartes, *Meditationes*, 173.

30. Newton, Review of *Commercium Epistolicum* published in *Philosophical Transactions*, Cambridge University Library, MS Add. 3968, fol. 101.

2

The Case of the Missing Author: The Title Page of Newton's *Opticks* (1704), with Notes on the Title Page of Huygens's *Traité de la lumière*

I. Bernard Cohen

The Missing Author

The title page of the first edition of Newton's *Opticks* (London, 1704) offers a puzzle. As an examination will show at once (see figure 2.1), it bears no author's name. Was this omission an oversight? That is, did the author's name get lost during the printing process? Or was this omission intentional? And if so, why would Newton have suppressed his name?

These questions are of more than purely bibliographic interest. The quest for answers leads us to a study of Newton's evaluation of his work in optics and of his attitude toward publication, and it reveals some publishing practices of the seventeenth and eighteenth centuries. For as we shall see, Newton's *Opticks* was not the only work on optical science of those days to appear without the author's full name; Huygens's *Traité de la lumière* was in this same category. Furthermore, the evidence leaves no doubt that Newton was familiar with Huygens's book and thus would have taken it as a model.

Although the title page of Newton's *Opticks* bears no author's name, the identity of the author was not in any real sense hidden from the reader. First of all, the contents are in part an enlargement of the presentation of Newton's experiments and ideas as published in the *Philosophical Transactions* in 1672.[1] Additionally, in the text of the *Opticks*, Newton writes in the first person about these very experiments and some new ones that he has performed, for example, "I took a black oblong stiff Paper" (p. 13), "I erected a glass Lens" (p. 15), "I placed a glass prism" (p. 18).[2] Hence any reader familiar with the subject of optics would have known the author was Newton. This was not the work of someone else reporting on what Newton had done.

Another sure indication of the identity of the author is offered in the preface, which is signed "I. N." (See figure 2.2.) In this preface, furthermore, Newton refers (writing in the first person) to his report to the Royal Society in 1675.[3] Finally, the last paragraph of the preface positively

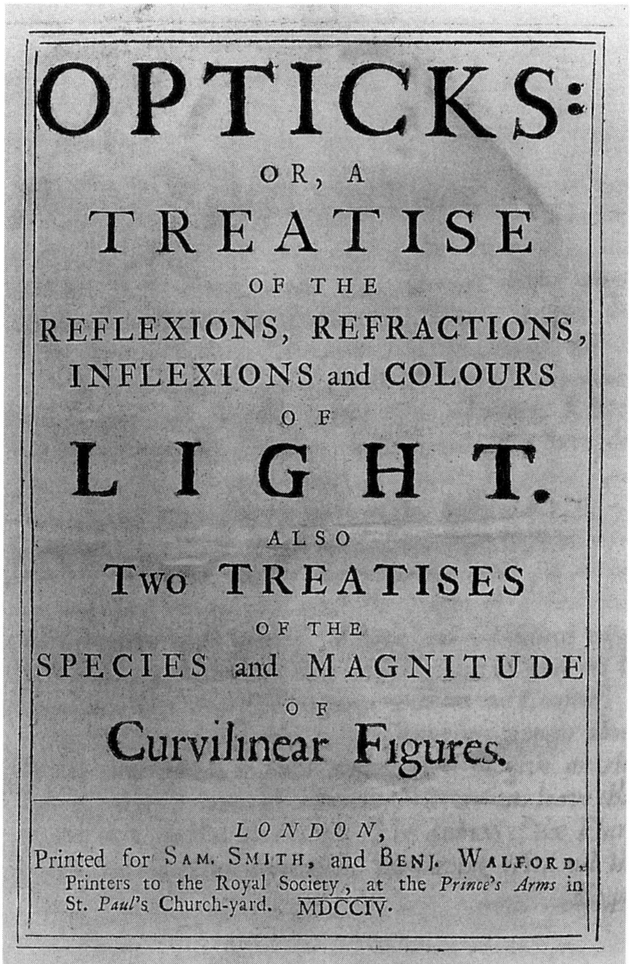

OPTICKS:
OR, A
TREATISE
OF THE
REFLEXIONS, REFRACTIONS,
INFLEXIONS and COLOURS
OF
LIGHT.
ALSO
Two TREATISES
OF THE
SPECIES and MAGNITUDE
OF
Curvilinear Figures.

LONDON,
Printed for Sam. Smith, and Benj. Walford,
Printers to the Royal Society, at the *Prince's Arms* in
St. *Paul's* Church-yard. MDCCIV.

Figure 2.1
Title-page of Newton's *Opticks* (1704), from the copy in the Grace K. Babson collection of Newtoniana in the Burndy Library, Massachusetts Institute of Technology. The words "OPTICKS," "LIGHT," "Curvilinear Figures," "SAM. SMITH," and "BENJ. WALFORD" are printed in red.

Experiments which I intended when I was about theſe Matters, nor repeated ſome of thoſe which I did try, until I had ſatisfied my ſelf about all their Circumſtances. To communicate what I have tried, and leave the reſt to others for further Enquiry, is all my Deſign in publiſhing theſe Papers.

*In a Letter written to Mr.*Leibnitz *in the Year* 1676. *and publiſhed by Dr.* Wallis, *I mentioned a Method by which I had found ſome general Theorems about ſquaring Curvilinear Figures, or comparing them with the Conic Sections, or other the ſimpleſt Figures with which they may be compared. And ſome Years ago I lent out a Manuſcript containing ſuch Theorems, and having ſince met with ſome Things copied out of it, I have on this Occaſion made it publick, prefixing to it an* Introduction *and ſubjoyning a* Scholium *concerning that Method. And I have joined with it another ſmall Tract concerning the Curvilinear Figures of the Second Kind, which was alſo written many Years ago, and made known to ſome Friends, who have ſolicited the making it publick.*

I. N.

Figure 2.2
Second page of the preface to the *Opticks* (1704), from the copy in the Grace K. Babson collection of Newtoniana in the Burndy Library, Massachusetts Institute of Technology.

identifies the author, as Newton, again in the first person, mentions his correspondence with Leibniz about the calculus and the reasons why he is now publishing two tracts on mathematics as supplements to the *Opticks* (the *Enumeratio Linearum Tertio Ordinis* and the *Tractatus de Quadratura Curvarum*).[4] They mark the first appearance in print of a tract by Newton on the method of fluxions or the calculus.[5]

Plainly, neither Newton nor his publishers were in any way hiding the author's identity. Furthermore, after the volume had been printed, it was found that a pair of paragraphs had been inverted; the leaf in question was cut out and replaced by a cancel pasted on the stub of the original. Surely if the author's name had been omitted inadvertently, this would also have been corrected by a similar replacement of the faulty title page.

There must, then, have been some reason or reasons why the author's name was not placed on the title page. In what follows, I shall present some grounds for believing that the omission of the author's name was not accidental. Additionally, I have found new evidence that reinforces this conclusion and that helps to understand why the title page bears no author's name. I believe that now we may write *finis* to the mystery of the title page and conclude the case of the missing author.

Newton's Optical Science: The Imperfect Science of the *Opticks*

It is well known that after the controversies that followed the publication of Newton's celebrated paper on light and colors in 1675,[6] he was reluctant to publish anything more on optical subjects and natural philosophy in general. Indeed, he even vowed that he would "bid adew to it [natural philosophy] eternally, excepting what I do for my privat satisfaction or leave to come out after me."[7] As the years went by, Newton was urged again and again to produce a book on optics, one that would contain the early work plus the results of his further researches. As we shall note in the next section, Newton actually agreed to a publication which would include a reprint, with copious annotations, of some of his optical papers that had appeared in the *Philosophical Transactions*. He apparently changed his mind about this edition and stopped the publication after a number of pages had been set up in type. In the late 1680s (after the publication of the *Principia*) and in the 1690s, he began to produce a treatise on optics in Latin to which he proudly gave the title *Fundamenta Opticae*. During the next years, this work went through various stages of development, of which the last was a rather different treatise—written in English rather than Latin—the *Opticks*, published in 1704.

The researches of Alan Shapiro, reinforced by the detailed analyses by Rupert Hall, make it possible to produce a chronology of Newton's stages of composition. The Latin treatise of the 1690s was to have contained four "books" or parts, of which the first was based on his "Lectiones Opticae," while the second contained his report to the Royal Society on thin films, augmented by some new experimental results. In retrospect the proposed Book IV may have been the most daring because in it Newton planned to marshal evidence from optical phenomena to support the concept of forces acting at a distance. This portion of the treatise was never completed, and our knowledge of it comes from Newton's outlines or sketches.

By 1694, Newton had shifted from a Latin to an English text and had produced a work much like the 1704 *Opticks*. In a memorandum of that year, 1694, David Gregory recorded that he had seen a book by Newton on optics, written in English and consisting of three "books," all but ready for publication. He did note, however, that the fourth book (on diffraction) was "not yet complete." It was in reference to this work that John Wallis wrote to Newton a year later that the book should be published. Newton replied that he dared not publish this work since doing so might cause him "some trouble."

The researches of Alan Shapiro, however, have shown that what Gregory saw was hardly the complete text ready for printing. Shapiro, furthermore, has made use of Newton's extensive manuscripts to reconstruct the early stages of the text. Shapiro's research (see chapter 3, this volume), combining historical-physical insight with an extensive knowledge of Newton's MSS, has enabled us to understand why Newton never completed the proposed Latin treatise. He was unable to produce a satisfactory mathematical theory that would explain the phenomena of diffraction.

It has long been generally supposed that a major reason why Newton finally changed his mind and in 1704 permitted the publication of a work on optics was that Hooke had just died, the last of those who had criticized his earlier publication on light and colors. In fact, however, Hooke died in March 1703, four months after Gregory recorded (in November 1702) a promise by Newton to publish the *Opticks*. At that time, however, Newton no longer needed to be concerned about a possible critique from Hooke since Hooke was then—in Rupert Hall's words—"wretchedly ill and incapable." November 1702 was also the month of Newton's election as President of the Royal Society, an event that perhaps made him feel secure enough to venture once again into print.

Most readers would have guessed that Newton was referring to Hooke when, in the preface to the 1704 *Opticks*, he wrote that he had "delayed the printing" of this work in order to "avoid being engaged in Disputes concerning these Matters." About a decade earlier, in 1693, Newton wrote to Leibniz that he would publish no more books since he feared that "disputes and controversies may be raised against me by ignoramuses." He was evidently referring, as in the preface to the *Opticks*, to some critics of his celebrated paper of 1672 on light and colors. As Westfall has noted, the *Opticks* did not appear in print until the last of such "ignoramuses," Robert Hooke, was dead.[8]

In 1937 there surfaced a memorandum (dated 1 March 1703/04) in which David Gregory recorded yet another impetus to Newton's final willingness to publish the *Opticks*: the appearance of George Cheyne's

Fluxionum Methodus Inversa (London, 1703). According to Gregory, "Mr. Newton was provoked by Dr. Cheyns book to publish his Quadratures, and with it, his Light & Colours, &c."[9] Cheyne's book was a presentation of Newton's method of "inverse fluxions" or integration. According to Gregory, Newton was so dissatisfied by Cheyne's presentation that he was goaded into publishing his own tract on this subject, together with a tract on "lines" (that is, curves) of the third order. These two tracts, both in Latin, do indeed appear as a kind of supplement to the English text of the 1704 *Opticks.*

In assessing the actual role of Cheyne's book, we must keep in mind that it was published in 1703 and that a year earlier, in November 1702, according to a memorandum by Gregory, Newton had already "promised Mr. Robarts, Mr. Fatio, Capt. Hally & me to publish his Quadratures, his treatise of Light, & his treatise of the Curves of the 2^{d} Genre." Note that even before Newton had seen Cheyne's book, he had agreed to publish his own treatise on his method of quadratures, together with the tract on curves of the second order and the treatise on optics. Accordingly, the only possible role of Cheyne's book would have been to cause Newton finally to implement his earlier decision to allow these works to be printed.

In the 1704 *Opticks* Newton himself gave two reasons why he had been reluctant to publish this work. Both are aspects of what Newton evidently considered to be shortcomings of this area of his research and thinking. The first is his awareness that he had not been successful in his studies of diffraction (which he called "inflexion"). In the third "book" of the *Opticks*, he apologized for this shortcoming. "When I made the foregoing Observations," he wrote, "I designed to repeat most of them with more care and exactness." He also intended to make "some new" observations "for determining the manner how the rays of Light are bent in their passage by Bodies, for making the fringes of Colours with the dark lines between them."[10] Unfortunately, as he explained, "I was then interrupted, and cannot now think of taking these things into further consideration." Since "I have not finished this Part of my Design," he concluded, he would end the *Opticks* by "proposing only some Queries, to a farther search to be made by others." To conclude a scientific treatise with a group of questions is surely a sign that the research on which the book was based had not been carried to its maximum potential. But Newton does not specify the nature of his failure, his inability to produce a mathematical solution to the problems of diffraction. As I have mentioned above, Alan Shapiro (in his chapter in the present volume) explores this subject in an original and profound way, showing from Newton's MSS the degree to which Newton was dissatisfied with his mathematical theory of diffraction. He was reluctant to publish an account of his work on this subject that had not achieved the level of perfection he usually demanded of his own work.

A second reason why Newton was reluctant to publish a book containing his researches in optics is given by the very fact, mentioned above, that the *Opticks* concludes with a series of unanswered questions rather than some positive assertions. Newton thus admitted openly that there were a number of aspects of optics and related scientific topics that he had not been able to explore as fully as he would have liked. It was not even clear to Newton that the solutions to these problems could be found by direct experiment. In any event, to have his treatise conclude with a set of problems needing solution gave emphasis to the state of imperfection or incompleteness of Newton's experiments and theoretical explanations.

Let us take note, however, that the Queries themselves have had an important place in the development of science. The first group of Queries, numbered 1 to 16, published in the original English edition of 1704, deal primarily with topics rather strictly in the domain of optical science, but later ones, added in subsequent editions, extended Newton's speculations into the domains of heat, gravity, the structure of matter, electricity, chemical action, the transmission of nerve impulses, methods of doing science (including resolution and composition), the creation of the world, the nature of God's "sensorium," and even the improvement of morals.[11]

These later Queries aroused enormous interest during the eighteenth century, rivaling in importance the rather strictly optical text of the *Opticks*, and they even helped to establish a variety of Newtonian philosophy different from that of the *Principia*.[12] They set forth a research program for the scientific world. After all, who would know better than Newton what questions should be further explored?

These Queries, furthermore, were not simple questions in the ordinary sense. Some, especially the later ones, were so lengthy that they occupy many pages of printed text. Almost all were written in the negative, as a rhetorical statement in the interrogative form: "Is not fire a Body heated so hot as to emit Light copiously?" Since Newton, in most cases, then proceeded to present a mass of evidence to support his point of view, these were hardly questions of the usual sort. In fact, Newtonians did not apparently take the question form seriously as a form of genuine interrogation. J. T. Desaguliers, for example, held that the Queries contain "an excellent body of philosophy," and that "upon examination" the Queries "appear to be true." Newton, he said, had adopted the question form only from "modesty." He had observations aplenty to satisfy himself that the things he had proposed "by way of queries" were "true." Another Newtonian, Stephen Hales, founder of quantitative plant physiology, simply wrote that in this part of the *Opticks*, Newton had "explained by Quaere" a number of varied physical phenomena.[13]

In any event, since Newton concluded the *Opticks* in a set of Queries, coupled with his admission that he had not completed the research, there can be little or no doubt that he wanted to make certain that the reader would be aware of the unfinished or incomplete nature of his presentation of this subject. This aspect of his work in the optical domain must have been an important factor in his long refusal to put his manuscript into a form suitable for publication as a book and then to submit this version for publication. That the book in question was a work on the subject of optics—a discussion of light and color—surely aggravated his reluctance, since his earlier venture into publishing in this field had incited such strong adverse criticism.

There was, however, another and very different reason why Newton was dissatisfied with his efforts to understand the phenomena of light and color: his awareness of his failure to construct a mathematical theory of optics on a par with the physics of the *Principia*. Alan Shapiro and Zev Bechler[14] have documented Newton's abortive attempts to produce a mathematical theory of light. In 1665 he boasted that his research on optics implied that this was a subject "more proper for mathematicians than naturalists."[15]

By 1704, the failure to produce a mathematical theory of optics was no longer a primary concern of Newton's. By this time, although he still believed that the theory of diffraction would yield to mathematical expression, he had abandoned the hope that the whole science of light and color could be mathematical in the sense in which that adjective was used in relation to the *Principia*. Nevertheless, he and many fellow members of the Royal Society shared a point of view in which the mathematical or exact sciences represented the highest type of science, whereas the experimental or observational sciences were inferior.[16] From this point of view the *Opticks* must have seemed to be on a different level of the scale of values than the mathematical *Principia*.

Of course, the *Opticks* is not a purely qualitative work. It is based on measurement and calculation. The later portion (Book II and Book III) uses quite sophisticated mathematics. But Book I validates its propositions by a "Proof by Experiments." The *Opticks* does not proceed, in the manner of the *Principia*, by proving its propositions by using mathematical techniques (algebra and geometry, theory of limits, infinite series, and fluxions or the calculus).

Readers of the *Opticks* would quickly become aware that this work was not a mathematical treatise in that sense in which the word "mathematical" characterizes the *Principia*. That Newton presented his text in the manner of an experimental record, in ordinary English prose, accentuated this difference. In this feature the *Opticks* differed from the *Principia*, of which the full title (*Philosophiae Naturalis Principia Mathematica*) declared its mathematical nature. The *Principia* was written and published in Latin, in the manner of Newton's mature mathematical works, such as the two tracts that were presented as appendages to the *Opticks*.

These factors, differentiating the *Opticks* from the *Principia*, seem to have been primarily responsible for the fact that the *Opticks* was published in English[17] and without the author's name emblazoned proudly on the title page.

NEWTON'S AMBIVALENCE ABOUT PUBLICATION: AN ABORTED EDITION OF NEWTON'S OPTICAL PAPERS

Newton was always very ambivalent about publishing his writings. He vowed, as we have seen, not to publish anything more in his lifetime and yet he did go on to compose and to send in to the Royal Society a long "Hypothesis" for explaining optical and other phenomena. Despite the criticism he received after publishing his paper on light or colors, he did at one time contemplate a reedition of that paper and those that followed. The existence of this abortive effort to reprint the papers or letters on optics from the *Philosophical Transactions* (and perhaps others as well) came to light in a curious way.

The evidence concerning this proposed edition was discovered by the late Derek Price, who had bought an old church library.[18] Among the waste sheets inside the decayed binding of one of these books, Price found two sets of proof sheets containing the text of a portion of Newton's famous letter of 1672. Price recognized that they might be of real importance for Newton scholarship and turned these over to me for study and for eventual publication if I should find anything interesting in them.[19]

Although at first glance, these sheets seemed to be either tear sheets or waste from the original publication in the *Philosophical Transactions*, a careful examination showed two differentiating features. One was that the page numbers differ markedly from those in the *Philosophical Transactions.* The other is that there are extensive notes, notes which were not mere glosses on the text but discussions of some very important philosophical issues raised by Newton's original publication, issues that had been the focus of argument by his critics. These notes were written in the first person and were clearly Newton's own commentary on his earlier papers.

A further examination of these sheets showed that they were not merely reprintings of the original pages in the Royal Society's *Philosophical Transactions.* Rather, they were set in type some time after the original type had been distributed and possibly were even composed in a different printing shop altogether. The evidence for this resetting lies in the fact that the paragraphing is somewhat different, as is the division of the text into lines of type.[20]

What is most interesting about these proof sheets, other than their serving as evidence for an abortive edition of Newton's papers on optics, is that they contain glosses on the text written in the first person. Evidently Newton had intended to provide extensive annotations. The sheets contain only portions of Newton's original communication to the Royal Society (1672) on light and color. Presumably there were to be similar annotations throughout the volume. The surviving sheets contain only three notes. These are keyed "c," "d," and "e." Hence there must have been further notes (at least) keyed to "a" and "b."

The first of these notes is very important because in it Newton clarifies what he means by his discussion of "whether Light be a Body." He also addresses what he calls an "improper" distinction between light being considered as a "body" and as "the action of a body." This leads him to a distinction between "the Peripatetick and Mechanick Philosophy" on the matter of "whether light be a quality or body":

> c) Through an improper distinction which some make of mechanical Hypotheses, into those where light is put a body, and those where it is put the action of a body, understanding the first of bodies trajected through a medium, the last of motion or pression propagated through it, this place may be by some unwarily understood of the former: Whereas light is equally a body or the action of a body in both cases. If you call its rays the bodies trajected in the former case, then in the latter case they are the bodies which propagate motion from one to another in right lines till the last strike the sense. The only difference is, that in one case a ray is but one body, in the other many. So in the latter case, if you call the rays motion propagated through bodies, in the former it will be motion continued in the same bodies. The bodies in both cases

> must cause vision by their motion. Now in this place my design being to touch upon the notion of the Peripateticks I took not body in opposition to motion as in the said distinction, but in opposition to a Peripatetick quality, stating the question between the Peripatetick and Mechanick Philosophy by inquiring whether light be a quality or body[,] Which that it was my meaning may appear by my joyning this question with others hitherto disputed between the two Philosophies; and using in respect of one side the Peripatetick terms *Quality*, *Subject*, *Substance*, *Sensible qualities*; in respect of the other the Mechanick ones *Body*, *Modes*, *Actions*; and leaving undetermined the kinds of those actions, (suppose whether they be pressions, strokes, or other dashings,) by which light may produce in our minds the phantasms of colours.

The second note is also of interest because in it Newton makes a distinction between the traditional views (those of "the Peripatetick Philosophy") and his own. He makes the important point that he does not believe colors to be "the qualities of light" or of "anything else without us," but rather "modes of sensation excited in our minds by light." As he expressed this at the end of the note and again in the *Opticks*, one should accordingly not talk of light as being of a certain color, say red, but rather as "rubriform" or red-producing:

> d) Understand therefore these expressions to be used here in respect of the Peripatetick Philosophy. For I do not my self esteem colours the qualities of light, or of any thing else without us, but modes of sensation excited in our minds by light. Yet because they are generally attributed to things without us, to comply in some measure with this notion, I have in other places in these letters, attributed them to the rays rather then to bodies, calling the rays from their effects on the sense, red, yellow, &c. whereas they might be more properly called rubriform, flaviform, &c.

The third note appears on the last page of these proofs and is incomplete. All that remains is:

> e) To determin after what manner light is a body, or whether it be a body more then by

The loss of the rest of Newton's note is all the more to be regretted in that the topic under discussion was one of the major points in the criticism of Newton's original publication on light and colors, in which he referred to his belief that light is "a body." In his reply to the criticism, Newton defended himself by saying that he had indeed made such a statement, but that he had done so "without any positiveness," as his use of the word "perhaps" would indicate.[21]

One last feature of these proof sheets merits being noted. At the end of the second extract and in the third extract, the word "than" is spelled "then." This idiosyncratic spelling affords yet another indication that it was indeed Newton who had made the comments printed in the margins, since he customarily spelled the word in this way.

The existence of these proof sheets leaves no doubt that Newton had at one time agreed to the publication of an edition of his early published papers and had added a kind of commentary. He must have changed his mind in midstream, however, and abandoned the project. Knowledge of this abortive edition of a work by Newton on optics would never have surfaced but for Derek Price's extraordinary discovery. The fact that this abortive edition contained Newton's own comments on his published articles is evidence that Newton had, at least at first, been party to this new presentation of his work. As we shall see below, Newton at one time consented to an edition of his *Arithmetica Universalis* and then, after the work was set in type, decided to cancel the edition. In this case, the only way out would have been to buy up the whole edition, since the work had gone beyond the proof stage and had already been printed. It is a matter of record that, whatever Newton's reluctance to publish the results of his research and thinking, he did in fact complete and publish two major treatises, the *Principia* and the *Opticks*, plus various mathematical works.

The *Principia* came into being as a result of Edmund Halley's urging after his visit to Newton in Cambridge in the summer of 1684. Newton at first agreed only to submit his tract *De Motu* to the Royal Society, not to be published in print, but only to be entered into the record to guarantee his priority. In an age of growing acrimonious controversy over priority of discovery, it was important to establish one's priority. We can surmise that Newton must have been torn between his wish to guarantee his priority and his fear of exposing himself by open publication. I believe that he adopted a somewhat similar compromise when he decided to publish the *Opticks* in English rather than in Latin and to issue this work without his name on the title page, identifying himself by the initials "I. N." at the end of the preface.

Newton as Perfectionist Author

Newton, perhaps because of a basic ambivalence between wanting his discoveries to be known and his fear of criticism, tended to be fussy about his publications. His first published book,[22] the 1687 *Principia,* contained a whole page of "Errata Sensum turbantia sic Emenda." After the 1706 Latin edition of the *Opticks* had been printed, a page was cut out and replaced by a cancel; Newton had opened himself up to criticism by unwisely writing about space as the "sensorium" of God. The substituted page would have him say "tanquam" or "as if" in God's sensorium.[23]

In 1712, after Book II of the *Principia* had been printed, Newton's attention was brought to an error in proposition 10.[24] The eventual repair required the replacement of pages 232–239, involving the introduction of a new signature of four leaves plus a single leaf of an adjacent signature.[25] The replacement page (or cancel) of the single leaf (pp. 231–232) was pasted onto the stub of the original (or cancellandum). This edition has also a page of "Errata" to be corrected by the reader.

While the *Arithmetica Universalis* (1707) was being printed under the editorship of William Whiston, Newton regretted that he had given permission to Whiston, his successor as Lucasian Professor, to put this work into print. As we shall see below, Whiston had not acted in a proper editorial capacity but had slavishly followed a manuscript written several decades earlier and deposited in the University Library. This behavior annoyed Newton. As in the case of the 1704 *Opticks*, Newton's response was to have a title page without any name of the author.

In the light of this history, we should expect that after the 1704 *Opticks* had been set in type and printed, Newton would have given it the same kind of critical reading that had produced lists of errata in his other works. And, indeed, as in those works, there is a final printed page of "errata" to be corrected by the reader. Some of these refer to Books I and II of the *Opticks*, whereas others are errors in the printing of the two mathematical tracts. Additionally, an offensive leaf contained an error that could not be easily be corrected simply by a statement in the errata. Accordingly, the offending leaf was cut out of the printed work and a cancel was substituted for the cancellandum, being pasted onto the stub of the original page.

The leaf in question is Q2, containing pages 123 and 124. There are two types of differences between the original text (on the cancellandum) and the corrected text (on the cancel). The most important of these is the order of the two paragraphs at the top of page 124, which were reversed. In the amended version, the final paragraph of this pair concludes with a reference to a variation of the "Experiment as follows." What follows at once is the next "Experiment," listed as "Exper. XVI." In the uncorrected version, the penultimate paragraph refers to the "Experiment as follows," but what follows is not the new experiment but another paragraph that is part of the previous discussion.

The second type of difference is textual. There are very few differences in the actual wording of the text on pages 123 and 124. The textual changes made on page 123 may be seen from the following extracts:

Cancellandum	Cancel
If a black object be incompassed with a white one ...	If a black object be encom passed with a white one ...
... therefore they appear in a contrary order to that, when a white Object is surrounded with black.	... therefore they appear in a contrary order to that, in which they appear when a white Object is surrounded with black.

The whole passage, in the final version reads:

> If a black Object be encompassed with a white one, the Colours which appear through the Prism are to be derived from the Light of the white one, spreading into the Regions of the black, and therefore they appear in a contrary order to that, in which they appear when a white Object is surrounded with black.

It would hardly seem likely that Newton would have ordered that the leaf be cancelled because of a misspelling of "encompassed." Furthermore, an entry in the errata could easily have corrected the omission of the phrase "in which they appear."

Accordingly, the reason for the cancellation must be in the text of page 124. There is only one textual change here. As may be seen in the following extracts, there was an omission of a phrase "with crooked sides."

Cancellandum	Cancel
... the Object-glass, or that part of it or of the Eye which is not covered, must be considered as a Wedge, and every Wedge of Glass or other pellucid substance has the effect of a Prism ...	... the Object-glass, or that part of it or of the Eye which is not covered, may be considered as a Wedge with crooked sides, and every Wedge of Glass, or other pellucid Substance, has the effect of a Prism ...

The text of the improved version has two new two commas, changes "must" to "may," and adds the important technical consideration that the object under discussion "may be considered as a Wedge with crooked sides," not merely a "Wedge." Here Newton uses "crooked sides" where we would say "curved sides."

A reading of these extracts shows that there can be little doubt that the reason for the cancelling of this leaf was the reversal of the two paragraphs on page 124; all other corrections could easily have been inserted into the errata. As a matter of fact, it would even have been possible to have a note in the errata alerting the reader to the inversion of these two paragraphs. But Newton[26] evidently was so fussy that he did not feel it adequate to list in the errata that two paragraphs had been reversed and believed an excision of the original leaf was required, even though this surgery required that every copy of the printed book should be altered, one by one.

In fact, as is so often the case for such corrections, a number of copies were never altered, and at least two of them have both the cancel and the cancellandum. The Grace K. Babson Collection of Newtoniana in the Burndy Library has two copies of the 1704 *Opticks*. One has the cancel, pasted onto an easily visible stub of the original leaf, but the other has both the cancel and cancellandum, the cancel being tipped in next to the leaf that should have been cancelled but was not.[27] The situation is much the same as in the Latin *Optice* of 1706, for which there exist some copies with the cancellandum rather than the cancel and at least one copy has both.[28]

The cancelling of a leaf in the 1704 *Opticks* differs significantly from that in the 1706 Latin *Optice*. In the latter case, Newton was making a significant textual alteration, eliminating a statement that might (and actually did) give rise to serious derogatory criticism. But in the case of the 1704 *Opticks*, the offending leaf does not contain a discussion of either the nature of space or God's sensorium. Rather, the changes made by Newton are those of a perfectionist author. I doubt whether many—if, indeed, any—readers would have noticed the fault in the cancelled page[29] and if they did, they would almost surely not deem it to be so great as to warrant the excision of a whole leaf.

I believe the conclusion is inescapable that Newton exhibited his usual fussiness after the 1704 *Opticks* had been printed, excising a page that did not fully meet the high standards he set himself. We may therefore be certain that if the omission of his name from the title page had been an error, he would have ordered the printer to cancel this page also and to substitute a "corrected" page with his name displayed on it. We shall see below that there is in fact a copy of the 1704 *Opticks* with just such a title page.

Another Missing Author: A Model for Newton

The presentation thus far would advance some plausible reasons for assuming that the omission of the author's name on the title page was a deliberate act of suppression on Newton's part. But even if we conclude that Newton was not entirely happy about the state of his science of optics, how can we be sure that Newton would indeed have gone to such extremes? In exploring this aspect of the topic, two sources of evidence would support the notion of a purposeful omission of the author's name.

The first line of evidence takes us to another treatise on optics of this same period by a different author: Christiaan Huygens's *Traité de la lumière* (Paris 1690). An examination of the usual title page of this work (see figure 2.3) shows that, like the *Opticks*, it does not declare the author's name. As in the case of the *Opticks*, the author is identified only by his initials, "C.H.D.Z.," standing for Christiaan Huygens de Zeelhem. The only difference between the two in this regard is that Huygens's initials appear on the title page, whereas (as we have seen above) Newton's initials appear at the end of the preface. (See figure 2.2.)

There are many other points of similarity between the two works. The subject matter of both is physical rather than traditional geometric optics, the properties of light rather than the traditional dioptrics or catoptrics. In the preface to the *Traité*, Huygens refers to his presentations of the subject to the French Academy of Sciences some years earlier, just as Newton refers to his presentations to the Royal Society. Both authors are thus making explicit their claims to priority of discovery. Even more important, the *Traité*, like the *Opticks*, contains both an explanation for the delay and an apology for publishing so imperfect a work.[30]

In the preface, Huygens explains that he had written this work "rather carelessly," hoping to rewrite and polish it later on. He finally judged it to be "better worth while to publish this writing, such as it is, rather than run the risk, by waiting longer, of its remaining lost." He then explains, in language that almost perfectly fits Newton's *Opticks*, that this work contains demonstrations of a kind "which do not produce as great a certitude as those of Geometry." He also refers specifically to the problem of "principles" that are "assumed [to] correspond perfectly to the phenomena."[31] This line of thought could easily be supposed to have been an expression of the point of view of Newton's *Opticks*.

Finally, we should note that Huygens's *Traité* is written in a vernacular language, French, rather than in Latin, the language usually used by Huygens in his mathematical publications and in his monumental treatise

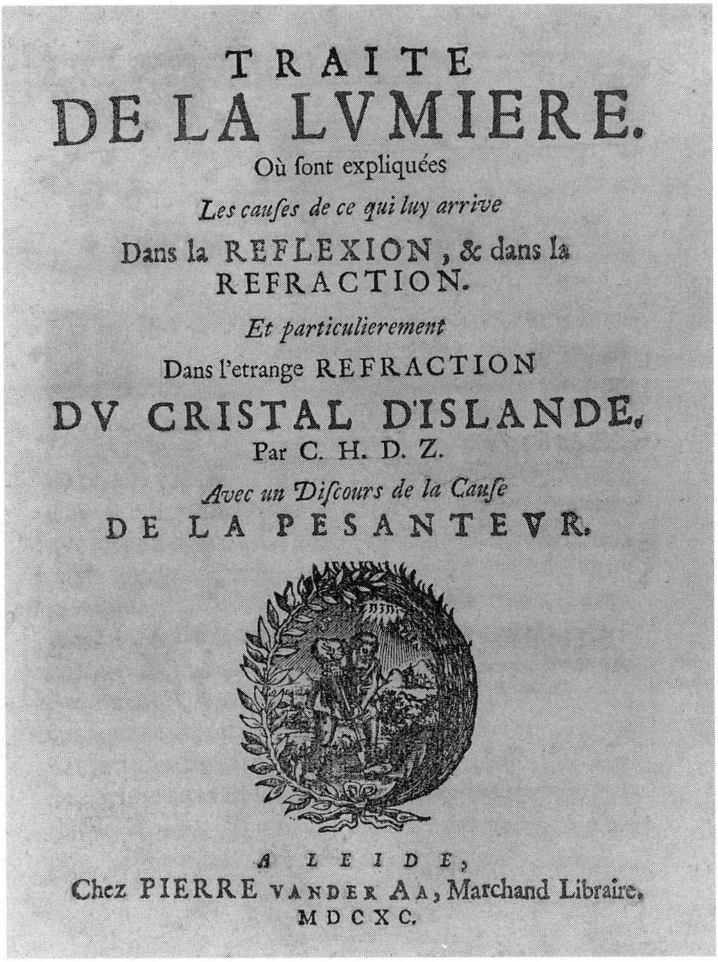

TRAITE
DE LA LVMIERE.
Où ſont expliquées
Les cauſes de ce qui luy arrive
Dans la REFLEXION, & dans la
REFRACTION.
Et particulierement
Dans l'etrange REFRACTION
DV CRISTAL D'ISLANDE,
Par C. H. D. Z.
Avec un Diſcours de la Cauſe
DE LA PESANTEVR.

A LEIDE,
Chez PIERRE VANDER AA, Marchand Libraire.
MDCXC.

Figure 2.3
Title page of Huygens's *Traité de la lumière* (1690), from the copy in the Grace K. Babson collection of Newtoniana in the Burndy Library, Massachusetts Institute of Technology. The words "DE LA LVMIERE," "Dans la REFLEXION, & dans la," "DV CRISTAL D'ISLANDE," "DE LA PESENTEVR," and "Chez PIERRE VANDER AA, Marchand Libraire" are printed in red.

on motion and on the principles of pendulum motion, the *Horologium Oscillatorium* of 1672. From the statement in the preface to the *Traité*, it would appear that Huygens deliberately chose to publish this work in French, rather than in Latin, because he considered it to be an imperfect work ("written carelessly") needing polishing and revision. Although Newton's *Opticks* was hardly "written carelessly," it too was stated by its author to be an imperfect work. We should note, however, that Huygens's *Traité* does not show signs of having been "written carelessly"; this disclaimer was a rhetorical flourish intended to disarm potential critics. Huygens would never have published anything that had truly been "written carelessly."

Certainly one reason why Huygens published his *Traité*, however imperfect he thought it, was (as has been mentioned) to guarantee his priority. After all, this work presents his original treatment of wave motion and is a foundational treatise in the theory of optics. Huygens resolved his ambivalence in the same manner Newton later adopted: he published the *Traité* in the vernacular, rather than making the contents universally available in the sense that a Latin version would have done.

WAS HUYGENS'S *TRAITÉ* A MODEL FOR NEWTON?

When Newton was completing his *Opticks* for publication, he was familiar with Huygens's *Traité*. In fact, Newton owned two copies of this work.[32] One was a presentation copy from the author and Newton wrote on the fly leaf, "Is. Newton Donum Nobilissimi Authoris."[33] We know that Newton read this work with great care, since he dog-eared some of the pages in the way in which he marked passages of special interest. Newton also owned a second copy of the *Traité*,[34] which also shows his characteristic dog-earing. He referred to Huygens's *Traité* later on, in Query 25 of the *Opticks*, which appeared for the first time in the English edition of 1717/1718.

This evidence leaves no doubt that Newton had a model before him of a work on optics declared by the author to be imperfect, a work that was published in the vernacular and not in Latin, a work that carried the identification of the author by his initials rather than by his full name. The parallelism goes even further for, like the *Opticks*, Huygens's *Traité* is devoted to the subject of physical optics rather than the classical or older geometric sciences of dioptrics and catoptrics.

There is yet another point of parallelism between the two works in that each contained a technical supplement. The supplement to the

Opticks was a pair of mathematical tracts; the supplement to the *Traité de la lumière* was a "Discours sur la cause de la pesanteur," an argument against Newton's theory of universal gravity.[35]

Newton, it must be remembered, held Huygens in very high esteem. It was to Huygens that Newton applied the greatest compliment he could imagine, calling him "summus."[36] Newton even said that he had invented the name "vis centripeta" or "centripetal force" in honor of Huygens.[37] Thus, in following the example of Huygens, Newton was imitating not just some ordinary contemporary scientist but rather the scientist whom he esteemed in the highest degree.

The Two Forms of the Title Page of Huygens's Treatise on Light

The omission of Huygens's name from the title page of his book differs in one very important respect from the treatment of the title page of the *Opticks*. A number of copies of the *Traité* issued by the publisher have the name of the author spelled out in full on the title page.

One such copy, formerly in the collection of the late Harrison D. Horblit,[38] is now in the library of the University of Natal, Durban, South Africa. This copy of the *Traité* has an interesting provenance: it was formerly in the personal library of the nineteenth-century astronomer and philosopher of science, Sir William Herschel, as indicated by a rubber stamp on the title page.[39] A second copy, in the Houghton Library, Harvard University, was a gift from David P. Wheatland, known for his pioneering work in the collection and preservation of old scientific instruments. In both of these copies, the title-page, printed in the same two colors as the usual ones, bears the full name of the author, "Par Monsieur Christian Huygens, Seigneur de Zeelhem." In both copies, the title page does not appear to be a cancel, pasted on the stub of the usual title page. (See figure 2.4.) The OCLC lists a total of fourteen libraries that hold a copy of the *Traité* with the author's name spelled in full.[40] In none of them does the title page appear to be a cancel.

These copies of the *Traité* are all printed on ordinary paper, of the same size as most other copies, about 19.7 × 15.9 centimeters. Huygens also had some special copies printed in a large-paper edition. In these, the type size and format of the printed text is the same as in the ordinary copies. All the ones I have seen have the author's initials on the title page rather than the full name. These copies have very large margins, producing a page size of about 25.4 × 21.6 centimeters. Huygens gave such a

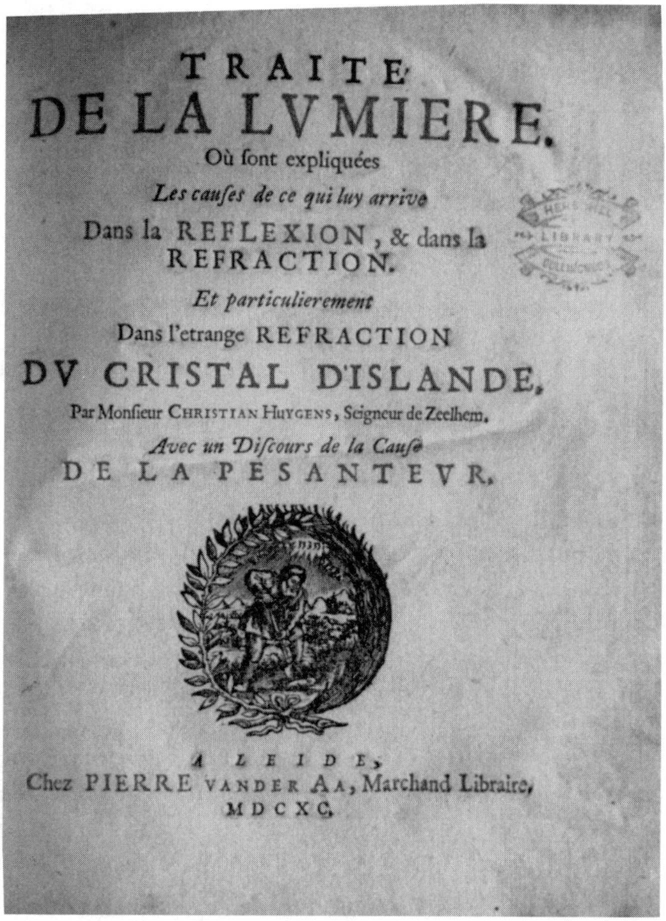

TRAITE
DE LA LVMIERE.
Où sont expliquées
Les causes de ce qui luy arrive
Dans la REFLEXION, & dans la
REFRACTION.
Et particulierement
Dans l'etrange REFRACTION
DV CRISTAL D'ISLANDE,
Par Monsieur CHRISTIAN HUYGENS, Seigneur de Zeelhem.
Avec un Discours de la Cause
DE LA PESANTEVR.

A LEIDE,
Chez PIERRE VANDER AA, Marchand Libraire,
MDCXC.

Figure 2.4
Title page of Huygens's *Traité de la lumière* (1690), the cancellandum with Huygens's name spelled out in full; from the copy in the University of Natal, Durban, South Africa. The words printed in red in the ordinary copies are also printed here in red.

copy, in the large-paper format, to Newton (and a similar large-paper copy to the philosopher John Locke).[41]

The fact that all the copies of Huygens's *Traité* with his name spelled out in full have a title page that is not a cancel but is part of the first gathering provides evidence that the title page was altered just after the printing had been begun. In this feature, Huygens's *Traité* differs from Newton's *Opticks*. There is no doubt, however, that any reader of the *Traité* familiar with the subject matter or with science in general would have had little or no doubt about the author's identity. Additionally, the initials "C. H." would give a clue to the name of the celebrated author, just as the initials "I. N." would give a positive identification to the author of the *Opticks*. In the case of Huygens's *Traité*, however, there might have been some uncertainty about the subsequent "D. Z." In any event, a little detective work could readily identify the person who had given lectures on this topic some years earlier at the Paris Academy of Sciences.

Although it is of bibliographic interest to note that some copies of the *Traité* bear the author's name and not just his initials, this has no bearing on whether Newton was following Huygens's example in presenting his *Opticks* with the author identified by his intials rather than by his name, since the two copies of the *Traité* in Newton's library did not contain the author's full name. He seems to have had no doubt, however, concerning the identity of C. H. D. Z.

ANOTHER BOOK OF NEWTON'S WITHOUT THE AUTHOR'S NAME ON THE TITLE PAGE

The *Opticks* was not the only book of Newton's to be printed without the author's name. The first edition of the *Arithmetica Universalis* (Cambridge 1707), said by D. T. Whiteside to be Newton's "most often read and republished mathematical work," was similarly issued with no author listed on the title page. Apparently, this omission resulted from Newton's wish to dissociate himself from final approval of this publishing venture. The case was somewhat similar to, but far from identical to, Newton's posture with regard to the *Opticks*.

The text of the *Arithmetica* had been deposited in the University Library in fulfillment of the requirements of the Lucasian Professorship.[42] There is no real evidence that this text, apparently deposited in the library in 1683, actually corresponded to his professorial lectures as read. In any event, Newton was aware that this work needed a competent and creative editor to transform it into a truly acceptable book.

At that time, Newton was established in London and his chair of Lucasian Professor was occupied by William Whiston. Whiston undertook to produce the edition of *Arithmetica Universalis*, perhaps with some urging by Richard Bentley, but evidently not wholly with Newton's approval.

The degree of Newton's original acquiescence to this publication and the degree of his involvement is far from certain. A memorandum by David Gregory declares that Newton did not even remember exactly what was in the text he had deposited in "the public Library according to the Statute [of his professorship].[43] According to Gregory, furthermore, in the coming summer, "He [Newton] intends to goe down to Cambridge . . . & see it, and if it doe not please him, to buy up the Coppyes." When Newton finally did see the work in print, he was disappointed, but it was apparently too late (or too costly) to purchase all of the copies before they could be distributed. Had Newton been successful, this would have been an aborted edition in a category similar to that of the aborted edition of the optical papers.

There were a number of reasons for Newton's dissatisfaction. Very likely, since this was an early work—dating from some time in the 1680s or earlier—Newton may have wished to introduce some revisions.[44] Additionally, as we may learn from his personal marked-up copy, Newton objected to a number of features of this edition, as follows. First of all, Newton found the running heads introduced by Whiston to be inappropriate and so, in his own copy, he introduced substitutes. On page 279 there was a half title that did not belong there; in his own copy, Newton marked this half title for deletion. Additionally, he indicated in his own copy what Whiteside has called "a large-scale reordering of the seventy-one geometrical problems" (comprising the "central portion"), "seeking to grade them into a more logical sequence and in increasing levels of difficulty." In the "concluding section on the 'curvilinear' construction of equations he pared away all not directly needed flesh reducing it to two skeletal conchoidal neuses," now "denuded of their proof."[45] Newton apparently did not approve of Whiston's insertion into the volume of an essay by Halley giving his own "method of finding the roots of equations arithmetically." This would be deleted in any future edition.

Whiteside concludes, "This savage butchery apart, all these improvements were incorporated in the Latin revise" that Newton "himself brought to publication in 1722." Newton also ordered the deletion of Halley's essay. This edition (London 1722) resembled the earlier one in that it too did not bear the author's name on the title page or anywhere

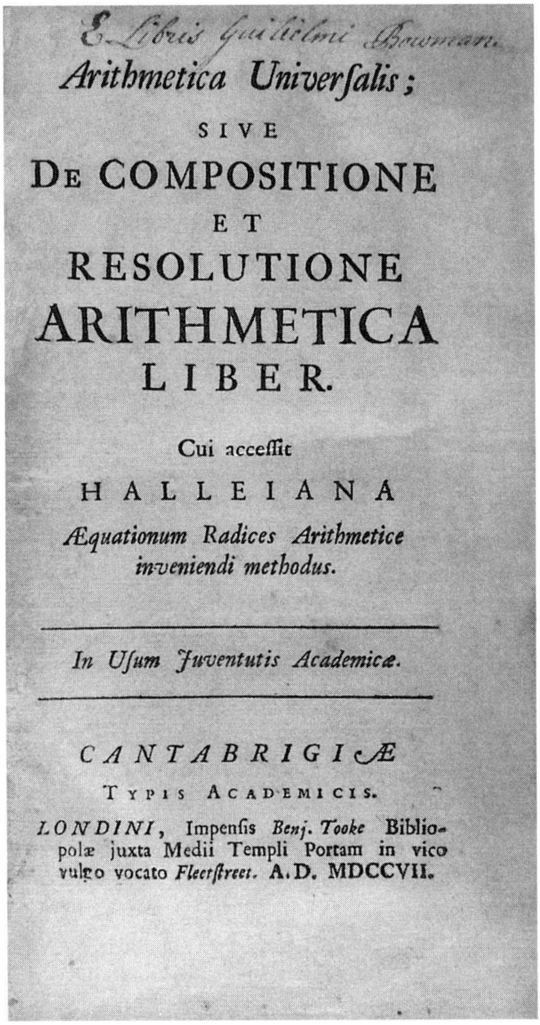

E Libris Gulielmi Bowman.

Arithmetica Univerſalis;

SIVE

DE COMPOSITIONE

ET

RESOLUTIONE

ARITHMETICA

LIBER.

Cui acceſſit

HALLEIANA

Æquationum Radices Arithmetice inveniendi methodus.

In Uſum Juventutis Academicæ.

CANTABRIGIÆ

TYPIS ACADEMICIS.

LONDINI, Impenſis *Benj. Tooke* Bibliopolæ juxta Medii Templi Portam in vico vulgo vocato *Fleetſtreet.* A.D. MDCCVII.

Figure 2.5
Title page of Newton's *Arithmetica Universalis* (1707), from the copy in the Grace K. Babson collection of Newtoniana in the Burndy Library, Massachusetts Institute of Technology.

else. In both editions, however, the editorial introduction would leave no doubt concerning the identity of the illustrious author. Furthermore, Newton took care, as Whiteside has said, "not to let knowledge of his participation [in the second edition] circulate beyond a small privy group."[46]

Clearly, this first edition of the *Arithmetica Universalis* did not present Newton's mathematics in a manner that met his standards. His final compromise was to follow the pattern established by the 1704 *Opticks* and have the work printed without his name on the title page. (See figure 2.5.)

The similarity between the *Arithmetica Universalis* and the 1704 *Opticks* lies in the fact that both volumes appeared without an author's name on the title page. But the *Arithmetica Universalis* went a step further toward disownment by the author than did the 1704 *Opticks*. The latter identified the author by the initials "I. N.," just as the *Traité de la lumière* identified its author by the initials "C. H. D. Z." But the *Arithmetica Universalis* nowhere contained either the name or the initials of Isaac Newton. Even so, his authorship was apparently "common knowledge from the day of publication."[47] Although the reasons are very different in the two cases, the absence of the author's name on the title page in both is a sign of the author's not having been fully satisfied with the contents of these works.

A Title Page of the *Opticks* Containing the Author's Name

There is yet another parallelism between the lack of author's name on the title page of Newton's *Opticks* and of Huygens's *Traité de la lumière*. One existing copy of Newton's *Opticks* has a title page bearing his name in full, just as some copies of Huygens's *Traité* have his name in full. But there is a difference between these two cases because, as we shall see, this special title page of Newton's *Opticks* is a cancel, whereas, as noted above, those carrying Huygens's full name do not appear to be.

This special copy of the *Opticks* is in the library of the British Optical Association, now part of the library of the British College of Optometrists in London. An examination of this special copy shows that, apart from the title page, it is identical in text to other copies. That is, it has the cancel of the leaf corresponding to pages 123/124. The only difference is in the title page.

This cancel title page was apparently produced very soon after the *Opticks* had been printed, that is, before the type had been distributed. This is evident from the fact that this title page uses the same type in the

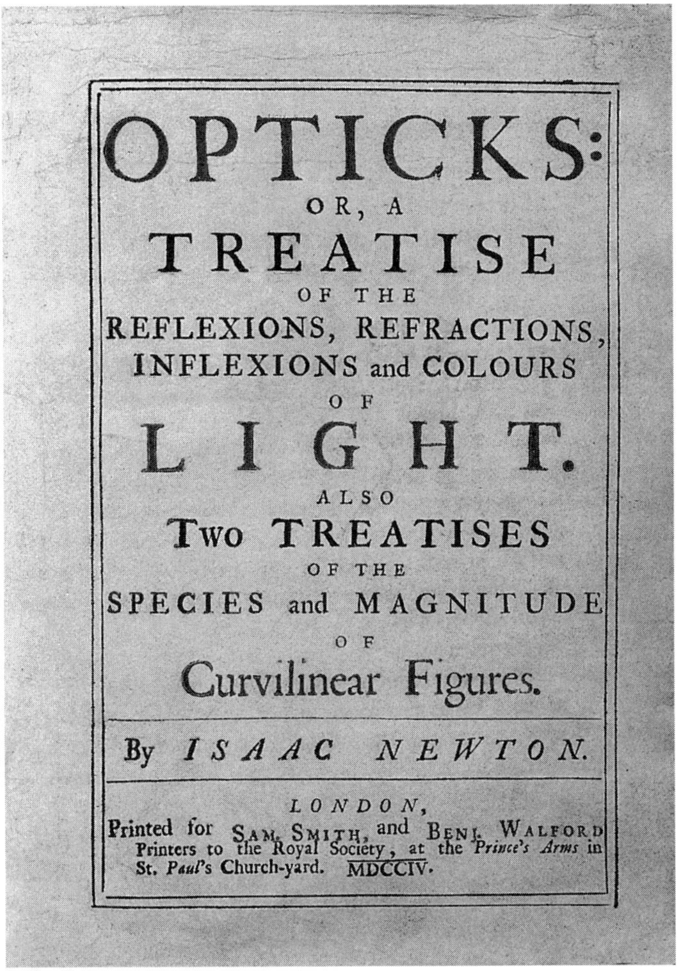

OPTICKS:

OR, A

TREATISE

OF THE

REFLEXIONS, REFRACTIONS, INFLEXIONS and COLOURS

OF

LIGHT.

ALSO

Two TREATISES

OF THE

SPECIES and MAGNITUDE

OF

Curvilinear Figures.

By *ISAAC NEWTON.*

LONDON,

Printed for Sam. Smith, and Benj. Walford Printers to the Royal Society, at the *Prince's Arms* in St. *Paul's* Church-yard. MDCCIV.

Figure 2.6
Title page of Newton's *Opticks* (1704), from the copy in the library of the British Optical Association (library of the British College of Optometrists), London. Unlike the title page of ordinary copies of the 1704 *Opticks*, this one is printed entirely in black.

same arrangement as in the normal title page, but with one difference. That is, the title page was not completely reset from scratch by the printer. Rather, the original typesetting was used with a repositioning of some of the lines of type. To make room for the author's name on this special title page (see figure 2.6), some lines of type of the usual title page had to be squeezed together, making space for the new line bearing the name of the author, Isaac Newton. As has been mentioned, this special title page is a cancel, pasted on the stub of the original. Evidently, the printer (or someone else) decided that there should be at least one copy of record, identifying the author by giving his name in full.

The existence of this copy of record provides evidence that the omission of Newton's name from the title page of the *Opticks* was not an accident that passed unnoticed. This omission did not arise from ignorance of the author's name on the part of the printer or publisher. Clearly, it was recognized that the *Opticks* lacked the author's name and for this reason at least one "corrected" title page was produced. The conscious omission of the author's name was recognized and "corrected," as we have just seen, soon after the publication of the book. Had the omission of the author's name been either the result of accident or of ignorance, then the error would have been corrected by printing a cancel of the "corrected" title page, which would have appeared in many, if not all, copies. We may assume that this special title page was made for record purposes and not for general distribution, since it is printed wholly in black ink and not in the black and red of ordinary copies.

In the case of the Latin edition (1706) of the *Opticks* and of the second edition of the *Principia*, as has been mentioned above, we can see that a fault was corrected after the text had been printed. No one has ever found a copy of the second edition of the *Principia* with the original printed and uncorrected text, probably because the fault was discovered before the whole treatise had been printed and offered for sale. In the case of the Latin version of the *Opticks,* however, the fault was apparently corrected only after the whole edition had been printed and partially distributed, with the result that quite a few copies contain the cancellandum rather than the cancel. As we have seen, the altered title page of the special copy of the *Opticks* was apparently printed and inserted into the volume very soon after the work had been printed.

The existence of at least one copy of the 1704 *Opticks* bearing the author's name on the title page provides firm evidence of the recognition, at the time of publication, that the author's name had been omitted. On at least this one copy, the omission was rectified. It is in any case out of all bounds of probability that Newton and the members of his circle could have been unaware that the title page lacks the author's name. Accordingly, we may be quite certain that this omission expressed the intention of the author. Someone, perhaps the printer or publisher, produced the special copy for record purposes.

The mystery is solved. Newton's omission of his name and his reference to himself only by his initials was evidently not the result of carelessness or a printer's error. Rather, this appears to have been a matter of intent and was modeled on Huygens's *Traité*, a work in many ways similar to Newton's. Additionally, the omission of the author's name, like the choice of English rather than Latin as the language of the text, would seem to be a kind of admission by Newton of the imperfect or incomplete nature of the *Opticks* and was to some degree an echo of the failure to produce a mathematical treatise or at least a treatise on optics based on mathematical principles.

Acknowledgments

I am grateful to the College of Optometrists, London, for permission to study their unique copy of the 1704 *Opticks* and especially to Ms. Jenny Taylor, the former librarian, for many kindnesses. This volume was originally in the library of the British Optical Association, donated to the College of Optometrists in 1980. This book is listed (and the title page is reproduced) in John H. Sutcliffe (ed.): *British Optical Association Library and Museum Catalogue* (London, 1932).

Notes

1. This letter is reproduced in facsimile from the original printing in I. B. Cohen and Robert E. Schofield, eds., *Isaac Newton's Papers & Letters on Natural Philosophy*, 2nd ed. (Cambridge: Harvard University Press, 1978); Newton's complete letter is printed in Herbert W. Turnbull, ed., *The Correspondence of Isaac Newton*, vol. 1 (Cambridge: Cambridge University Press, 1959). A detailed comparison of the two versions is given in *Newton's Papers & Letters*, 505ff.

2. *Opticks: or, A Treatise of the Reflexions, Refractions, Inflexions and Colours of Light* (London: Printed for Sam. Smith and Benj. Walford, 1704; facsimile reprint, Brussels: Culture et Civilisation, 1966); there is a convenient reprint of the fourth edition (New York: Dover Publications, 1952; rev. ed., 1979).

3. "Part of the ensuing Discourse about Light was written at the Desire of some Gentlemen of the Royal Society in the Year 1675, and then sent to their Secretary, and read at their Meetings."

4. Although the text of the *Opticks* is in English, these two tracts are in Latin.

5. Newton had already published a short introduction to the calculus in book 2, section 2, lemma 2 of the *Principia* (1687). In the 1704 publication, but not in the *Principia*, Newton uses the dot notation that he had invented during the 1690s.

6. The documents in the controversy are published in Cohen and Schofield, *Newton's Papers & Letters.*

7. Newton to Henry Oldenburg, 18 November 1676, in Turnbull, *Correspondence*, 2:182–3.

8. A. Rupert Hall's *All Was Light: An Introduction to Newton's Opticks* (Oxford: Clarendon Press, 1993) contains a detailed analysis of the stages of composition of the *Opticks* and a discussion of the various texts that preceded the *Opticks* as printed in 1704. A convenient account of the controversy with Hooke and others is given is R. S. Westfall: *Never at Rest* (Cambridge/London/New York: Cambridge University Press, 1980), ch. 7. Westfall gives a brief outline of the stages of composition of the *Opticks* on pp. 638–641.

Among Alan Shapiro's writings, the following (in addition to chapter 3 of the present volume) are most significant in relation to the present chapter:

"Newton's Achromatic Dispersion Law: Theoretical Background and Experimental Evidence," *Archive for History of Exact Sciences*, 21 (1979): 91–128.

"Beyond the Dating Game: Watermark Clusters and the Composition of Newton's *Opticks*," in P. M. Harmon and Alan E. Shapiro, eds., *The Investigation of Difficult Things: Essays on Newton and the History of the Exact Sciences in Honor of D. T. Whiteside* (Cambridge: Cambridge University Press, 1992), 181–227.

Fits, Passions, and Paroxysms: Physicks, Method, and Chemistry and Newton's Theories of Colored Bodies and Fits of Easy Reflection (Cambridge: Cambridge University Press, 1993), 4.1, 138–150.

9. W. G. Hiscock, ed., *David Gregory, Isaac Newton and Their Circle: Extracts from David Gregory's Memoranda 1677–1708* (Oxford: Printed for the Editor, 1937).

10. Newton, *Opticks*, Dover edition, 338; 1704 edition, 132 (second enumeration).

11. A first group of Queries was added in the Latin edition (1706); others appeared first in the English edition of 1717/18.

12. My *Franklin and Newton* (Philadelphia: The American Philosophical Society, 1956; reprint, Cambridge: Harvard University Press, 1966) presents in detail the Newtonian tradition in experimental natural philosophy. See Robert E. Schofield, *Mechanism and Materialism: British Natural Philosophy in an Age of Reason* (Princeton: Princeton University Press, 1950).

13. J. T. Desaguliers, *A Course of Experimental Philosophy*, eleventh annotation (London: printed for J. Senex ..., 1734–1744); cf. Cohen, *Franklin and Newton*, chaps. 6 and 7 (secs. 4 and 6).

14. Alan E. Shapiro, "Experiment and Mathematiucs in Newton's Theory of Color," *Physics Today* 27 (1984): 34-42; "The Evolving Structure of Newton's Theory of

White Light and Color," *Isis* 71 (1980): 211–35; Zev Bechler, "Newton's Law of Forces Which Are Inversely as the Mass: A Suggested Interpretation of His Later Efforts to Normalize a Mechanistic Model of Optical Dispersion," *Centaurus* 18 (1974): 184–222; "Newton's Search for a Mechanistic Model of Colour Dispersion," *Archive for History of Exact Sciences* 11 (1973): 1–37.

15. See Cohen and Schofield, *Newton's Papers and Letters*, 509; Turnbull, *Correspondence*, 1:389; also my *The Newtonian Revolution* (Cambridge: Cambridge University Press, 1980, 1983), 138–39.

16. On this hierarchy among the sciences, see chapter 4.

17. At the end of the first paragraph of the "Advertisement," Newton stated his wish that this work "may not be translated into another Language without my Consent." In the event, he decided that the contents should reach a wider international audience and not be restricted to the anglophone world. Accordingly, he commissioned Samuel Clarke (known today primarily for his epistolary exchange with Leibniz over the Newtonian natural philosophy) to produce a Latin translation, which was published in 1706.

18. Derek J. de Solla Price, "Newton in a Church Tower: The Discovery of an Unknown Book by Isaac Newton," *Yale University Library Gazette* 34 (1960): 124–26.

19. I. Bernard Cohen, "Versions of Isaac Newton's First Published Paper: With Remarks on the Question of Whether Newton Planned to Publish an Edition of His Early Papers on Light and Color," *Archives Internationales d'Histoire des Sciences* 11 (1958): 357–75.

20. These sheets may be dated as follows. They had to have been produced before 1703/1704, when the actual *Opticks* was being composed by the printer and publisher. It is also very likely that Newton decided to print this annotated edition once he had shifted into a publication frame of mind, that is, after 1687 and the publication of the *Principia*. Of course, it is possible that after the heat of the controversy of the 1670s, Newton might have wanted to reprint his communications with explanatory notes that would make his own position clear.

21. Newton's defense of his use of the word "body" appears in *Philosophical Transactions* no. 88, 18 November 1672, 5084–5103; cf. Cohen and Schofield, *Newton's Papers & Letters*, 116–35, esp. 118 (= p. 5086).

22. The *Principia* was the first published book that Newton had written. Earlier on, he had edited and published an edition of Bernard Varenius's *Geographia Universalis*; see Westfall, *Never at Rest*, 252; also, A. Rupert Hall, "Newton's First Book," *Archives Internationales d'Histoire des Sciences* 13 (1960): 39–61.

23. This episode in publishing history is discussed in Alexandre Koyré and I. B. Cohen, "The Case of the Missing Tanquam: Leibniz, Newton, and Clarke," *Isis* 52 (1961): 555–66.

24. On this emendation of the *Principia*, see the book length discussion by D. T. Whitseside in vol. 8 of his edition of *The Mathematical Papers of Isaac Newton* (8 vols.;

Cambridge: Cambridge University Press, 1967–1981). The results of Whiteside's research are summarized in the *Guide to the "Principia,"* accompanying the new translation of the *Principia* by I. B. Cohen and Anne Whitman (Berkeley and Los Angeles: University of California Press, 1999), chap. 7, sec. 4.

25. This surgery required a wholly new text for page 232 (the verso side of the last leaf of signature G) and pages 233–239 (occupying the first three leaves and the recto side of the fourth leaf of signature H).

26. There is no evidence concerning whether the printer discovered this fault or Newton brought it to his attention. It is not likely, however, that a printer would have proofread the whole book a second time, looking for proof errors. The *Opticks* was set in type in two separate sequences. As a result, the "Second Book" opens with a new pagination, starting from 1.

27. A note by Josiah Q. Bennett, "Mea Culpa: A Cancellation in the First Edition of Newton's *Opticks*," in *Serif* 7 (1970): 33–35, describes two copies of the 1704 *Opticks*. One has the usual cancel, but the other shows how the cancellation was done carelessly or incorrectly. In this copy, leaf R2 has been excised rather than leaf Q2 and the cancel was then attached to the stub of R2, leaving leaf Q2 "in place undisturbed."

28. Some data on copies with and without the cancelled leaf are given in Koyré and Cohen, "The Case of the Missing Tanquam." The copy of the *Optice* with both the cancel and the cancellandum is in the University Library, Cambridge.

29. Probably they would have been more puzzled by the statement about the lens of the eye rather than by the inversion of the paragraphs. After all, the reference to the "following" experiment makes perfect sense even if what "follows" does so after another short paragraph.

30. Christiaan Huygens, *Traité de la lumière* (Leyden: Chez Pierre vander Aa, 1690).

31. These extracts are quoted from the English translation made by Silvanus P. Thomson, *Treatise on Light, in Which Are Explained the Causes of That Which Occurs in Reflexion & in Refraction* ... (London: Macmillan, 1912).

32. Both are listed in John Harrison, *The Library of Isaac Newton* (Cambridge: Cambridge University Press, 1978), items 822, 823.

33. Huygens sent this copy to Newton via Fatio de Duillier, 24 February 1689/1690; Turnbull, *Correspondence*, vol. 3, 390.

34. Harrison, *The Library of Isaac Newton,* item 823.

35. Huygens's "Discourse on the Cause of Gravity," translated by Karen Bailey and annotated by Karen Bailey and George Smith, is scheduled to appear in a volume entitled *A Measure in Evidence: Huygens's Determination of the Strength of Surface Gravity* (in process).

36. Newton's exact phrase was "vir summus Hugenius"; Newton to Gottfried Wilhelm Leibniz, 16 October 1693, in Turnbull, *Correspondence of Isaac Newton*, 3:285.

37. U.L.C., MS Add, 3968, fol. 415*v*, transcribed in I. B. Cohen, *Introduction*, p. 296: "Mr Newton in honour of that author [Mr Hygens] retained the name [vis centrifuga] & called the contrary force vis centripeta."

38. Harrison D. Horblit, *One Hundred Books Famous in Science: Based on an Exhibition Held at the Grolier Club* (New York: The Grolier Club, 1964), no. 54.

39. The catalogue cited in n. 38 contains a full-page reproduction in color of both the original title page and the special title page with Huygens's name spelled out in full. It should be noted, however, that in reproducing the special title page, the rubber stamp indicating the previous ownership by Sir William Herschel has been eliminated.

40. These are the libraries of the University of Chicago, the University of Pennsylvania, the University of Western Ontario, the University of Edinburgh, the University of California (Berkeley), the University of Nebraska (Lincoln), Harvard University, Stanford University, the University of Iceland, the John Crearar Library (Chicago), and the library of Saint Luke's-Roosevelt Hospital Center (New York).

I have examined only two of these copies, the one formerly belonging to Harrison Horblit and now in the University of Durban (which I saw when it was still in Harrison Horblit's collection) and the one in the Houghton Library at Harvard University. Information concerning the other copies has been supplied by the rare book librarians at the various libraries holding copies.

41. This large-paper copy, inscribed by Huygens to Locke, was also part of the Harrison F. Horblit collection and was sold by H. P. Kraus of New York.

42. On the terms of this chair, and Newton's deposited lectures, see my *Introduction*, suppl. 3.

43. According to a memorandum by David Gregory (21 July 1706), Newton was "forced seemingly to allow of it, about 14 months ago, when he stood for Parliament-man at that University." Furthermore, "he has not seen a sheet of it, nor knows ... how many sheets it will make, nor does he well remember the contents of it." See Hiscock, *David Gregory, Isaac Newton and Their Circle*, p. 36.

44. For details, see the commentary by D. T. Whiteside in *Mathematical Papers of Isaac Newton*, 5:14.

45. The provenance of this copy of the *Arithmetica* is given in Harrison, *The Library of Isaac Newton*, item 1254. It ended up in the possession of William John Greenstreet (1865–1950) and was offered for sale by Bernard Quaritch (Catalogue 1164). In 1998, this copy was offered for sale by an American bookseller.

46. According to Westfall's *Never at Rest*, 649, Newton's reaction to this edition was a case of "finding faults for the sake of finding faults." It is hard to understand why Westfall belittled Newton's complaints. Why would Newton have refused to have his name appear anywhere in the volume if there were no real grounds for his objections?

47. Ibid.

3

Newton's Experiments on Diffraction and the Delayed Publication of the *Opticks*

Alan E. Shapiro

When Newton was deciding how to complete the *Opticks* sometime around 1690, he planned to end it with an account of diffraction or, as he called it, inflexion. This required launching a serious experimental investigation of the phenomenon, for ever since he first took note of diffraction in 1675, he had offered only vague, qualitative, and often physically impossible descriptions together with some speculations on its cause. He soon undertook a series of experiments in the hope of uncovering quantitative laws governing the phenomenon and evidence for the existence of short-range forces between light and matter. When Newton wrote the *Principia* just a few years earlier his knowledge of diffraction was still based on casual observations, but in the third book of the *Opticks*, which contains his experimental investigation of diffraction, we can see that he had gained some comprehension of its vagaries. Yet it is also clear that he was unable either to develop a coherent physical model or to deduce a series of physical principles or phenomenological laws as he had done in the rest of the *Opticks*. It is not that Newton did not try, and indeed for a few months he thought that he had succeeded. Some time around the fall of 1691 he had put the finishing touches on the *Opticks* and written up his experimental research on diffraction as Book IV, Part II, but by February 1692 he had removed those pages from what he had just a few months earlier judged to be the completed manuscript of the *Opticks*.[1] Newton's inability to conclude the part on diffraction satisfactorily was a principal reason that the publication of the *Opticks* was delayed for twelve years, until 1704.

The surviving papers for the *Opticks* allow us to retrace his investigations from 1690 to 1692 and see what caused the problem. And the problem is clear: Newton had developed a model of diffraction that assumed that the paths of the fringes were identical to, or coincided with, the rectilinear paths of the rays that produced them. When he discovered that this could not possibly be true, he recognized that he had to all but start over in his effort to uncover the laws underlying diffraction. At this stage of his life—the end of his active scientific career, as it turned out—he was either unwilling or unable to commence such a task, and the *Opticks* languished incomplete until late 1702 or 1703, when he finally rewrote the brief section on diffraction and added the Queries. Newton's unpublished papers on diffraction also provide an unusually detailed look into the way he carried out an experimental investigation and utilized his data and calculations to deduce and reject laws.

Francisco Maria Grimaldi's discovery of diffraction—the bending of very narrow beams of light into and away from shadows, thereby appearing to violate the law of rectilinear propagation, and the formation of alternating colored and dark bands or fringes alongside and within these shadows—was published posthumously in 1665, but the work was then little known in England. It is doubtful that Newton ever read it. Newton learned of diffraction from Robert Hooke at a meeting of the Royal Society on 18 March 1675 and from Honoré Fabri's *Dialogi Physici*. Though Fabri's book was sent to Newton in 1672, it does not seem that he did anything more than glance at it until he heard Hooke's lecture. The two authors presented different aspects of the phenomenon. Following Hooke, Newton initially stressed the bending of light into the shadow and, following Fabri, the appearance of the colored fringes. Newton's first public account of diffraction in the "Hypothesis" in December 1675 was not based on his own observations and, as Stuewer and Hall have pointed out, is terribly confused.[2] He notes, for example, the existence of six colored fringes, three of which are in the shadow, when light passes by a sharp edge. Later he correctly noted the existence of fringes only outside of the shadow. He referred to diffraction a number of times before the *Principia*, but these do not exhibit any deeper understanding. In the *Principia* (Book I, Proposition 96, Scholium) he tells us that he made his own observations, but his comments are mainly directed to justifying the analogy between various optical phenomena and the motion of light corpuscles subject to forces acting at a distance, and not to elucidating diffraction itself.

Newton began to compose the *Opticks* in about 1687, and he completed the portions on color and much of that on the colors of thin films by around 1690.[3] Up to this time, and even when he began to sketch his new part on diffraction, he still did not adequately describe the most basic phenomena. For example, in the "Fundamentum Opticae," which is really the first draft of Book I of the *Opticks* when Newton was planning to write it in Latin, he has fringes produced by a knife edge within the shadow, as can be seen in his drawing (figure 3.1). And in his

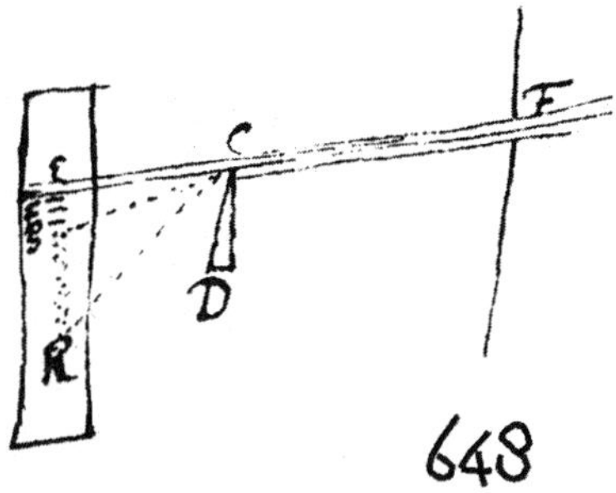

Figure 3.1
Diffraction by a knife edge; from the "Fundamentum opticae," Add. 3970, f. 648. (By permission of the Syndics of Cambridge University Library.)

first attempt at the section on diffraction for the *Opticks*, he still drew six fringes, although the text mentions only three; and in the following observation he explicitly put the fringes in the shadow before deleting the passage.[4]

When Newton became serious about investigating some facet of nature, he typically set out to measure it and to subject it to laws (ideally mathematical). This is precisely what he did with diffraction. The earliest drafts for the *Opticks* reflect genuine observations and experiments, but there was no serious attempt at measurements. Newton clearly intended to make them, for he left gaps to insert numbers. As real numbers appear, so does a firmer grasp of the phenomenon.

Newton first attempted to write the part on diffraction after he had completed what were then Books I and II with his theory of color and had made his initial revision of his observations on the colors of thin plates which he had submitted to the Royal Society in 1675, and which then formed the three parts of Book III. The new part on diffraction was immediately to follow the latter, but Newton discovered an entirely new phenomenon, the colors of thick plates, and launched an investigation of it. As his work on thick plates and diffraction seemed to be proceeding fruitfully, Newton decided to put them together in the last book of the *Opticks*. The part on thick plates formed Book IV, Part I, "Observations Concerning the Reflexions & Colours of Thick Transparent Polished Plates," and that on diffraction Part II.[5] I estimate that these investigations occupied Newton from the beginning of 1691 through the summer or fall of that year, when he completed Book IV, although it is possible that he began this work a few months earlier. When Newton first considered concluding the *Opticks* with diffraction, all of his accounts of it involved a

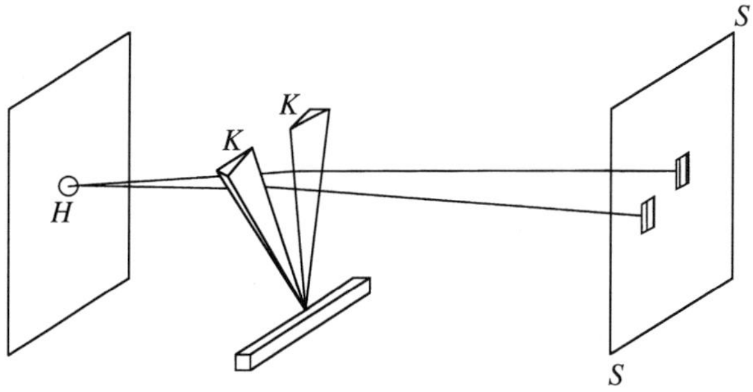

Figure 3.2
A schematic diagram of Newton's experiment to observe the diffraction pattern produced by two knife blades *K*, *K* making a small angle with one another to form a V.

single experiment (described in Hooke's lecture), that of a narrow beam of light passing by a knife edge. In his first draft on diffraction he at last added an original experiment on the fringes formed by two parallel knife edges, and he concluded with an innovation that allowed him to determine the distance at which the rays forming each fringe pass the knife edge.

Newton took two knives with straight edges *K* and stuck their points in a board so that the blades formed a V making an angle of about half a degree (figure 3.2). He placed the knives in a narrow beam of sunlight that passed through a hole *H* with a diameter of 1/42 inch, and he observed a pattern of fringes on a screen *S*. Three fringes parallel to each knife edge appeared (as in figure 3.3 from the published *Opticks*), and as the blades came closer to one another the fringes intersected. (Newton eventually concluded that the fringes formed hyperbolas.) By measuring the height at which the rays passed between the knives, the space between the blades was easily determined; and for rays that passed through the middle of the slit, the distance of the rays from the edge was simply half that space. Newton argued that when each of the three fringes intersect, the rays pass through the center of the slit. By measuring the heights at which various colors of various fringes intersected, he was able to come up with a law: "the distances of the edge of the knife from y^e rays wch go to y^e same colours of y^e several fasciae seem to be in arithmetical proportion of y^e number[s] 3 2 & 1."[6] If we let *s* be the distance of the rays from the edge of the diffracting body, then we can represent this simple, linear law as

$$s \propto 1, 2, 3.$$

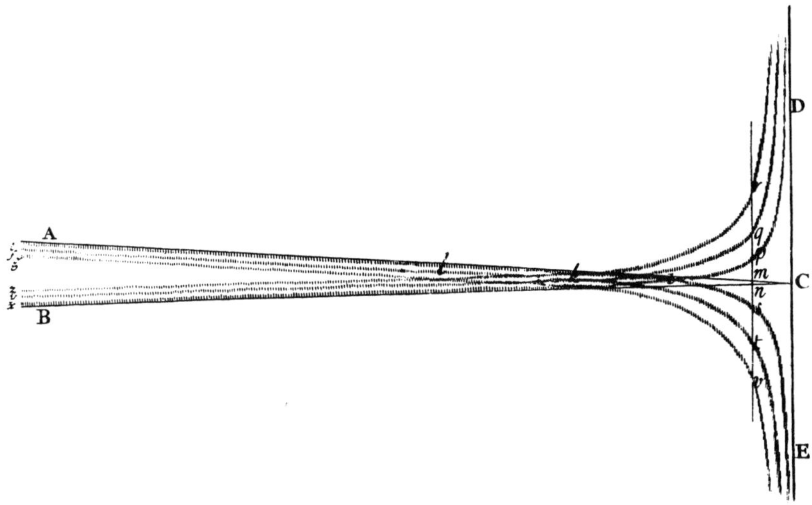

Figure 3.3
The diffraction fringes formed by light passing through two knives whose edges form a V. The solid line represents the shadow of the knife edges. From the *Opticks*, Book III, Figure 3.

Many of the essential features of Newton's work on diffraction for the *Opticks* can be seen here. The first, and most fundamental for the suppressed version, is the assumption that the paths of the rays and the fringes are identical, rectilinear paths. His experimental rejection of this assumption would eventually bring his work on diffraction to a halt. The second is the replacement of his earlier qualitative descriptions with measurements, and third is his quest to describe the phenomenon mathematically. Even though Newton left gaps in his manuscript for his measurements, I have no doubt that he had already made them and that he later intended to enter refined ones. Still more refined, different measurements would soon lead him to adopt a new law and abandon the simple linear law. The quest for mathematical laws underlying phenomena—and their continual refinement—is characteristic of all Newton's optical work. Finally, we can see that, as he had suggested in the *Principia*, he assumes that diffraction is an instance of action at a distance. In a somewhat later draft he formulated this assumption as a proposition:

> The surfaces of bodies act regularly upon the rays of Light at a distance on both sides so as to inflect them variously according to their several degrees of refrangibility, & distances from the surface.

> The truth of this Proposition will appear by the following observations.[7]

I do not intend to examine here Newton's attempt to find a physical explanation for diffraction, but I can indicate the problems that continually frustrated him, namely, explaining what caused dark bands to appear between the bright ones and what happened to the light from those dark bands. To explain the deflection of the light into or away from the shadow was in principle a simple matter for Newton. Before the *Principia* he invoked variations in the density of the aether, which would cause the light rays to change direction or bend, and in the *Principia* and afterward he replaced the aether by attractive and repulsive forces. To explain the alternating bright and dark fringes, it was natural for Newton to attempt to relate them to the similar alternating bright and dark rings he had observed in thin films and the vibrations or fits that he used to explain them. He suggested this mechanism in 1675 in his first discussion of diffraction in his "An Hypothesis Explaining the Properties of Light," and again in about 1690 in drafts of the "Fourth Book," which are early sketches for the conclusion of the *Opticks*.[8] He soon abandoned this idea, presumably because he never had any evidence that periodicity was involved in diffraction, and also because once he had formally specified the properties of the fits to explain the colors of thin and thick plates, they were incompatible with diffraction. For example, the fits acted in the direction of propagation, whereas for diffraction he needed a cause that would act transverse to that direction. Newton simply could not explain the cause of the dark fringes and what happened to that light. For a while he thought that the rays from the dark bands are inflected into the shadow "so as faintly to illuminate all y[e] dark space behind the body," but he had to abandon that idea too.[9] In the *Opticks* all he could do was lamely ask whether the rays are not "in passing by the edges and sides of Bodies, bent several times backwards and forwards, with a motion like that of an Eel? And do not the three fringes of coloured Light above-mentioned, arise from three such bendings?"[10] That Newton had no physical explanation for the cause of diffraction did not hinder him from pursuing his investigation of diffraction, for he never required a causal explanation of phenomena whose laws he had described.

Returning to Newton's experimental investigations, we see that in his next draft he began a new, promising series of observations of the diffraction pattern formed by a hair. When he illuminated a hair with a

Table 3.1

	At the distance of	
	half a foot	nine feet
1. The breadth of y^e^ shadow	$\frac{1}{54}$ digit.	$\frac{10}{85} = \frac{2}{17}$ $\frac{1}{8\frac{1}{2}}$
2. The breadth between y^e^ brightest yellow of y^e^ innermost fasciae on either side the shadow.	$\frac{1}{40}$ $\frac{1}{39}$	$\frac{10}{67}$ $\frac{7}{50}$ $\frac{3}{20}$
3. The breadth between y^e^ brightest parts of y^e^ middlemost fasciae on either side y^e^ shadow.	$\frac{1}{23\frac{1}{3}}$ $\frac{1}{23}$	$\frac{4}{17}$
4. The breadth between y^e^ brightest parts of the outmost fasciae on either side y^e^ shadow	$\frac{1}{18}$	$\frac{10}{33\frac{1}{3}} = \frac{3}{10}$
5. The breadth of y^e^ luminous part (green yellow & red) of y^e^ first fascia	$\frac{1}{180}$	$\frac{1}{35}$
6. The distance between y^e^ luminous parts of y^e^ first & 2^d^ fascia	$\frac{1}{240}$	$\frac{1}{42}$
7. The breadth of y^e^ luminous part of the 2^d^ fascia	$\frac{1}{260}$	$\frac{1}{52}$
8. The distance between y^e^ luminous parts of y^e^ second & 3^d^ fascia	$\frac{1}{360}$	$\frac{1}{63}$
9. The breadth of y^e^ luminous part of y^e^ 3^d^ fascia	$\frac{1}{420}$	$\frac{1}{80}$

narrow beam of light admitted through a small hole about twelve feet away, he observed three colored fringes on each side of a central shadow that was much larger than it should be according to the laws of geometric optics. This observation, incidentally, marks Newton's recognition that shadows are larger than they should be, for previously (following Hooke) he had emphasized the bending of light into the shadow. After a cursory description of the fringes, Newton abruptly ended this draft (Add. 3970, ff. 375–76) and undertook systematic observation and measurement of the fringes. He then expounded the results of his investigation in another draft devoted primarily to diffraction by a hair (Add. 3970, ff. 377–78, 328–29). He found the outermost fringes "so faint as not to be easily visible." The colors of all the fringes were difficult to distinguish unless he let them fall obliquely on to a screen so that they were much broader than when they fell perpendicularly. Table 3.1 from Newton's manuscript gives his measurements of the fringes in fractions of an inch when the screen was placed half a foot and nine feet from the hair.[11]

Let me first note that Newton switched from the term "fascia" to "fringe" (which is still used in optics) only during revision for publication in about 1703. Second, he was not directly able to measure 1/360 or 1/420 of an inch, as the table seems to imply. By letting the fringes fall obliquely on a ruler or scale that was divided into sixteenths of an inch, he observed magnitudes that were actually twelve times larger than the values in the table, that is, he was able to observe an interval of about one-half of one-sixteenth of an inch.[12] The multiple values for the entries in lines 2 and 3 indicate Newton's inability to distinguish between them in this early draft. The differences vary from 0.75% to 7%, but even in the table in the published *Opticks* (see table 3.3) Newton still had multiple values. Since the largest difference is 2.5%, we can take this as a reasonable estimate of the smallest difference that Newton could distinguish by his measurements.

From these measurements Newton "gathered" that, "y^{e} rays in the most luminous part of the first fascia passed by the hair at y^{e} distance of 0.00767 = 1/130 part of an inch & in their passage by it were bent & turned outwards from y^{e} shadow so as to contein with the line produced in wch they came from the hole to y^{e} hair, an angle of ^{min}."[13] Values were also given for the rays of the next two fringes as well as for the edge of the shadow, and in each case he left a space for the angle to be inserted. From the phrasing ("ye rays in the most luminous part of the first fascia passed by") we can see that Newton has assumed the identity of the paths of the rays and the fringes. This will become perfectly evident from his calculation of these distances and angles. Though Newton does not explain how he determined these values, in his papers there are a number of worksheets that contain his actual measurements and calculations.

I shall first present my reconstruction of Newton's model of diffraction and then show how he actually used this model in his calculations. He assumed (figure 3.4) that the incident rays coming from the point H at the center of the hole are bent away from the hair at a point a short distance $f/2$ away from the center of the hair. The rays then proceed in a straight line and depict a fringe of a given order. If the diffracted rays on each side of the hair are projected backward, they intersect at a point E on the axis of symmetry. Newton, in fact, believed that the rays were gradually deflected as they passed the hair and followed a curved path, but for the purpose of measurement and calculation he assumed the deflection to occur at a point, just as he did in all his work on refraction, where he also held that the paths are curved. It is a matter of simple geometry to calculate the distance f for each fringe (figure 3.5). Let c be the distance between the middle of the fringes on each side of the shadow

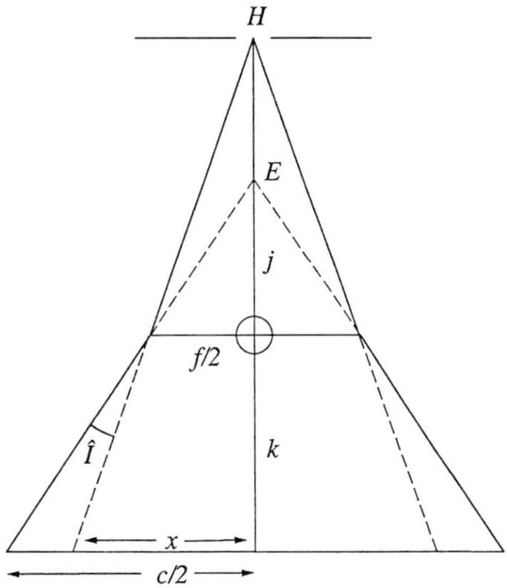

Figure 3.4
Newton's model for diffraction by a hair that assumes the paths of the rays and fringes coincide and propagate in straight lines.

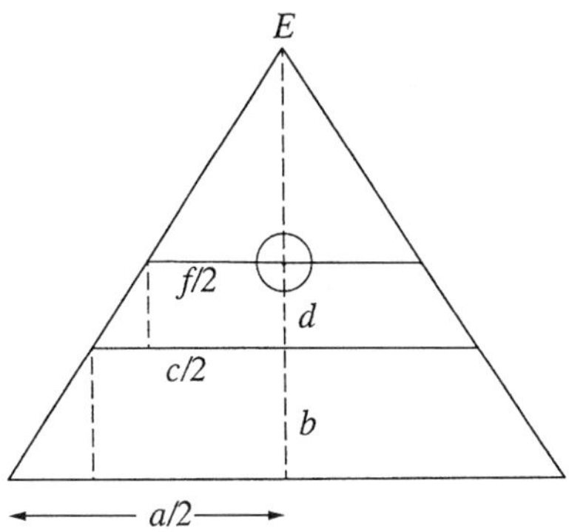

Figure 3.5
Determining the points of inflection according to Newton's linear-propagation model for diffraction by a hair.

of the hair, when it is observed on a screen at a distance d of one-half foot from the hair; and let a be the distance between the middle of the fringes when the screen is 9 foot from the hair, so that b, the distance of the first screen to the second, will be $8\frac{1}{2}$ feet. From similar triangles we have

$$\frac{\frac{1}{2}(c-f)}{\frac{1}{2}(a-c)} = \frac{d}{b},$$

which yields

$$f = c - \frac{d}{b}(a-c). \tag{3.1}$$

If we let h be the diameter of the hair, and s the distance of the point of inflection from the surface of the hair, then

$$s = \frac{f-h}{2}. \tag{3.2}$$

At the beginning of his investigation Newton determined the diameter of the hair to be 1/280 in., which is within the range of the size of a human hair, and he never altered that value. He does not explain how he measured that diameter.[14]

One of Newton's work sheets for diffraction is filled with calculations and sketches for diffraction from a hair and knife edges.[15] One series of numbers on f. 646r (6.a in figure 3.6) turns out to be his calculation of the inflection distance from the hair as related in the draft on f. 378v that I recently quoted, and we shall see that he used the proportion that I just derived to determine f and s as defined by equations (3.1) and (3.2). In table 3.2 I have transcribed Newton's sequence of numbers and added the headings and the right-hand column, which explains the steps of the calculation. Except for rounding off, these are precisely the numbers found in Newton's text. Thus, Newton built his calculation of the inflection distance on the assumption that the rays and fringes propagate along identical rectilinear paths. If any ray is intercepted anywhere along its path, it always depicts a fringe of the same order, that is, the rays do not cross or intersect one another.

Newton, we should recall, had left gaps in the same text in which to insert his calculation of the "angle of inflection," which he defines as the line between the incident ray produced and the inflected ray. His worksheets show that he used the same linear-propagation model as in the calculation of the inflection distances. To determine the angle of inflection, he first calculates the distance x from the axis where the extension of

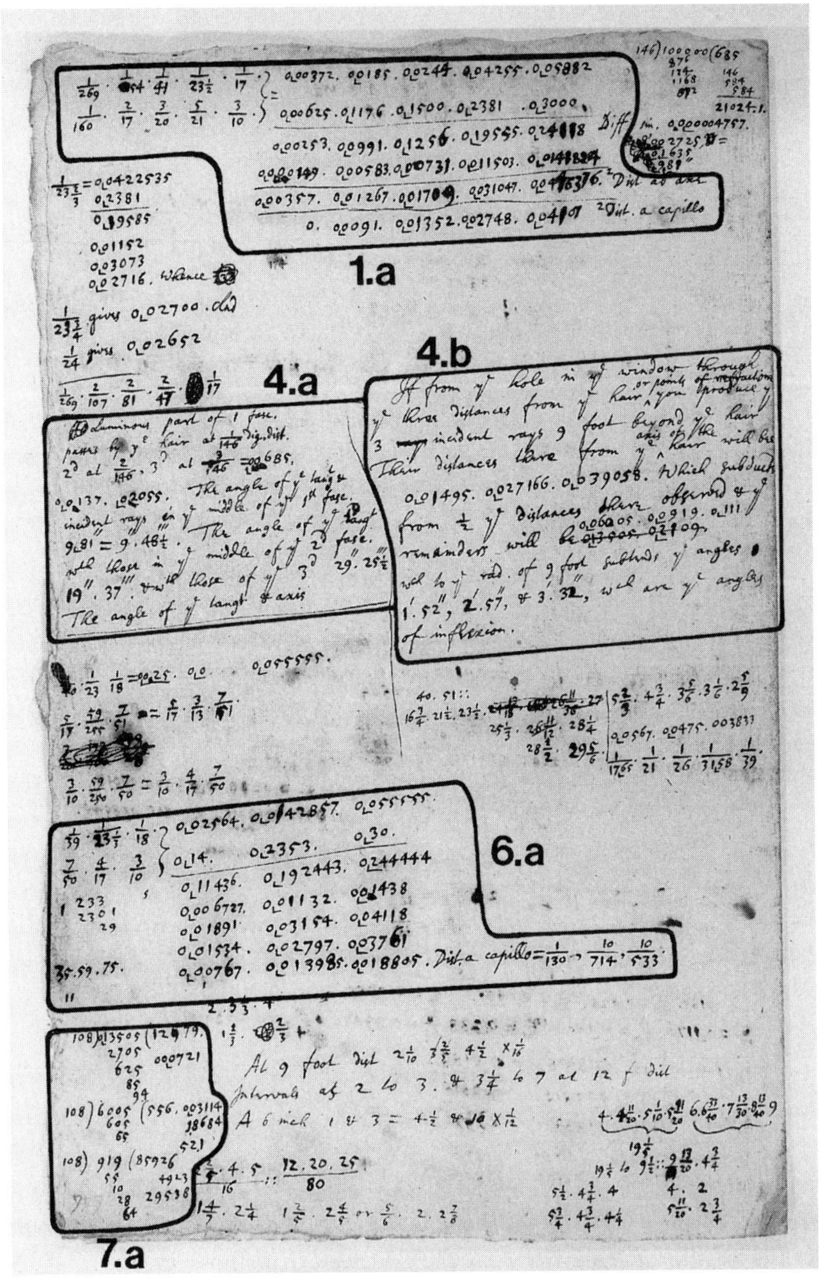

Figure 3.6
One of Newton's worksheets on diffraction; Add. 3970, f. 646r. (By permission of the Syndics of Cambridge University Library.)

Table 3.2

Distance between innermost fringes	Distance between middle fringes	Distance between outermost fringes	
$\frac{1}{39} = 0.02564$	$\frac{1}{23\frac{1}{3}} = 0.042857$	$\frac{1}{18} = 0.055555$	c
$\frac{7}{50} = 0.14$	$\frac{4}{17} = 0.2353$	$\frac{3}{10} = 0.30$	a
0.11436	0.192443	0.244444	$\mathrm{a} - \mathrm{c}$
0.006727	0.01132	0.01438	$\frac{\mathrm{d}}{\mathrm{b}}\left[= \frac{6\ \mathrm{in}}{8.5\ \mathrm{ft}} = \frac{1}{17}\right](\mathrm{a} - \mathrm{c})$
0.01891	0.03154	0.04118	$\mathrm{f} = \mathrm{c} - \frac{\mathrm{d}}{\mathrm{b}}(\mathrm{a} - \mathrm{c})$
0.01534	0.02797	0.03761	$\mathrm{f} - \mathrm{h}\left[= \frac{1}{280} = .00357\right]$
0.00767	0.013985	0.018805	$\frac{\mathrm{f} - \mathrm{h}}{2} = \mathrm{s}$
$\frac{1}{130}$	$\frac{10}{714}$	$\frac{10}{533}$	Distance from hair

the incident ray intersects the screen (see figure 3.4). From similar triangles we have

$$\frac{\frac{1}{2}([f - h] + h)}{j} = \frac{x}{j + k},$$

where j and k, the distances of the hair from the hole and screen respectively, are known. Since the distances f of the inflection point from the hair were already calculated, x is readily determined for each fringe. Newton then subtracts these distances from one-half the distance between the fringes c and calculates the angle of inflection $\hat{I}$ (in radians),

$$\hat{I} = \frac{\frac{c}{2} - x}{k}.$$

I reconstructed Newton's method of calculating the angle of inflection from two entries on his worksheets that are determined from different values of the distance between the fringes c than those in table 3.2. A passage on his worksheet (see 4.b in figure 3.6) is unusually informative for Newton actually explains the nature of the calculation,[16] while an actual calculation on another worksheet (Add. 3970, f. 357r) illuminates the procedure. I shall spare you the details of the calculation, but it further confirms my reconstruction of Newton's model. Let us now look at a physical implication of Newton's model and his experimental confirmation of it.

Newton introduced a new empirical result in the draft that I have been considering. When he observed the fringes on a screen from their first appearance at about one-quarter of an inch from the hair out to nine feet, he found that they "kept very nearely the same proportion of their bredths & intervalls wch they had at their first appearing." For example, he found that the ratio of the distance between the middle of the first fringes to that between the third fringes was as "nine to nineteen & by some of my observations neare the hair, as six to thirteen."[17] Since these two ratios differ by about 2.5%, and another one by nearly 7%, this gives us another estimate for what Newton considered to be acceptable accuracy in his work on diffraction. From his calculations and comparisons of his measurements with his laws, I have found that is about the range of what Newton considered a "reasonable" or "acceptable" result, though generally the upper bound for error is about 5% or 6%.

Was this simply an empirical law that Newton discovered, or was it a consequence of his model? I am reasonably confident it was the latter. In pondering the model, some questions naturally arise about the disposition of the sets of rays that are inflected at different distances from the hair and then proceed to depict the three fringes. Are they parallel, or do they diverge, and if they diverge, do they do so from a common point? If the last assumption is made, together with that of the rectilinear propagation of the rays and fringes, it immediately leads to the conclusion that the proportion of the fringes is constant at all distances (figure 3.7). Thus, Newton's identification of the paths of the rays and fringes was proving to be quite fruitful.

Newton's commitment to empirical confirmation becomes apparent from the vanishing of his simple linear law for the distance of the point of inflection from the diffracting edge. In the draft I have been considering he makes no statements about any law for these distances. It is evident from his worksheets, however, that he had abandoned the old law, because his improved measurements did not confirm it. Using his earliest hair measurements, he did initially adopt it, for he jotted down "Luminous part of 1 fasc. passes by y^{e} hair at 1/146 dig. dist.$_{[,]}$ 2nd at 2/146, 3^{d} at 3/146 = .00685, 0.0137, .02055."[18] If this linear law is tested against the improved values adopted in this draft, it differs by 9% for the second fringe and 18% for the third. These are well beyond what I estimated Newton considered an acceptable error in his diffraction experiments, so

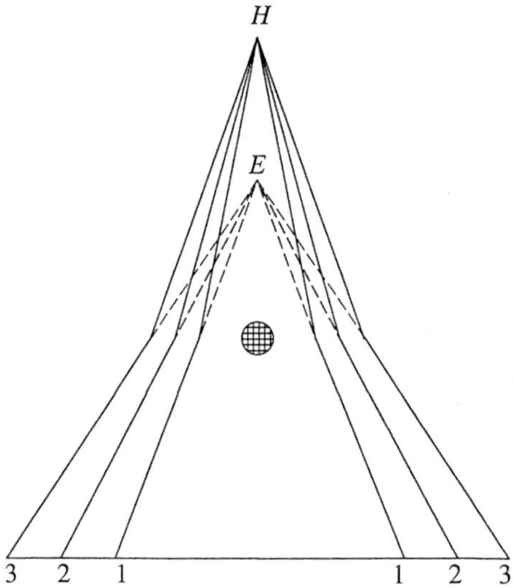

Figure 3.7
The three pairs of fringes formed in diffraction by a hair, based on Newton's assumptions that the fringes propagate in the same straight lines as the rays and all the diffracted rays and fringes diverge from a common point.

we can appreciate why he abandoned it. We should also note how quickly Newton generalized his results, both from a limited number of measurements and also from particular cases. He did not, for example, test his model with observations at some distance besides those at 6 inches and 9 feet, the two to determine the model's parameters; nor did he try another set of measurements with a hair of a different diameter. Similarly, Newton deduced the linear law from experiments with a knife edge and then extended it to the hair, so that he considered it to be a general property of diffraction. He then rejected it for both a hair and knife edges because of its predictive failure with refined measurements with a hair.

Probably not long after Newton composed the draft I have been considering he wrote out a fair copy of "Observations Concerning the Inflexions of the Rays of Light in their Passage by the Surfaces of Bodies at a Distance," then Book IV, Part II (Add. 3970, ff. 79–90) to conclude the *Opticks*. This final section of the *Opticks* was not one of its more ambitious parts. I think Newton's aim was simply to get a handle on some of the basic features of diffraction and to show that they were consistent with his physical theories, especially on color. Most of the observations are devoted either to diffraction by a hair or various arrangements of knife edges. The observations on the hair are largely a revision of the earlier

Table 3.3

	At the distance of	
	half a foot	nine feet
1. The breadth of the shaddow	$\frac{1}{54}$ dig.	$\frac{1}{9}$ dig.
2. The breadth between the middles of the brightest light of y^{e} innermost fasciae on either side y^{e} shaddow	$\frac{1}{38}$ or $\frac{1}{39}$	$\frac{7}{50}$
3. The breadth between the middles of the brightest light of the middlemost fasciae on either side y^{e} shaddow	$\frac{1}{23\frac{1}{2}}$	$\frac{4}{17}$
4. The breadth between y^{e} middles of y^{e} brightest light of y^{e} outmost fasciae on either side y^{e} shaddow	$\frac{1}{18}$ or $\frac{1}{18\frac{1}{2}}$	$\frac{3}{10}$
5. The distance between the middles of the brightest light of the first & second fascia	$\frac{1}{120}$	$\frac{1}{21}$
6. The distance between the middles of y^{e} brightest light of the second & third fascia.	$\frac{1}{170}$	$\frac{1}{31}$
7. The breadth of the luminous part (green white yellow & red) of the first fascia	$\frac{1}{170}$	$\frac{1}{32}$
8. The breadth of the darker space between the first & second fascia	$\frac{1}{240}$	$\frac{1}{45}$
9. The breadth of y^{e} luminous part of the second fascia	$\frac{1}{290}$	$\frac{1}{55}$
10. The breadth of the darker space between y^{e} 2^{d} & 3^{d} fascia	$\frac{1}{340}$	$\frac{1}{63}$

draft. Newton, however, recorded some new distances, in lines 8 and 10, and eliminated one, the breadth of the third fringe, line 9 in table 3.1. As can be seen from the new table of measurements (table 3.3), he changed a number of the values.[19]

For the distance between the middles of the three fringes (lines 2–4 in each table) the changes are not great; they are either nil or on the order of 2 to 2.5%. For the more delicate measurements in lines 7 and 9 (corresponding to lines 5 and 7 in table 3.1), the changes were larger, ranging from 5.5% to 11.5%.[20] As in the preceding draft, Newton calculated the distances of the inflection point from the hair and also the angle of inflection. On the basis of these more refined values, he was able to announce new laws for both the distances and the angles of inflection, namely, that "the squares of these distances are in the arithmetical progression of y^{e} odd numbers 1, 3, 5, wthout any sensible errors & the squares of these angles of inflexion are also in the same progression." Thus the new laws are

$$s^2 \propto 1, 3, 5, \text{ and } \hat{I}^2 \propto 1, 3, 5,$$

or

$$s \propto \sqrt{1}, \sqrt{3}, \sqrt{5}, \text{ or } \hat{I} \propto \sqrt{1}, \sqrt{3}, \sqrt{5}.$$

This new law differs significantly from the old, linear law for the distances that Newton first proposed for diffraction from a knife edge. No measurements at all survive from this earliest period, so the basis of that law must have been very casual observations. Newton's new measurements agreed with this new law with an error ranging from about 0.5% to 4%, which according to my assessment was within the limit of his tolerance.[21]

In the next observation Newton returns to the proportionality of the breadth and intervals of the fringes, and introduces a greater number of measurable proportions. The differences between the law and the measurements are mostly around 1.5%, but all remain well within his acceptable range. In an addition written in the margin of the manuscript, he added yet another law:

> the breadths of the fasciae seemed to be in the progression of the numbers 1, $\sqrt{\frac{1}{3}}$, $\sqrt{\frac{1}{5}}$, & their intervalls to be in the same progression wth them, that is the fasciae & their intervalls together to be in the continuall progression of the numbers[s] 1, $\sqrt{\frac{1}{2}}$, $\sqrt{\frac{1}{3}}$, $\sqrt{\frac{1}{4}}$, $\sqrt{\frac{1}{5}}$, or there abouts.[22]

This new law agrees remarkably well with the data—all measurements and predicted values within 1.5%—and fits the data the best of all Newton's diffraction laws. The measurements in the preceding draft, table 3.1, do not lead to this law, since they differ from it by about 5% to 17%. The values in lines 5 and 6 (corresponding to lines 6 and 8 in table 3.1) were changed by a factor of 2, and I have no doubt that the earlier values were caused by an error in reading or conversion. Almost no one, let alone Newton, makes experimental errors of that order. He does fiddle with his measurements (usually by no more than a few percent), but I do not believe that his aim here was to get better agreement with his law. Since he added the law to the manuscript after it was written out, it appears that he had not yet come upon the law when he entered these new values. Moreover, these values were not subsequently altered in the manuscript, as are many others in the table. This law does not follow from his model, as the simple proportion does, and was no doubt derived inductively. It is clear, though, that he was searching for some regularity of this sort, because of the new measurements in lines 8 and 10.

What do I mean when I say that Newton "tinkered" with his measurements? The words "fiddling" or "tinkering" were suggested to me by the manuscripts where one sees a range of numbers differing from one another by small amounts, from which Newton finally chose one. Newton, it seems, was trying out slightly different values to find which yielded better agreement with his laws or a more evident law. These are all measured values or within the range of his measured values. For example, in 1.a in figure 3.6 Newton carried out his calculation with $23\frac{1}{2}$, but to the left and below that we can see him trying out what he would get with $23\frac{2}{3}$, $23\frac{3}{4}$, and 24. He had determined a range for his measurements of the second fringe (as well as the others), as can be seen in row 3 in tables 3.1 and 3.3. Newton seems to have been playing with his numbers within what he judged to be his experimental error in order to get a better fit with his various laws.[23] It is not apparent how he finally decided which value to use. This process of massaging the data—as contemporary scientists call it—may improve agreement with the laws, but it is risky. We shall encounter an example of the risky side in Newton's renunciation of the linear-propagation model that he added to Observation 1 during the final revision of the manuscript of the *Opticks*.

It is important to recognize that in this first, and soon to be suppressed, state of the book on diffraction Newton never expounded his linear-propagation model. He left unexplained the way in which he had determined the distances and angles of inflection, s and $\hat{I}$, that appeared in his laws describing diffraction. I had to uncover their physical foundation, the linear-propagation model, from his work sheets and calculations. Newton's methodology compelled him to suppress his model, because it was a hypothesis imposed on the observations and not derived from them. He believed that hypotheses or conjectures could not be mixed with the more certain part of science, the phenomena or principles derived from the phenomena. To do so would compromise the quest for a more certain science.[24] Thus, the linear-propagation model lay hidden in the background, underlying his data and guiding his search for the laws governing diffraction.

Having completed his account of diffraction by a hair, Newton then turned to diffraction by knife blades. If we recall, in his earliest observations on diffraction he found that the edge of a single knife forms three fringes, and that when he placed the knives so that they formed a V (figure

3.2), he found two sets of intersecting fringes (figure 3.3). While he was carrying out his investigation of diffraction by a hair, to which I have devoted so much attention, he also continued his study of diffraction from knife edges. In Observation 8 he more carefully repeated the experiment in which two knife edges formed a V. They now made an angle of 1° 54′, and they were placed 10 or 15 feet from the hole, which was 1/42 inches in diameter. Let me first present a brief overview of Newton's description of the fringes formed by the V-edges, in which I shall identify the paths of the fringes and rays, just as he did. The fringes are observed on a screen (a white ruler) that is gradually moved farther away from the two knives. When the incident rays pass each edge, they are bent away from it and form three fringes parallel to that edge; as they move farther away from the two knives, the two sets of fringes/rays cross one another. Moreover, the rays that pass through the V closer to the intersection of the blades, that is, where the edges are closer to one another, cross one another nearer to the two knives than those higher up. Thus, if the screen were placed farther from the knives then they were in Newton's figure, the intersections of the fringes would shift to the left in his figure.

Now let us turn to Newton's own account. When he let the light fall on a white ruler about an inch beyond the knife blades, he

> saw the fasciae made by ye two edges run along the edges of the shadows [of the knives] in lines parallel to those edges without growing sensibly broader, till they met in angles equal to the angle of the edges, & where they met & joyned they ended without crossing one another. But if the ruler was held at a much greater distance from ye paper they became something broader as they approached one another & after they met they crossed & then became much broader then before.

From these observations he concluded:

> Whence I gather that the distances at wch the fasciae pass by ye knives are not encreased nor altered by ye approach of the knives but the angles in wch ye rays are there bent are much increased by that approach. And that ye knife wch is nearest any ray determins wch way the ray shall be bent & ye other knife increases the bent.[25]

The first conclusion follows from the fact that the fringes are parallel to the knife edges. I cannot resist noting that the second conclusion, which also satisfies observation, is physically odd, for it implies that the edge closer to the rays bends them away, while the opposite one "attracts" them.

In Observation 9 Newton related that at 1/3 of an inch from the knives, he found that the shadows between the first two fringes crossed one another 1/5 of an inch above the intersection of the blades. He readily calculated that the distance between the blades was 1/160 of an inch, so that the "dark intervalls of the first & second fascia meeting in y^e middle of this distance are in their passage beetween the knives distant from their edges $1/320^{th}$ part of an inch." Now he invokes the "law" of Observation 3—which was derived from measurements on the *hair* and assumed identical rectilinear paths of propagation for the rays and fringes—that the "the bright fascias & their intervalls pass by y^e edge of y^e knife at distances wch are in y^e progression of y^e numbers $\sqrt{1}.\sqrt{2}.\sqrt{3}$ &c."[26] With this law he calculated that the three bright fringes passed by the edge of the *knife* at 1/452, 1/261, and 1/202 of an inch. This implies that the light from the second and third fringes comes from the half of the light closer to the opposite edge, which seems to contradict his observation that the light along the closer edge produces the fringes. Another hidden implication—and this would be the fatal one—is that if he had made his measurement at any other position, he would have found that the light forming each of the fringes passes the edges at different distances. Newton, however, was for the time being content with these results, which form the first completed state of the *Opticks*, and he then proceeded to demonstrate that the fringes form hyperbolas.

Sometime between about the fall of 1691 and February 1692 Newton became dissatisfied with his newly completed Book IV and dismembered it. He revised Part I on the colors of thick plates and made it Part IV of the preceding book; he also devised the theory of fits and added a number of propositions on fits at the end of Book III, Part III. He had, however, become sufficiently disheartened with Part II on diffraction that he removed it from the manuscript of the *Opticks* without adding a revised replacement. By comparing the suppressed version with the published one that Newton wrote around 1703, it is apparent that a new set of observations that he had made on the V-knives had undermined his model for diffraction and forced him to recognize that the paths of the rays and the fringes are distinct. The experiment dates from the period around February 1692, when he removed the just completed part on diffraction from the *Opticks*, for it is written on the top of a single folio that is immediately followed by an account of halos observed in June 1692. Newton decided to measure the distance between the knife edges when the intersection of the first dark lines fell on a white paper placed at different distances from the knives (table 3.4).[27]

Table 3.4

Distances of the Paper from the Knives in Inches.	Distances between the edges of the Knives in millesimal parts of an Inch.
$1\frac{1}{2}$	0'012
$3\frac{1}{3}$	0'020
$8\frac{3}{5}$	0'034
32	0'057
96	0'081
131	0'087

These measurements show that the light forming the *same* fringe or dark interval, when it is observed at various distances from the knives, comes from different distances from the edges and is deflected at different angles. Thus, when the paper was $1\frac{1}{2}$ incher from the knives, the light passed at most .006 of an inch from each edge, whereas at 131 inches it was at most .043 of an inch away. Moreover, the angle at which the rays are diffracted (assuming that they are diffracted at the edges of the knives) varies from nearly $30'$ at the former to just $2'$ at the latter. Since light propagates in straight lines after it has passed through the blades and been deflected, it cannot be the same light that forms the fringes at different places, that is, the fringes do not propagate rectilinearly. Newton's record of this experiment from 1692 contains no comments on its implication. For the *Opticks* he tersely noted:

> And hence I gather that the Light which makes the fringes upon the Paper is not the same Light at all distances of the Paper from the Knives, but when the Paper is held near the Knives, the fringes are made by Light which passes by the edges of the Knives at a less distance, and is more bent than when the Paper is held at a greater distance from the Knives.[28]

This conclusion, as important as it is, nonetheless rendered much of Newton's work on diffraction invalid, since he had carried out many of his investigations and derived many of his conclusions based on the assumption that the rays and fringes followed identical rectilinear paths.[29] Newton no doubt recognized the implications of this experiment as soon as he performed it. He removed the part on diffraction from Book IV of the *Opticks*, added the theory of fits to Book III, and shifted its revised companion on the colors of thick plates there as well.

About two years later, when David Gregory visited Newton in Cambridge in May 1694 and was allowed to study the manuscript of the *Opticks*, the first three books were finished, but Newton had done nothing about restoring the part on diffraction. Gregory noted that he "saw *Three Books of Opticks*. . . . The fourth, about what happens to rays in passing near the corners of bodies, is not yet complete. Nonetheless, the first three make a most complete work." While Newton had demonstrated his model's untenability, there was no intrinsic reason that his work on diffraction had to come to a halt. Earlier in his career he had rejected his own initial approaches and then gone on to develop new ones. Newton did intend to start over, as he recounts in the conclusion he added to the published *Opticks*:

> When I made the foregoing Observations, I designed to repeat most of them with more care and exactness, and to make some new ones for determining the manner how the rays of Light are bent in their passage by Bodies for making the fringes of Colours with the dark lines between them. But I was then interrupted, and cannot now think of taking these things into further consideration.

Newton was nearing the end of his active scientific career, and, despite his intentions in the early 1690s, he never returned to experimental research on diffraction. In April 1695 he told John Wallis, who was trying to convince him to publish the *Opticks*, that the last part of it was still incomplete, and a year later he had moved to London to assume the post of warden of the mint.[30]

After Newton decided to publish the *Opticks* in late 1702, he chose to revise the final book on diffraction rather than to eliminate it altogether.[31] Most of the revision entailed removing passages based on the linear-propagation model, for example, the calculations of the distance of the points of inflection from the diffracting edge and the laws governing them. In Observation 9 he added the measurements that had invalidated that model.[32] Since it was in the observations on the hair that Newton developed his model of the linear propagation of the fringes, it was only fitting that he renounced that model in a major addition that he made to Observation 1. He introduced an illustration of diffraction by a hair (figure 3.8) that clearly showed that the rays crossed one another, so that the paths of the fringes were not the same as that of the rays. He explained that

> the Hair acts upon the rays of Light at a good distance in their passing by it. But the action is strongest on the rays which pass by at least dis-

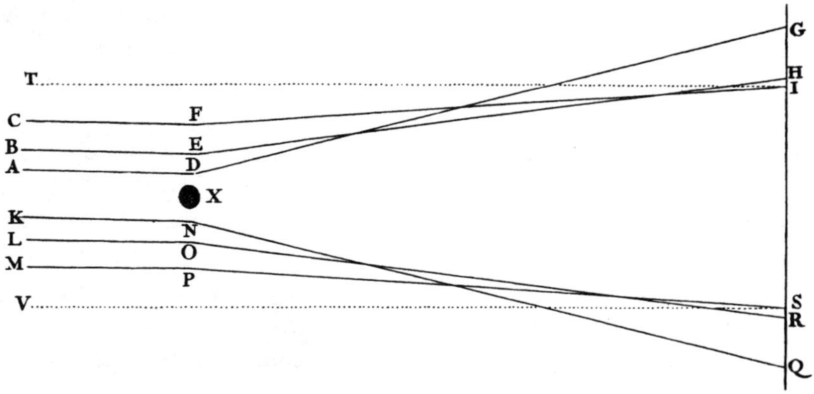

Figure 3.8
Diffraction by a hair; from the *Opticks*, Book III, Figure 1.

> tances, and grows weaker and weaker accordingly as the rays pass by at distances greater and greater, as is represented in the Scheme: For thence it comes to pass, that the shadow of the Hair is much broader in proportion to the distance of the Paper from the Hair, when the Paper is nearer the Hair than when it is at a great distance from it.[33]

Newton writes as if the nonlinear propagation of the shadow and, hence, the crossing of the rays is obvious and follows directly from observation. On the contrary, his data do not demonstrate that the rays must cross or that the linear-propagation model must be abandoned. Newton, as I believe I have shown, came upon the crossing of the rays during the course of his observations with knife edges and not the hair. That the data do not support Newton's claim further supports this interpretation. When Newton revised his manuscript of the part on diffraction for publication and added this new paragraph to Observation 1, he did not present any new experimental evidence. Rather, he let the old observations stand. He had reported three measurements of the breadth of the shadow—at 4 inches, and 2 and 10 feet—in Observation 1, but he had also included two measurements of the shadow along with his measurements of the fringes in Observation 3 (see line 1 in table 3.3). Newton justified the crossing of the diffracted rays (which implies the curvilinear propagation of the fringes) by arguing that the shadow is disproportionally broader when it is measured closer to the hair than further away. The measurements in Observation 1 do support this claim, but when those in Observation 3 are added it becomes a bit ambiguous.[34] Nonetheless, if the disproportion is accepted as valid, it may still be readily accounted for by his original linear-propagation model, where the diffracted rays diverge from a common point of intersection (see figure 3.7). If one compares Newton's measurements of the breadth of the shadow at various distances from the hair with the values calculated according to his original model, they differ only by an undetectable fraction of a percent to 2.5%.[35] Thus, the rays need not cross.

If Newton's claim that the rays cross one another because the shadow is disproportionally broader when it is observed closer to the hair than further away is correct (and it is), we can ask why his data does not support that claim. In the first place, it is because he discovered the crossing of the rays in another experiment with knife blades and then chose to announce this in Observation 1 at the beginning of his account of diffraction. In the second place, it may be strongly suspected that Newton had earlier fiddled with his observations so that they would agree better with the linear-propagation model that he was then using.[36] The shadow has to propagate according to the same laws as the rays, since its boundary is defined by the rays (namely, by the innermost edges of the first fringe). Newton was certainly aware of this, and in a number of his calculations for the linear-propagation model, similar to the one that I presented in table 3.2, he included the shadow.[37] Since the shadow is not at all well defined (the intensity of the light decreases gradually from the first fringe), it is difficult to measure. Newton thus had a fair amount of leeway for exercising his judgment in measuring the shadow's breadth and forging agreement with his linear-propagation model. A reasonable conjecture, then, to explain the incongruence between Newton's data and his interpretation of them is that after he performed his experiment with the knife blades and found that the rays cross one another, he then went back to Observation 1 and reinterpreted his original, "massaged" data.

If my account of Newton's investigation of diffraction for the *Opticks* is valid, then the traditional explanation for the delay of publication, namely, that he was waiting for Hooke to pass from the scene (he died in March 1703) is not altogether sufficient. For much of the period between the purported completion of the *Opticks* and its publication in 1704, there was no *Opticks* to publish. It was incomplete. I suspect that at least until 1697, and perhaps for another few years, Newton intended to return to his experiments on diffraction. Until he fully accepted that he would not take them up again and would instead edit and revise the original completed part, there was no *Opticks* to delay.[38] There is no doubt that Newton's wish to avoid controversy—especially with his nemesis, Hooke—was as much a factor in his holding back the publication of the *Opticks* as its incompleteness, as he told John Wallis in 1695, but to place the entire burden for its long delay on Hooke is unfair.[39]

It could be objected that the threat of Hooke still delayed the publication of the *Opticks* because Hooke had introduced Newton to the phenomenon and had challenged him with it. Consequently, Newton had to meet this challenge.[40] Hooke, however, had not challenged Newton to explain diffraction, as he had done with the colors of thin films, but rather Newton's interpretation of the rectilinear propagation of light. Newton's principal argument against a wave theory of light—in particular, Hooke's—was that wave motions, as sound shows, do not obey rectilinear propagation. Hooke invoked diffraction to show that light in fact does not always propagate in straight lines, so that Newton's objection is besides the point. Newton responded that diffraction is actually a species of refraction, an inflection or deflection, that occurs in the region of a diffracting body, and that the light rays afterward propagate in straight lines.[41] That Newton dreaded yet another confrontation with Hooke is not the issue, but whether he withheld the *Opticks* because of that. He withheld the *Opticks* for at least five or six years because the section on diffraction was not ready. Newton, quite properly, I believe, judged that by the 1690s any comprehensive treatise on optics, such as he was writing, had to include an account of diffraction.

Newton's investigation of diffraction may seem to some to be too slipshod to be worthy of the great man, as he fiddled with his numbers, derived imaginary laws, and developed an untenable physical model. In fact, I find that these are characteristics of all of his optical research. Newton—in a way that I do not yet fully comprehend—plays off his measurements against his mathematical descriptions, which allows him both to change his measurements and to revise his laws. He did not change them at will to get perfect agreement, but controlled them by an awareness of his experimental error and by additional experiments. The only unusual feature of Newton's investigation of diffraction is that he did not unrelentingly pursue it until he had control over the phenomenon as he did earlier with chromatic dispersion and the colors of thin plates, and—nearly simultaneous with his work on diffraction—the colors of thick plates. Far from being slipshod, Newton's critical standards and rigorous methodology are in fact what led him to reject and withdraw his own initial investigation of diffraction. By continually striving to quantify phenomena, developing laws that could be tested, devising new experiments to extend and test his results, and setting realistic measures of experimental error, Newton controlled his fertile imagination by experiment, which, in turn, stimulated his imagination. By peering in Newton's "workshop" to examine his notes, drafts, data, and worksheets, we can begin to grasp how he actually worked at his science.

Acknowledgement

I thank Michael Nauenberg for a stimulating exchange in spring 1997 on the accuracy of Newton's measurements of diffraction patterns. Using the wave theory of light, Nauenberg derived equations for the diffraction fringes cast by a hair that can be applied under the conditions of Newton's observations. To our knowledge such a calculation has not been carried out before. At various points in the notes I will refer to his results, which he intends to publish in Michael Nauenberg, ed., *The Foundations of Newtonian Scholarship* (River Edge, NJ: World Scientific, forthcoming).

Notes

1. Until shortly before publication the books in the manuscript of the *Opticks* were numbered differently than those of the published edition. Parts I and II of Book I of the published *Opticks* were Books I and II in the manuscript (which is now in Cambridge University Library), so that the manuscript had four books rather than the three of the published edition. I use the published numbering except in some references to the manuscript. See Alan E. Shapiro, *Fits, Passions, and Paroxysms: Physics, Method, and Chemistry and Newton's Theories of Colored Bodies and Fits of Easy Reflection* (Cambridge: Cambridge University Press, 1993), 139, 149–50.

2. Grimaldi, *Physico-Mathesis de Lumine, Coloribus, et Iride, Aliisque Adnexis Libri Duo* (1665; reprint, Bologna: Arnaldo Forni, 1963); Fabri, *Dialogi Physici Quorum Primus Est de Lumine ...* (Lyon, 1669). For Hooke's lecture, see Thomas Birch, ed., *The History of the Royal Society of London, for Improving of Natural Knowledge, from Its First Rise*, 4 vols. (1756–57; reprint, Brussels: Culture et Civilisation, 1968), 3:194–95; and Richard Waller, ed. *The Posthumous Works of Robert Hooke, M.D. S.R.S., Geom. Prof. Gresh. &c. Containing His Cutlerian Lectures, and Other Discourses, Read at the Meetings of the Illustrious Royal Society*, (1705; reprint, New York: Johnson Reprint, 1969), 186–90. Roger H. Stuewer, "A Critical Analysis of Newton's Work on Diffraction," *Isis* 61 (1970): 188–205; and A. Rupert Hall, "Beyond the Fringe: Diffraction as Seen by Grimaldi, Fabri, Hooke and Newton," *Notes and Records of the Royal Society of London* 44 (1990): 13–23.

3. Alan E. Shapiro, "Beyond the Dating Game: Watermark Clusters and the Composition of Newton's *Opticks*," in P. M. Harman and Alan E. Shapiro, eds., *The Investigation of Difficult Things: Essays on Newton and the History of the Exact Sciences in Honour of D. T. Whiteside* (Cambridge: Cambridge University Press, 1992), 181–227; and *Fits, Passions, and Paroxysms*, 138–50.

4. Cambridge University Library, MS Add. 3970 (hereafter "Add. 3970"), ff. 371r, 372r. For clarity and simplicity, in my quotations from Newton's manuscripts I will not indicate the many textual changes, unless they are of significance for this study.

5. See Shapiro, *Fits, Passions, and Paroxysms*, sect. 4.1.

6. Add. 3970, f. 372v. Newton also included this law in two drafts of the "Fourth Book," Prop. 2, ff. 335r, 338v; on the "Fourth Book," see Shapiro, *Fits, Passions, and Paroxysms*, 141–43. On the hyperbolic fringes, see Newton, *Opticks: Or, a Treatise of the Reflexions, Refractions, Inflexions and Colours of Light* (1704; reprint, Brussels: Culture et Civilisation, 1966), Book III, Observation 10; and for a modern theoretical and experimental account of the experiments with the V-knives as presented in the *Opticks*, see M. P. Silverman and Wayne Strange, "The Newton two-knife experiment: Intricacies of wedge diffraction," *American Journal of Physics* 64 (1996): 773–787.

7. Add. 3970, Prop. 3, f. 377r. We can see that Newton initially intended to follow the format of Book I with propositions that are proved by experiment, but by the time he had composed (the suppressed) Book IV, Part II he had adopted the style of Book II, that of a sequence of observations with principles deduced from them.

8. "Hypothesis," H. W. Turnbull, J. F. Scott, A. Rupert Hall, and Laura Tilling, eds., *The Correspondence of Isaac Newton*, 7 vols. (Cambridge: Cambridge University Press, 1959–1977), 1:384–85. "Fourth Book," in Add. 3970, Prop. 13, f. 335v; and Prop. 10, f. 337r,v.

9. "Fourth Book," Add. 3970, Prop. 3, f. 338v.

10. *Opticks*, Query 3, 133.

11. Add. 3970, f. 378r,v. I have added the numbers in the left hand column and edited the manuscript slightly. Newton subsequently crossed out the observation in line 9. There is a (faint) bright fringe at the center of the shadow of the hair that Grimaldi had observed but which Newton failed to detect; see Stuewer, "A critical analysis."

12. Both Grimaldi, *Physico-Mathesis de Lumine*, 3, and Fabri, *Dialogi Physici*, 11, used the term "fascia." It is evident from a sheet on which Newton recorded his measurements that his scale was divided into sixteenths of an inch; Add. 3970, f. 373r. Since the intervals were increased by a factor of twelve, Newton had inclined his ruler at an angle of 5° to the beam, that is, nearly parallel to it.

13. Add. 3970, f. 378v.

14. Peter Spargo, University of Cape Town, has proposed a way, using a method available to Newton, in which he could have measured the diameter of his hair; " 'Y^e exactest measure I could make . . .': A possible explanation of Newton's determination of the thickness of a hair," *Transactions of the Royal Society of South Africa*, 50 (1995): 165–68.

15. This worksheet, Add. 3970, ff. 645–46, is written on a folded sheet that has an undated draft of a letter to the Commissioners of Taxes on one side. Newton used all four sides for his calculations. The letter is tentatively dated to the early 1670s in

Turnbull et al., *Correspondence of Isaac Newton*, 7: 371, and to about 1675 by Richard S. Westfall, *Never at Rest: A Biography of Isaac Newton* (Cambridge: Cambridge University Press, 1980), 209, n. 89. This dating must be revised. The diffraction data and calculations as well as the watermark argue for the early 1690s. The watermark that I designate "GLC" was used in other parts of the manuscript of the *Opticks*, for most of the drafts on diffraction for the *Opticks*, and the suppressed part on diffraction; see Shapiro, "Beyond the Dating Game," 202, table 1.

16. "If from ye hole in ye window through ye three distances from ye hair or points of refraction you produce y^{e} 3 incident rays 9 foot beyond y^{e} hair Their distances there from y^{e} axis of the hair will bee 0.01495. 0.027166. 0.039058. Which subducte from 1/2 y^{e} distances there observed & y^{e} remainders will be 0.06005. 0.0919. 0.111 wch to y^{e} rad[ius]. of 9 foot subtends y^{e} angles $1'.52''$, $2'.57''$, & $3'.32''$, wch are y^{e} angles of inflexion" (Add. 3970, f. 646r). In all of Newton's calculations, $j = 12$ ft., $k = 9$ ft., and $h = 1/280$ in. He uses the values for c and f that are in the calculation on the top of figure 3.6 at 1.a, and another calculation is in the bottom left corner at 7.a.

17. Add. 3970, ff. 378v, 328r.

18. Ibid., f. 646r; see 4.a. in figure 3.6.

19. Ibid., f. 81r. I have added the numbers in the left column.

20. In order to compare Newton's measurements with modern or Fresnel theory, Michael Nauenberg derived an equation for the distance between the fringes for diffraction by a hair under Newton's conditions of observation (personal communication 6 April 1997). Two factors may seem to hinder a valid comparison: (1) determining the particular wavelength or part of the spectrum to use in the calculation, since Newton worked with sunlight, which is a mixture of wavelengths; and (2) the exact value of the diameter of the hair, for although Newton's measurement may have been a remarkable achievement, it surely is not an accurate value. Nauenberg chose 5.5×10^{-5} cm for the wavelength, which is a reasonable value, but this leaves an arbitrary factor of a few percent for wavelengths close to that one. The calculation of the distance between the fringes, however, turns out to be independent of the hair's width under Newton's conditions of observation, so that that factor becomes irrelevant. When Nauenberg compared Newton's measurements of the distance between the three fringes with that calculated according to modern theory, he found Newton's measurement at one half foot differed from the calculated values by 1.5 to 4.5%, while at 9 feet they ranged from 1.7 to 2.4%. This shows that Newton had achieved a high degree of accuracy with his measurements.

21. Ibid., f. 82r. Newton had confidently written that "if they could be measured more accurately it's without any sensible errors," before he decided to delete it and let the reader make his own assessment.

22. Ibid., f. 83r.

23. Another example of Newton trying out a set of values can be found in his fringe measurements in Add. 3970, ff. 377v, 378r (table 1). On the sheet facing the table of his draft he has four more sets of values for the key first four lines, some of which

differ by 7% or 8%; he published the third of these. We should recall that these measurements stand up quite well against the predictions of modern theory (see note 20).

24. On Newton's quest to construct a hypothesis-free science and how it affected the formulation of his theory of fits, see Shapiro, *Fits, Passions, and Paroxysms*, chapters 1.2 and 4.3.

25. Ibid., Observation 8, ff. 87r-88r. This passage is essentially unchanged (except for the switch from "fasciae" to "fringes") in the published *Opticks*, Book III, Observation 7, 125.

26. Add. 3970, f. 88r.

27. The values in the *Opticks* (Book III, Observation 9, p. 127) scarcely differ from the original measurements on Add. 3970, f. 334r, and I give those from the *Opticks*. The observation of the halo was subsequently incorporated in Book II, Part IV, Observation 13. The watermark on f. 334 is consistent with the early 1690s, since it is the same as that used for the definitions that begin Book I. I have designated this watermark "None" in the two tables in "Beyond the dating game," pp. 202–203.

28. *Opticks*, Book III, Observation 9, 127. There is a draft of this passage from around 1703 in Add. 3970, f. 477v.

29. David Brewster judged this to be one of Newton's "two new and remarkable results" on diffraction; *Memoirs of the Life, Writings, and Discoveries of Sir Isaac Newton* (1855; reprint, New York: Johnson Reprint, 1965) 1:200.

30. Gregory, *Notae in Newtoni Principia*, Royal Society, MS 210, insert between pp. 55, 56, my translation; for the Latin see Shapiro, *Fits, Passions, and Paroxysms*, 148. *Opticks*, Book III, 132. Wallis to Newton, 30 April 1695, in Turnbull et al., *Correspondence of Isaac Newton*, 4:117.

31. According to a memorandum by David Gregory on 15 November 1702: "He [Newton] promised Mr. Robarts, Mr. Fatio, Capt. Hally & me to publish his Quadratures, his treatise of Light, & his treatise of the Curves of the 2^{d} Genre"; W. G. Hiscock, ed., *David Gregory, Isaac Newton and Their Circle: Extracts from David Gregory's Memoranda 1677–1708* (Oxford: Printed for the Editor, 1937), 14.

32. In the Advertisement to the *Opticks*, pp. [iii–iv], Newton felt it necessary to warn the reader that "the Subject of the Third Book I have also left imperfect, not having tried all the Experiments which I intended when I was about these Matters, nor repeated some of those which I did try, until I had satisfied my self about all their Circumstances." During revision he added the concluding Observation 11, which describes diffraction in monochromatic light, but this observation goes back to one of his earliest drafts on diffraction from the early 1690s, Add. 3970, ff. 324–25. I cannot now explain why Newton omitted it from the first completed state of the part on diffraction, Book I, Part II.

33. *Opticks*, Book III, Observation 1, 115–16.

34. The breadth of the shadow at distances of 4 in., 6 in., 2 ft, 9 ft., and 10 ft. from the hair were 1/60, 1/54, 1/28, 1/9, and 1/8 in. The ratios of the breadth to the dis-

tance are then $1/240 > 1/324 > 1/672 > 1/972 < 1/960$. Since the difference in the last two ratios is not great, one could attribute the discrepancy to experimental error.

35. Using equation (1) I was able to calculate from these measurements the distance from the center of the hair, $f/2$, at which the rays at the edge of the shadow are inflected. It was then a simple matter to compare the breadth of the shadow predicted by the linear-propagation model with the observed values.

36. Michael Nauenberg first called my attention to the "obvious" fact that Newton's observations had to be inconsistent with Fresnel's theory, and he estimated that some of his observations err by as much as 25% (personal communications, 14 and 15 April 1997). I had already recognized that Newton's data were inconsistent with the claim that the rays must cross, or that the linear-propagation model was incapable of explaining the disproportion.

37. See, for example 1.a in figure 3.6, where the first column treats the calculated geometrical shadow, the second the measured "physical" shadow, and the next three the three fringes; see also Add. 3970, f. 646v.

38. Gregory's memoranda of his visit to Newton in May 1694 includes a passage concerning the publication of the *Opticks* that has received no notice. Gregory states that the *Opticks* "will be published within five years after he resigns from the University"; Turnbull et al., *Correspondence of Isaac Newton*, 3:336. This seems to imply that Newton would publish the *Opticks* after he left Cambridge. This makes sense, since he was already looking for a position outside the university. Gregory does not suggest any reason why Newton proposed a five-year delay. One possibility is that Newton thought that he could revise the *Opticks* within that period. The next sentence, however, confuses the issue: "If [it is published] while at the university, [it will be] translated into Latin, which task he would do willingly." This implies that even if Newton remained at Cambridge, he would still publish the *Opticks*, though in Latin, and no delay is mentioned. In the event, Gregory's memorandum is not far off the mark: Newton left Cambridge in 1696, but did not resign his professorship until 1701; the *Opticks* was published in English in 1704, and Newton arranged for a Latin translation to appear in 1706.

39. Wallis to Newton, 30 April 1695, in Turnbull et al., *Correspondence of Isaac Newton*, 4:116–17; Newton's letter to Wallis is lost, but his reasons for not publishing the *Opticks* can be readily reconstructed from Wallis's counterarguments. Neither Gregory, nor Wallis, nor any other contemporary source that I am aware of mentions Newton's fear of Hooke's criticism as being responsible for delay of the publication of the *Opticks*. So far I have traced the claim that Newton had delayed the publication of the *Opticks* until after Hooke's death only as far back as Jean-Baptiste Biot's biography of Newton in the *Biographie universelle*, vol. 31 (Paris: L. G. Michaud, 1821), 170; reprinted in Biot, *Mélanges scientifiques et littéraires*, vol. 1 (Paris: Michel Lévy Frères, 1858), 196. See also Westfall, *Never at Rest*, 638–39; and A. Rupert Hall, *All Was Light: An Introduction to Newton's "Opticks"* (Oxford: Clarendon Press, 1993), 92.

40. A. I. Sabra raised this point when I delivered the paper on which this chapter is based at the Dibner Institute conference, and I thank him for his useful comments.

41. Hooke's clearest statement of his position is in a letter to Lord Brouncker (?), ca. June 1672, although there is no evidence that Newton received this document; Turnbull et al., *Correspondence of Isaac Newton*, 1:200–201. Hooke was replying to Newton's charge in his letter to Oldenburg for Hooke, 11 June 1672, ibid., 174–76. From Newton's account of his exchange with Hooke on diffraction at a Royal Society meeting in March 1675, it appears that this issue was then discussed; see Newton, "Hypothesis," ibid., 383–85. In the *Principia*, Book I, Proposition 96, Scholium, Newton most clearly treats diffraction as a form of refraction. He returned to the problem of rectilinear propagation and the wave theory in the *Opticks*, Query 28, which was added in the Latin translation in 1706; see Stuewer, "A Critical Analysis," 202–205.

4

Mathematicians and Naturalists: Sir Isaac Newton and the Royal Society

Mordechai Feingold

Following the death of Sir Isaac Newton in March 1727, a bitter contest for the succession of the presidency of the Royal Society ensued. For months the supporters of the acting president, Sir Hans Sloane, battled the champions of Martin Folkes, Newton's heir presumptive, and as the November elections drew nearer, the wrangling intensified. "You must have heard of the Philomats' putting up Mr. Fowlks for President of the Royall Society, in opposition to Sir Hans Sloane," wrote Richard Richardson to botanist William Sherard. "They were positive in their success, but lost it on Thursday last. It has been the whole talke of the town; and there has been much canvassing and intrigue made use of, as if the fate of the Kingdome depended on it. The Society will suffer by it, which I am sorry for."[1] Although they were defeated, the ringleaders remained recalcitrant. James Jurin, prior to his being forced to relinquish the editorship of the *Philosophical Transactions*, managed to insert a defiant dedication of volume 34 to Folkes, in the course of which Newton himself was called on to pronounce upon the Society's misguided action:

> I shall not, I presume, need any other Apology for prefixing your Name to this *Thirty Fourth Volume* of *Philosophical Transactions*, when I declare, that the Motive of my doing so was the same, which induc'd the greatest Man that ever liv'd, to single you out to fill the Chair, and to preside in the Assemblies of the *Royal Society*, when the frequent Returns of his Indisposition would no longer permit him to attend them with his usual Assiduity. This Motive, Sir, we all know, was your uncommon Love to, and your singular Attainments in those noble and manly Sciences, to which the Glory of Sir *Isaac Newton*, and the Reputation of the *Royal Society* is solely and entirely owing. That Great Man was sensible, that something more than knowing the Name, the Shape and obvious Qualities of an Insect, a Pebble, a Plant, or a Shell, was requisite to form a Philosopher, even of the lowest rank, much more to qualifie one to sit at the Head of so great and learned a Body. We all of us remember that Saying so frequently in his Mouth, That *Natural History might indeed furnish Materials for Natural Philosophy; but,*

> *however, Natural History was not Natural Philosophy*; and it was easy for all his Friends to see, with what intent he so often us'd this remarkable Expression. We knew his Love to the *Royal Society*, and his Fears for it. It was not that he despis'd so useful a Branch of Learning as *Natural History*; he was too wise to do so: But still he judg'd that this humble Handmaid to Philosophy, tho' she might be well employ'd in amassing Implements and Materials for the Service of her Mistress, yet must very much forget her self, and the Meanness of her Station, if ever she should presume to claim the Throne, and arrogate to her self the Title of the *Queen of Sciences*.[2]

The thinly disguised slander of Sloane and his fellow naturalists was lost on no one.

My recounting this episode is intended not to serve as the climax of a survey of Newton's membership in, and presidency of, the Royal Society. Such an appraisal will be part of my forthcoming history of the Society, in which I shall seek to build on the important contributions to the subject made by Frank Manuel, Richard Westfall, and John Heilbron. Instead, in this chapter I wish to address one important facet of this relationship that has hitherto received little scholarly attention: namely, what were the consequences for the fortunes of the Society of Newton's uncompromising conviction concerning the primacy of mathematics in the domain of natural philosophy, on the one hand, and his condescending view of the natural history tradition, one the other? I argue that Newton's stance on these issues was responsible both for the antagonism that greeted his first paper on colors, especially from Hooke, and for the subsequent emergence of a mathematical "party," whose members openly perpetuated Newton's prejudices in their own works as well as at the meetings of the Royal Society. Together, they constituted a divisive clique that had little in common with the interests and concerns of the generality of membership, and on several occasions their haughty actions brought the Society to the brink of a rift.

To date, historians have tended to be relatively forgiving of the takeover of the Society by Newton and his party, for it has been viewed as an invigorating force vis-à-vis the Society's scientific activity. Prior to Newton's election as president, so the argument goes, the Society "had long since lost its original vigor and direction." Its weekly meetings were the forum for displaying rarities and antiquities, and the papers read by its members bordered on trifles, a state of affairs that presented the wits with ample material to mock the Society and its proceedings. Newton "delivered" the Society from its destructive self and under his dictatorial management "serious" scientific work was once again instituted and the experimental program reinstated to its former glory, at least for a season.[3] This characterization, I believe, is in need of revision. And although I obviously cannot tackle this here in its entirety, I hope that my highlighting one facet of this revision will generate discussion that, in turn, can be incorporated into the larger work in progress.

Let me begin by examining briefly the design and aims of the Royal Society on the eve of Newton's first contact with it. The Society's expressed mission when it was founded in 1660 was to promote "the knowledge of natural things, and useful Arts by Experiments." And whereas the extent to which the Society adopted the Baconian ideology is debatable, it is generally agreed that Thomas Sprat was just one voice among many urging the eschewing of theories in favor of the collection of observations and experiments, which were deemed the necessary foundation for a comprehensive new natural philosophy: "The *Society* has reduc'd its principal observations," wrote Sprat, "into one *common-stock*; and laid them up in publique *Registers*, to be nakedly transmitted to the next Generation of Men; and so from them, to their Successors. And as their purpose was, to heap up a mixt Mass of *Experiments*, without digesting them into any perfect model: so to this end, they confined themselves to no order of subjects; and whatever they have recorded, they have done it, not as compleat Schemes of opinions, but as bare unfinished Histories." One of the key figures of the early Royal Society, Sir Robert Moray, expounded in detail the road toward this ideal:

> In the mean time this Society will not own any Hypothesis, systeme, or doctrine of the principles of Naturall philosophy, proposed, or maintained by any Philosopher Auncient or Moderne, nor the explication of any phaenomenon, where recourse must be had to Originall causes ... Nor dogmatically define, nor fixe Axiomes of Scientificall things, but will question and canvas all opinions[,] adopting nor adhering to none, till by mature debate and clear arguments, chiefly such as are deduced from legittimate experiments, the trueth of such positions be demonstrated invincibly.—And till there be a sufficient collection made, of Experiments, Histories and observations, there are no debates to be held at the weekly meetings of the Society, concerning any Hypothesis or principle of philosophy, nor any discourse made for explicating any phenomena ... but the time of the Assemblyes is to be employed, in proposing and making Experiments, discoursing of the trueth, manner, grounds and use thereof; Reading and discoursing upon Letters, reports, and other papers concerning Philosophicall and mechanicall matters; Viewing and discoursing of curiosities of Nature and art.[4]

Robert Boyle's many programmatic statements drove home a similar message. In introducing his discussion of the elasticity of air, for example, he insisted that he did not intend "to assign the adequate cause of the spring of the air, but only to manifest, that the air hath a spring, and to relate some of its effects." Several years later an anonymous member of the council went to great lengths to argue that the Society's "foundation maxime" was the requirement to fetch materials "from Observation and Experiment, lying in the two Worlds of Nature and Art." The members were asked not to despise the patient process of collecting, no matter how trivial it may seem, for "the meanest observation of what is really existent in nature, is more valuable than the most illustrious, if ungrounded, phancy."[5]

Such an advocacy of "naive" empiricism and utilitarianism, joined with the incessant rebuke of theory, was fortified by the proscription against mathematics. Again, it was Bacon who offered the rationale for such an attitude:

> For it being plainly the nature of the human mind, certainly to the extreme prejudice of knowledge, to delight in the open plains (as it were) of generalities rather than in the woods and inclosures of particulars, the mathematics of all other knowledge were the goodliest fields to satisfy that appetite for expatiation and meditation. But though this be true, regarding as I do not only truth and order but also the advantage and convenience of mankind, I have thought it better to designate Mathematics, seeing that they are of so much importance both in Physics and Metaphysics and Mechanics and Magic, as appendices and auxiliaries to them all. Which indeed I am in a manner compelled to do, by reason of the daintiness and pride of mathematicians, who will needs have this science almost domineer over Physic. For it has come to pass, I know not how, that Mathematic and Logic, which ought to be but the handmaids of Physic, nevertheless presume on the strength of the certainty which they possess to exercise dominion over it.[6]

Equally emphatic in his stricture against mathematicians and other theoreticians was William Harvey. "The example of astronomy is not to be followed here; in this subject, merely from appearances and the actual fact, the causes and the reason why come up for investigation. But (as one seeking the cause of an eclipse would be placed above the moon to discover that cause by sensation and not by reckoning) with regard to sensible things, that is, the things that come under the senses, it will be impossible to bring forward any surer demonstration to induce belief than the actual sensation and seeing for oneself." For Harvey, even more than for Bacon, first-hand observations and experiments constituted the "way of the anatomists" and were the ultimate arbiters of the naturalists' knowledge.[7]

For his part, Boyle was willing to admit that a naturalist stood in need of "a competent knowledge of mathematicks." Yet he insisted, "the phaenomena which the mathematician concurs to exhibit, do really belong to the cognizance of the naturalist. For when matter comes once to be endowed with qualities, the consideration how it comes by them, is a question rather about the agent or efficient, than the nature of the body itself." Elsewhere he cautioned further: "we must not expect from mathematicians the same accurateness, when they deliver observations concerning such things, wherein it is not only quantity and figure, but matter, and its other affections, that must be considered." Sir Robert Moray made the message explicit when he confessed to Christiaan Huygens that "it is true that for myself, I like philosophical speculations much more than any other kind of study. But it is proper (and just) that the mathematical sciences should hold the second place."[8]

Of course not all the fellows assented. Already in 1654 Seth Ward opined that "it was a misfortune to the world, that my Lord *Bacon* was not skilled in Mathematicks, which made him jealous of their Assistance in naturall Enquiries; when the operations of nature shall be followed up to their Staticall (and mechanicall) cause, the use of Induction will cease, and sylogisme succeed in the place of it, in the interim we are to desire that men have patience not to lay aside Induction before they have reason." Ward's intimate friend, Isaac Barrow, firmly believed that "there is no Part of this [physics] which does not imply Quantity, or to which geometrical Theorems may not be applied, and consequently which is not some Way dependant on Geometry; I will not except even *Zoology* itself." Mathematics, he insisted, was "adequate and co-extended with *Physics*." And given the likelihood of attaining certitude in mathematical physics, Barrow proceeded to argue that "one Experiment will suffice (provided it be sufficiently clear and indubitable) to establish a true Hypothesis, to form a true Definition; and consequently to constitute true Principles. I own the Perfection of Sense is in some Measure required to establish the Truth of hypotheses, but the universality or Frequency of Observation is not so." As Alan Shapiro has recently pointed out, Barrow's conception of mathematical physics exerted considerable influence on the formation of Newton's own views on the subject.[9]

At times the issue was debated in the context of the Society's own work. Thus when in 1668 the Society embarked on an examination of

Huygens and Wren's rules of motion, John Wallis was asked to contribute his ideas as well. Henry Oldenburg, however, responding to some discussion at one of the meetings, requested in a subsequent letter a *non*-mathematical account. That request elicited a somewhat indignant response from the Oxford Professor of Geometry:

> I have this to adde in reference to one of your letters . . . where you tell me that the *Society in their present disquisitions have rather an Eye to the Physical causes of Motion, and the Principles thereof, than the Mathematical Rules of it.* It is this. That the Hypothesis I sent, is indeed of the *Physical* Laws of motion, but *Mathematically* demonstrated. For I do not take the Physical and Mathematical Hypothesis to contradict one another at all. But what is Physically performed, is Mathematically measured. And there is no other way to determine the Physical Laws of Motion exactly, but by applying the Mathematical measures and proportions to them.[10]

William Neile submitted an even more theoretical paper on the same subject, conferring on experiments the sole task of confirming his principles. For as he told Oldenburg, "to know a thing barely by experiment is good for use but it is not science or philosophye." Evidently, the reaction to his paper was critical, as Neile complained that "for my owne part if the principles I offer should prove never so much true[,] yet I could not wish for the name of a Philosopher if I could have it[,] it is so troublesome a name that[,] and the name of Mathematician is almost as dangerous as the name of a Poet . . . I desire everie bodie should presume mee to be ignorant unlesse in any particular matter they find mee not altogether so. I have no great time nor much affection to propound this business to people[;] if it be false it is in vain[,] if true I thinke truth need not much goe a begging."[11]

Given such sentiments, it is hardly surprising that Neile's contemporaneous proposal for reforming the Society was explicit about the need to proceed beyond mere experiments, which in "themselves are but a dry entertainment without the indagation of causes." Indeed, he continued, "it seems a litle belowe the name and dignity of Philosophers to sitt still with the bare registring of effects without an inquiry into their causes." As Michael Hunter observed, Neile's "Proposalls" reveal his "deep misgivings about the rather narrow experimentalism espoused by some of his colleagues and celebrated in Sprat's *History*, which eschewed causal enquiries and confined itself almost exclusively to the compiliation of facts through random experimentation." Several years later, Sir William Petty delivered a paper in which he urged members to "apply [their] mathematics to matter, for only by the rules of number might natural philosophy, and especially matter theory, free itself from the confusion of qualities and words."[12]

Nonetheless, at this stage the insistence on the primacy of mathematics and the elevated position of (informed) theory was still a minority position that could coexist alongside the largely tolerant and pluralistic activities of the naturalists. Those who attempted to narrow the Society's sphere of activity did so cautiously and apologetically, and there was no organized attempt to either challenge the public ideology presented by the Society or to denigrate the interests and activities of many of its members.

Against this background the reaction to Newton's first letter on colors, and, particularly, to the bombshell inserted in the middle of it, can be better appreciated:

> [A naturalist would scearce expect to see the science of those become mathematicall, and yet I dare affirm that there is as much certainty in it as in any other part of Opticks. For what I shall tell concerning them is not an Hypothesis but most rigid consequence, not conjenctured by barely inferring 'tis thus because not otherwise or because it satisfies all phaenomena (the Philosophers universall Topick,) but evinced by the mediation of experiments concluding directly and without any suspicion of doubt. To continue the historicall narration of these experiments would make a discourse too tedious and confused, and therefore] I shall rather lay down the *Doctrine* first, and then, for its examination, give you an instance or two of the *Experiments*, as a specimen of the rest.[13]

Oldenburg, I suspect, expurgated the passage in brackets before he read the paper at the Society's meeting; he certainly was careful to excise it from the version printed in the *Philosophical Transactions*. Consequently, the only other person to read the paragraph may have been Robert Hooke. And I believe that much of Hooke's criticism of Newton's paper was prompted not simply by what he regarded as dogmatic formulation of theory, but by what he considered to be, at least implicitly, disparaging of the naturalist and experimental tradition of the Society. For Hooke was particularly sensitive to the issue; his own works and programmatic statements concerning the Society's objectives indicate his vacillation between the two traditions. Indeed, even his quip that has long puzzled scholars—that he had "confirmed" Newton's experiments "having by many hundreds of tryalls found them soe"—should perhaps be understood as an admonishment of Newton's presumption of demonstrative truth from experiments. More than simply defending his own theory, Hooke's methodological message was that the experiments warranted neither Newton's "hypothesis" nor even his own, for he could explain the phenomena according to two or three *other* theories equally well.[14]

Space does not permit further consideration of the reaction to Newton's paper. I should only mention that Hooke and Newton continued to harp on the relation between theory and experiment, while Oldenburg proceeded to expurgate Newton's responses before publishing them in the *Philosophical Transactions*. Hooke's rebuttals did not matter, for Oldenburg never bothered to print them. Ultimately, stunned by the reaction, Newton withdrew. I would like to suggest, however, that he did so not, as some historians assume, on account of his distaste for controversies, but because he came to consider the Royal Society a forum inhospitable to his beliefs.[15] He refused to abide by its moratorium on theoretical pronouncements and, though he availed himself of much of the same language that people like Hooke used when broaching theories, was never apologetic. Nor was he concerned about the possible implications of his position for the public image of the Royal Society. Responding to Gaston-Ignace Pardies' objections to his first paper, for example, Newton wrote Oldenburg: "in the conclusion of which you may possibly apprehend me a little too positive, but I speak onely for my selfe." Three years later, when he sent Oldenburg his "hypothesis of light," Newton grudgingly consented to accommodate his critics by offering a corpuscular conception of light. At the same time, however, he decried the inability of the "great virtuosos" to "take my meaning when I spoke of the nature of light and colors abstractly."[16]

A decade later, though, Newton inadvertently became entangled in a more serious dissension among the fellows. From 1685 Edmond Halley, who was elected clerk of the Society in January 1686, was making a great fuss about the *Principia*. He had coerced the Society into publishing the book and, subsequently, cheerfully justified his neglect of the Society's business on the grounds that the publication of that "incomparable" work had required all his attention. As he wrote Martin Lister in 1687, "I have lately been very intent upon the publication of Mr Newtons book, which has made me forget my duty in regard of the Societies correspondants; but that book when published will I presume make you a sufficient amends for this neglect."[17] This story is well known. Less appreciated is the more general background. Halley's election as clerk occurred shortly after a major storm at the Society, as a result of which the two Secretaries, Francis Aston and Tancred Robinson, resigned in a huff within a week of the November elections. The exact nature of the dispute is not quite clear, but it obviously involved a showdown between the Society's naturalists, on the one hand, and the mathematicians on the other. Clearly, the former group lost. As a consequence, several fellows increasingly distanced themselves from the Society, including Lister, Aston, Robinson, and John Flamsteed, the last of whom locked horns with the mathematicians on somewhat different grounds. Furthermore, Halley's subsequent election as clerk was another indication of the party's strength, as he defeated Hans Sloane for the position. On 8 April 1686 William Molyneux wrote Halley that the latter's election reassured him that the Society was now on the right track:

> I thank you for the account you give me of the affairs of the Society: I had it before, but it was from a person concerned, whom I always thought to blame in this particular; for I found thereby, there was a party arising in the Society, that were for rejecting all kinds of useful knowledge except ranking and filing of shells, insects, fishes, birds, etc. under their several species and classes; and this they termed *Natural History*, and *Investigating Nature*, never attending to the uses and properties of these things for the advantage of mankind, and reckoning chemistry, astronomy, mathematics, and mechanics, as rubs in their course after nature. This indeed seemed to me something strange; and I must confess, I could not but laugh at it.

Molyneux's "concerned" source of information was undoubtedly John Flamsteed to whose letter (now lost) Molyneux responded on 20 February 1685/1686:

> I hear that at last the commotions in the Royal Society are something allayed. Indeed if Mr. H[alley] will be diligent, he may discharge that place with that great advantage; but I fear he may love his ease a little too much; and if so, all the correspondence fail of course. But you that are his friends must desire him to bestow himself briskly in this affair. But what I understand of those heats, they proceeded from mere passion, that all things are not carried according to the fancies of each party. I hear withall, that there are a certain set of Gentlemen in that Society, that would fain model the rest to their useless way of Philosophising; I mean, in search of shells, insects, etc. neglecting those parts of Philosophy, that may be really useful to mankind; and idly contemn Mathematics as an hindrance in investigating nature; whereas all sensi-

ble men will allow, that nothing solid or useful in Philosophy can be obtained without them."[18]

Just then, on 21 April, Halley informed the Society that Newton's book was nearly finished; a week later, following the presentation of Newton's manuscript to the Society, "It was ordered, that a letter of thanks be written to Mr. Newton; and that the printing of his book be referred to the consideration of the council; and that in the meantime the book be put into the hands of Mr. Halley, to make a report hereof to the council." Owing to the chaotic state of the Society at the time, Halley took advantage of the lull in council meetings and on 19 May managed to push through at a regular meeting of the Society a resolution to print the book. In the following month he also extracted from the president, Samuel Pepys, the Society's imprimatur.[19]

Halley's victory could also be interpreted as another manifestation of the triumph over the naturalists. After all, shortly before the palace coup, the Society had undertaken the publication of Francis Willughby's *Historia Piscium*, which proved a commercial failure of major proportions. In contrast, writing to various of the Society's correspondents, Halley praised the forthcoming *Principia* as a divine treatise. The fight, however, was far from over. On 14 February 1686, a month after Halley was elected clerk, Robert Plot wrote to Lister, who had applied for membership in the Oxford Philosophical Society, expressing the hope that other disaffected members of the Royal Society—including Aston, Flamsteed, Edward Tyson, and Robinson—would follow suit. In fact, Plot apparently had been contemplating having the Oxford Society replace the London one. He told Lister that before his death Bishop John Fell had "thought of a good designe of advancing the reputation of this [Oxford] Society beyond what [could] be imagined." He still hoped to implement the plan in 1686, "but of this say nothing, but bring in those men to follow if you can, and you shall know all in good time. Nevertheless I could not have you or any of the rest, throw yourselves out of the Royal Society, for we may live to have a pluck with them for the Authority there too another time."[20]

One such "pluck" occurred on 23 June, a week after Robinson, Aston, Flamsteed, and Tyson were elected members of the Oxford Society. Some members of the council (Flamsteed and Robinson, I suspect) challenged the legitimacy of Halley's position, but to no avail. Then, on 29 November 1686, a day before the annual elections, a no-confidence resolution against Halley continuing as clerk was nearly passed.

Five weeks later a committee was established to "investigate his performance of his duties," though it eventually exonerated him. Within this context we should also view the resumption by Robert Hooke of his lectures on earthquakes in December 1686. Hooke announced that he intended to defend the Society against those critics who charged it with devotion to collecting "a rude heap of unpolished and unshaped materials." Finally, not unrelated was the ensuing liaison between Halley and Wallis to refute the physical basis of Hooke's claims.[21]

The publication of the *Principia* marked a new phase in the relations between mathematicians and naturalists, not only because its immediate success contributed greatly to the widespread acquiescence to the claims of mathematicized physics (even by those who were unable to understand it), but also because the triumph of the Newtonian method appeared to sanction its application to other scientific domains. Newton himself led the way with the publication of *Opticks* in 1704, and even more importantly, with the enlarged Latin edition two years later. Toward the end of Query 31, Newton promulgated that "[a]s in Mathematicks, so in Natural Philosophy, the Investigation of difficult Things" ought to follow the method he utilized in both his works. And, he continued, "if natural philosophy in *all* its Parts, by pursuing this Method, shall at length be perfected, the Bounds of Moral Philosophy will be also enlarged."[22] A growing number of young converts immediately picked up his cue. As early as 1700 John Arbuthnot publicly sang the praise of the *Principia* and the forthcoming *Opticks*, while enumerating the great value of mathematics in all scientific domains: "A natural philosopher without mathematics," he thundered, "is a very odd sort of a person, that reasons about things that have bulk, figure, motion, number, weight, etc., without arithmetic, geometry, mechanics, statics, etc. I must needs say, I have the last contempt for those gentlemen that pretend to explain how the earth was framed, and yet can hardly measure an acre of ground upon the surface of it." For John Harris, mathematics was "the only solid Foundation on which a Useful Enquiry into Nature and all Physical Learning can possibly be built." All errors in natural philosophy, John Keill argued, "seem to spring from hence, that Men ignorant of Geometry presume to Philosophize, and to give the Causes of Natural Things. For what can we expect but Mistakes from such, as having neglected Geometry, the Foundation of all Philosophy, and being acquainted with the Forces of Nature, which can only be estimated by the means of Geometry, do yet attempt to explain its Operations, by a Method not at all agreeing with the Rules of Mechanicks?"[23]

It was primarily because of Archibald Pitcairne's influence that the Newtonian ebullience penetrated the domain of medicine so swiftly. In a series of provocative and highly influential dissertations, beginning in the late 1680s, Pitcairne insisted on the need to "mathematicize" medicine, effectively utilizing both Newtonian methodology and matter theory, and concluding along the way that knowledge claims based on experience are virtually indistinguishable from mere empiricism. To such an advocacy of reductionist theories, not to mention the haughty manner in which they were expressed, Pitcairne conjoined a willful subversion of Harvey's injunction to the physicians to follow the "way of the anatomists," rather than that of the mathematicians. Notwithstanding the perception to the contrary, Pitcairne wrote, Harvey's method was mathematical, not experimental. "Reasoning in Medicine ought to be founded upon the same Principles with those which are made use of by Astronomers," he asserted. And since "it is not allowable, either in Theory or Practice, to advance any thing into a Principle, which is Matter of Doubt amongst Mathematicians, and Men altogether clear of prejudice," it follows that "it behoves Physicians to follow rather the Example of Astronomers."[24]

Pitcairne's disciples were equally strident. "How it comes to pass," asked Richard Mead provocatively in 1704, "that, notwithstanding the considerable advances made in the study of nature by the moderns, especially in the last century, this useful art has not received those benefits, which might reasonably be expected from a surer method of reasoning, than men were formerly acquainted with?" The problem, according to Richard Mead, was that whereas such moderns as Kepler, Galileo, and Newton "have made vast improvements in natural philosophy, by joining mathematical reasoning to their inquiries into nature ... medicine still deals so much in conjecture, that it hardly deserves the name of a science." Hence "in order to prove, how beneficial the study of geometry must be to physicians, as well for investigating the causes of diseases, as for finding proper remedies for them; I have attempted to explain a very difficult question, *concerning the courses and returns of some distempers*; the nature of which is such, that it cannot be thoroughly well handled by any other means." John Quincy, Pitcairne's editor, argued as a matter of fact that inasmuch as the "study of Medicine has in all Ages been influenced by the Philosophy in vogue," ever "since the Introduction of Mathematical Reasoning, and the Application of mechanical Laws" to the study of medicine it has been discovered to be "no otherwise knowable."[25]

The younger generation of physicians continued to advocate both reductionism and quantification in medicine as well as to engage in speculative theories, ostensibly deriving their license from Newton. "Is not the Mechanism of the Body conducted by the same Laws that support the Motions of the great Orbs of the Universe?" asked Nicholas Robinson in the preface to his *A New Theory of Physick and Diseases, Founded on the Principles of the Newtonian Philosophy*, dedicated to Mead. He informed his readers that he had written his treatise "with an Intention to discover the Certainty of Theory, and the Advantages to be reaped from it in point of Practice." The book was published, he continued, "as an Essay towards reducing Physick to a Standard, by erecting the Basis upon Principles that can never fail us." The book's title made the source of these principles clear. Some who subsequently moderated their views had nevertheless articulated strong Newtonian sentiments at the outset of their career. Peter Shaw, for example, whom we shall encounter in a different context below, had initially advocated the applicability of Newtonian methodology to medicine, in the belief that it could enhance the elimination of disease: "It is such a Kind of Geometrical Method which appears to me the most proper to be observed in this Pursuit; that way of Reasoning from *Data* to *Quaesita*, which has done Wonders in Philosophy, Astronomy, and Mechanicks. But it happens most unfortunately, that instead of encouraging and pursuing this noble Method in Physic, we seem almost entirely to discountenance and contemn it."[26]

Small wonder, then, that the beleaguered naturalists responded in kind. Tancred Robinson, in his letters to Lister, often castigated the fashionable trends. "Our Physitians here," he wrote on 9 November 1708, "begin to put on all shapes, except the true one." Ten days later he censured James Keill, "who goes upon imaginary Lemmata and data" in his account of the quantity of the blood in the body.[27] As for the state of the Royal Society, Robinson believed it "may have great men in their number, but alas very litle souls, and narrow minds, so I despair of the further advancement of true Learning in this Kingdom." Sir Edward Eizat made a similar argument when criticizing Archibald Pitcairne's medical ideas: "The Mathematicks deserve their room in the World and common-wealth of Learning, and are very good Neighbours, while they keep within the bounds prescribed them by Nature, and do not wander beyond the limits of their own Orb: But if they come to make Incursions on the territories of another Vortex, they may chance to share with the Comets in their fate." With greater moderation John Ray summarized the naturalists' prevailing attitude: "But the greatest of all the particular *Phaenomena* is the Formation and Organization of the Bodies of Animals, consisting of such variety and curiosity; that these mechanick Philosophers being no way able to give an account thereof from the necessary motion of Matter ... [they should] prudently therefore break off their System there, when they should come to Animals, and so leave it altogether untoucht." Thomas Smith supported Lister's attempt to take on the Newtonians in similar terms: "You have bestowed some animadversions upon Keil, who with some others, under the shelter of geometry and other nice speculations, are setting up new hypotheses, which have no influence in the practical part of medicine, and leaving the beaten roads of Nature, the knowledge of which may bee more certainly attained by observation and experiment, pursue her in by-paths, where, they phansy, they shal overtake and command her. Their Mathematics may be of wonderfull use to these gentlemen, provided, that they lay not aside the old Methods, which direct to the true knowledge of things, especially the little world of man."[28]

The resentment spread more broadly to the scholarly community. Dedicating in 1700 his *Six Philosophical Essays* to Jeremy Collier, Samuel Parker, the young nonjuror Oxford scholar, remonstrated bitterly at the imperious claims of the mathematicians:

> I shall only inform you of the occasion of a remarkable Omission, I mean my neglect of Mathematical Arguments, of which the World is become most immoderately fond, looking upon every thing as trivial, that bears no relation to the Compasse, and establishing the most distant parts of Humane Knowledge; all Speculations, whether Physical, Logical, Ethical, Political, or any other upon the particular results of number and Magnitude. Nor is it to be questioned, but the Dominion of Number and Magnitude is very large. Must they therefore devour all Relations and properties whatsoever? 'Tis plainly unreasonable. In any other commonwealth but that of Learning, such attempts towards an absolute Monarchy would quickly meet with opposition. It may be a kind of Treason, perhaps, to intimate thus much; but who can any longer forbear, when he sees the most noble, and most usefull portions of Philosophy lie fallow and deserted for opportunities of learning how to prove the Whole bigger than the Part, etc.

Roger North, too, inveighed against the very pretensions of the Newtonians. He viewed the queries of the *Opticks* as "a parcel of dogmata ... designed to favour the attractionall against the corpuscular scheme," while he assailed the members of the "attractive sect" who insolently held that "everything is mathematically solvable." "This science of Naturall philosophy," he amplified elsewhere, "is Now under disgrace and contempt, and nothing applauded but Mathematics."[29]

The increasingly constricting atmosphere that pervaded the English scientific community at large was even more noticeable within the Royal Society, where the Newtonian party reigned supreme. Numerous objections to the "tyranny" of the philomats' management of the Society can be cited. Martin Lister thought it strange that mathematicians boasted so much in the face of the utter certainty that mathematics could not be as useful a tool in medicine. "I find they think themselves Lords of nature and masters of all natural sciences and stick not to run down all experiments and observations that stand in their way." Walter Moyle concurred. "I find that there is no room in Gresham College for Natural History," he demurred in a letter of 1719 to William Sherard: "mathematics have engrossed all; and one would think the Gentlemen of that Society had forgot that the chief end of their institution was the advancement of natural knowledge." Small wonder, then, that following the election of Hans Sloane as president, James West felt confident "that the natural history of our country will be more encouraged now."[30]

Many of the Fellows took exception to the mathematicians' narrowly defined conception of the kind of knowledge most worth pursuing, further objecting to altering the Society's weekly meetings and its journal to an almost exclusive vehicle for the defense and extension of Newton's work. The excessive preoccupation with selective topics of a privileged program, it was felt, allowed the (Newtonian) officers of the Society to disregard their duties with impunity. Richard Richardson put the problem most succinctly: "the rest of the Society [is] so enamoured with the mathematical sciences that often matters equally useful were neglected."[31] In 1721 John Woodward wrote Cotton Mather in similar terms: "As to what you Say of your *Curiosa Americana*, after I had carefully pursued them my Self, I delivered them to the Roy. Society; that they might print any thing out of them, that they judgd proper, in the Philosophical Transactions. But the Editors, since Mr. Wallers Death, are very neglectfull and partial; by which the Society suffers not a little; and indeed things are very low with them at present. For my own Part I have not been wanting in Doing you Justice: and makeing the Curious here sensible of your great Diligence there."[32]

Moreover, the naturalists, who were convinced that it was premature to apply the mathematical way to the biological and life sciences, also feared the consequences of the new hierarchy of knowledge for the venerable English observational and experimentalist tradition. Consequently, they valiantly attempted to resurrect the more traditional icons of English science, Bacon and Boyle. They "changed medicine from an art into a science," warned Lister, "when the degree of uncertainty in the subject made this impossible. In its existing poor and immature state, any attempt to give a universal theory was an excuse to avoid the industry of experiment and observation." He therefore resolved to shelter himself under the great patron of experimental philosophy, Lord Bacon, and be short in his reproofs of the men of Gresham College, "which yet they deserve who have outrun both ancient and modern discoveries, greatly pleasing themselves with masteries from the mathematics and an all-sufficiency to practise without study and useful application of the materia medica, which yet is all in all." Lister thus became determined to demonstrate that he could explain medicine without recourse to either mechanical explanations or even quantitative reasoning.[33] Lister returned to Bacon in 1711 when he wrote to (in all likelihood) Lord John Somers, Newton's immediate predecessor as president of the Royal Society, expressing his delight in Somers' favorable reception of his *De Humoribus*. That book, Lister emphasized, embodied "the Summa of [his] most serious thoughts about the Animal Oeconomie," after a lifetime of medical practice and careful comparison of ancient and modern learning. "I have also my L. Bacon on my side," he continued,

> who was the first restorer of this sort of Learning, and gave the first notice, of our wants in, and methods how to advance Natural Philosophie, upon the truth of Exp[erimen]ts and Observations, and not wild mathematical Hypotheses; such as Descartes and others since his time have and do dailie vented. I must owne I have been somewhat hard upon the late medical Geometricians, who having got the ball at their foot, by verie surprising methods gone about to overthrow all ancient Learning, and that upon verie precarious foundations.

A decade and a half later Peter Shaw set up Robert Boyle against the detractors of natural history, which, he perceived, "seems at present to lie under some Disgrace upon account of the small Benefit that it is presumed to arise from the Study of it ... But if any Man has a despicable opinion of Natural History in general, let him look upon it in that View wherein Mr. Boyle considered it; for here, as in every Thing else, our excellent Author has regarded Usefullness and the Benefit of Mankind. Natural History, as managed by him, has no superfluous Branches."[34]

Curiously, we encounter considerable deference to Newton himself, with most critics reluctant to ascribe blame to him. Like Gottfried Wilhelm Leibniz and John Flamsteed in other contexts, most critics tended to exculpate Newton from the sins committed by his overenthusiastic epigones. Even Martin Lister, though believing he needed to "animadvert" on Newton—because Newton's patronage encouraged all other mathematicians, and his *Opticks* was the source of all the pernicious new hypotheses in physics and philosophy—nonetheless was determined to treat him with "all imaginable candour and deference due to so great a man." After all, he rationalized, even the best mathematicians "when they go out of their lines are but like other men." His correspondent, Thomas Smith, approved. "You will do wel to shew," he wrote, "where our great and famous Mathematician has failed in his *new doctrine*, and I know, you will interpose your dissent from some of his Singularities with all possible respect due to his great name, wch is deservedly famous all christendome over." A few weeks later he was delighted by what he saw, even expressing his conviction that Newton himself would be appreciative: "I am extremely wel pleased with your respectfull treating of Sir Isaac N. who, I am persuaded, wil be so farre from quarrelling with you, upon the account of your reflexions, that upon your next interview, when the book is published, he wil thanke you not onely for your civility, but for the weight and strength of your reasonings."[35]

The Newtonians' ability to establish their hegemony over the Royal Society, and English science more generally, was not merely the natural outcome of acquiescence by an admiring community to a manifestly higher form of knowledge. At least in part it was an inevitable consequence of the bitter personal and professional disputes that had wreaked havoc among the naturalists at least since the 1680s. Fierce debates over cosmogonies and ruthless competition for rarities not only divided the community, but offered choice ammunition to the satirists, who wasted little time in making natural history look contemptible and its practitioners ridiculous. Unavoidably, the raillery attached itself to the Royal Society as well, to the chagrin of many of its members. On the eve of Newton's election as president, matters had deteriorated to such an extent that various fellows could be restrained only with difficulty from a public exchange of blows (or, in one case, the drawing of swords), the intended target being, more often than not, John Woodward. William Nicholson, dean of Carlisle, articulated the sentiments of many contemporaries. Reflecting in 1699 on the quarrels among the naturalists (as well as on the Richard Bentley vs. Charles Boyle debate) Nicholson remarked wryly: "These skirmishes are much more diverting (to me) than the late broils in Flanders; but we have now so many of them that this war is come to be almost as expensive as the other." The following year later another fellow informed Lister that "a Monstrous Faction [had] sprung up in the R. S.

occasioned by the Quarrells of Cowper, and Bidloo, of Woodward, Harris, Sloan, Pettiver and Plukenet, and the Severall Parties begin to proclaim their Warrs, which may end in the destruction of the R. S."[36]

The contrast with the rapidly forming mathematicians' party could not have been more striking. Unified around the undisputed leadership of the great Newton, its members shared a well-defined, and generally agreed upon, body of knowledge, as well as unanimity of purpose. Equally important, their outlook could be easily promoted as being intellectually superior and more morally edifying than that of the naturalists—and hence the proper business of the Royal Society—if only by virtue of the mystique that had already formed around the *Principia* and the *Opticks*, and the virtual immunity of Newtonian science, by reason of its recondite nature, to the satirists' barbs. Yet the Newtonians' propensity to make such claims warrants neither the assumption that they were correct, nor the conclusion that the Society became divided between "scientists" and "amateurs."[37] By the early eighteenth century the cleavage between a group comprised primarily of mathematicians, astronomers, and physicists, on the one hand, and naturalists, physicians, and general scholars, on the other, was indicative of taste, not competence. Members of the latter group, representing the lion's share of the Society's fellowship, were both serious and proficient in their respective fields, though not in the exact sciences. Most were incapable of or unwilling to follow, let alone contribute to, the mathematical sciences. Analogously, many mathematicians lacked the inclination, not to mention the talent and patience, to partake in the naturalists' great painstaking enterprise. Thus, we see for the first time the ripening of two distinct traditions among members of the Royal Society and the English scientific community at large. Again, the intrinsic tension between naturalists and mathematicians was not new. But in the early decades of the Society's existence many of the foremost naturalists (and active fellows) were themselves proficient mathematicians, whose broad range of interests ensured that all branches of scientific knowledge received proper recognition, and that none were allowed to reign supreme. Between the 1650s and 1670s individuals such as Sir Charles Scarburgh, Jonathan Goddard, William Croone, Francis Willughby, and William Petty contributed to contemporary research in astronomy, optics, physics, and mathematics, at the same time participating in the biological, medical, and botanical work of the Society and its members. And at least until the mid-1680s, a near consensus prevailed that the Society was broad enough in aim to encompass—and respect—the entire range of scientific activity. In contrast, whereas by the turn of the eighteenth century most naturalists still upheld that open-mindedness, most mathematicians did not. And following Newton's assumption of power, the latter were in a position to act on their beliefs.

The ramifications of the Newtonians' attempted takeover, however, went beyond a derision of natural history. At least from the naturalists' perspective, the mathematicians undermined the very cornerstone upon which the English empiricist scientific tradition stood: experiments and observations. The strength of this conviction for Lister and others has already been cited. Even more vehement was Flamsteed's reaction. As far as he was concerned, Newton's contemptuous attitude toward a lengthy and systematic program of observations was both destructive to science and iniquitous to the work of a lifetime. For Flamsteed, theories were only as exact as the observations on which they were based. As he wrote in 1695, Newton "adds a great many words to persuade me that to have the theory of the moon published with my observations, would be a great proof of their accuracy; whereas theories do not commend observations; but are to be tried by them." Newton, of course, subscribed to a diametrically opposed view of the proper relations between theory and observations. When a friend once quoted to him Flamsteed's boasting that Newton merely "worked with the oar [observations] that he had dug," Newton's rebuttal was crushing: "if Flamsteed dug the oar he [Newton] had made the golden ring." Equally revealing of Newton's outlook concerning the strict subordination of empirical work to the construction of theories was the recorded exchange between him and John Machin. When the latter lamented that "it was a pity that when people had geometrical demonstrations they should trust their senses," Newton promptly quipped "that he first proved his inventions by geometry and only made use of experiments to make them intelligible and to convince the vulgar."[38]

Not that politicking and cabals had ever been lacking in the Society. Quite the contrary. As I have described elsewhere, the Society was plagued almost from the start with intense personal rivalries and party intrigues that on more than one occasion brought it to the brink of disaster.[39] What distinguished the discord generated by the advent of the Newtonian party was the fastening of an explicit ideological component to the predictable human one. Under the new regime it was not enough to be loyal to Newton and his handpicked policies; those permitted to assume office were expected not only to be proficient mathematicians, but to uphold the president's exacting perception of the hierarchy of knowledge. Accordingly, Richard Waller's conscientious performance of his duties as secretary was insufficient to ensure his position, and in 1709 he was ousted. The other secretary, Hans Sloane, who threatened to resign in protest (an empty threat, as it turned out), was forced to resign four years later, notwithstanding his unconditional loyalty to Newton for almost a decade, a devotion that alienated many of Sloane's fellow naturalists.[40] Loyal Newtonians were invariably elected. John Harris replaced Waller, and Edmond Halley succeeded Sloane. The pattern continued for the remainder of Newton's presidency. Not even Newton's great admirer, William Stukeley, was granted permission to stand as a candidate for secretary, and when he did anyway, he not only lost the election, but "Sir Isaac showed a coolness toward [him] for 2 or 3 years," and it was only thanks to Stukeley's perseverance that Newton "began to be friendly again."[41]

Ultimately, however, the success of the Newtonians was contingent on Newton's presence. Given that his principles and policy went against the grain of English science, only his continued presence could ensure the prevailing of a minority opinion. And just as Newton's mystique failed to rub off on any of his disciples, so, too, did his ability to impose the "truth" on an intractable community. Hence the great contest over the presidency of the Royal Society in 1727 marked both the culmination and the demise of the mathematicians' hegemony. And a battle royal it was. The warring parties mobilized the entire body of members, as well as their patrons, and each fellow was obliged to account for his actions. Cambridge botanist Richard Bradley, for example, following the election recited for Sloane all the canvassing he had done on his behalf. Even James Brydges, duke of Chandos, felt compelled in the election's aftermath to send a very apologetic letter to Lord Cardigan, explaining why he had failed to attend the meeting and cast his vote on behalf of his "good friend" Sloane. The many favors shown him by Archbishop Wake (Folkes's uncle), he explained, "would not suffer [him] to be against a Gentleman who was so near a Relation of His Grace and whom he was so desirous of seeing at the Head of the Society." But since he discovered that Cardogan was on Sloane's side, he "forbore" from appearing.[42]

Sloane won the election, and Folkes, according to Stukeley's narrative, "being baffled . . . went to Rome with his wife and dau[ghte]rs, dog, cat, parrot, and monkey. There his wife grew religiously mad." After Sloane's death Folkes finally became president. And although Stukeley accused him of high-handed rule and indifference to natural history, by then Folkes's passion for antiquities was far stronger than any desire to promote a mathematical party, so much so that the Society's great critic,

John Hill, complained that under Folkes the Society's meetings allowed "personages acting the importants ... to trifle away time in empty forms and grave grimaces," and the *Philosophical Transactions* contained "a greater proportion of trifling and puerile papers than are anywhere else to be found."[43]

The quarrels continued into the second half of the eighteenth century and beyond. The mathematicians kept up their challenge for control of the Society, invariably invoking Newton's name in the process of criticizing the Society's activities. On 19 November 1772, for example, Sir William Browne delivered an impassioned speech at a meeting of the Royal Society, "recommending Mathematics, as the Paramount Qualification for their Chair." After citing Jurin's 1727 dedication at length, Browne proceeded to add his own denunciation of the naturalists: "Let the Natural Historian horizontally range the whole globe in search of 'an insect, a pebble, a plant or a shell,' " he thundered, "but let him not look up so high above his level or element, as even so much as to dream of ascending or clinging to the Chair of Natural Knowledge." Mathematics, he continued, was "the only key, capable of opening the doors of such vast researches" and "this key the immortal Sir Isaac Newton has, indeed, completed, and made a master-key." And since none but a mathematician should "either judge himself, or be judged by others, qualified to take the chair of Natural Knowledge," Browne concluded by recommending that "as so many members are Mathematicians, ten of the most deserving may be nominated for the New Council, out of whom the most eminent may, both receive himself, and do the Society the honour of becoming their President."[44]

The mathematicians fared no better than they had half a century earlier, however, and physician Sir John Pringle was elected president. Yet they remained unrelenting and in the name of "serious" science they continued to challenge "dilettante" presidents. In 1784 there was an attempt to unseat Sir Joseph Banks from the presidency. Samuel Horsley, the editor of Newton's works, went so far as to threaten to divide the Society:

> But, Sir, I am united with a respectable and numerous band embracing, I believe, a majority of the scientific part of those who do its scientific business. Sir, we shall have one remedy in our power when all others fail. If other remedies should fail, we can at last SECEDE. Sir, when the hour of secession comes, the President will be left, with his train of feeble *Amateurs* and that Toy [the Mace] upon the table, the Ghost of

> that Society in which Philosophy once reigned and Newton presided as her minister.[45]

Once again the coup failed by a three-to-one margin. But the acrimonious disputes over the kind of natural knowledge most worth pursuing continued to plague the Society, and would become a cornerstone of Charles Babbage's attack on the Society half a century later.

ACKNOWLEDGMENTS

I would like to thank the rector, fellows, and wonderful library staff of the Wissenschaftskolleg, Berlin, for providing me with the time and support necessary for writing this chapter, and to Jed Buchwald, Raine Daston, Michael Hunter, and Rob Iliffe for helpful comments on an earlier draft.

NOTES

1. *Extracts from the Literary and Scientific Correspondence of Richard Richardson* (Yarmouth: S. Sloman, 1835), 284.

2. *Philosophical Transactions*, dedication to vol. 34 (1727).

3. John L. Heilbron, *Physics at the Royal Society during Newton's Presidency* (Los Angeles: William Andrews Clark Memorial Library, 1983), 1, 5–15; Richard S. Westfall, *Never at Rest: A Biography of Isaac Newton* (Cambridge and New York: Cambridge University Press, 1980), 627–28. For an attempt at reevaluation see David P. Miller, " 'Into the Valley of Darkness': Reflections on the Royal Society in the Eighteenth Century," *History of Science* 27 (1989): 155–66.

4. Thomas Sprat, *History of the Royal Society*, ed. Jackson I. Cope and Harold W. Jones (St. Louis and London, Routledge, 1959), 115; Michael Hunter, *Science and the Shape of Orthodoxy* (Woodbridge: Boydell Press, 1995), 172.

5. Robert Boyle, *New Experiments Physico-Mechanical Touching the Spring of the Air*, in Thomas Birch, ed., *The Works of the Honourable Robert Boyle*, 6 vols. (London, 1772; reprint, Hildesheim: Olms, 1965) 1:12; Michael Hunter, *Establishing the New Science* (Woodbridge: Boydell Press, 1989), 226.

6. James Spedding, Robert L. Ellis and Douglas D. Heath, eds., *The Works of Francis Bacon*, 14 vols. (London 1857–1874; reprint, Stuttgart: Olms, 1962) 4:370. In the *Novum Organum* he likewise insisted that mathematics "ought only to give definiteness to natural philosophy, not to generate or give it birth." Ibid., 4:93.

7. William Harvey, *The Circulation of the Blood and Other Writings*, trans. Kenneth J. Franklin (London: Dent, 1963), 159. A few pages later Harvey reiterated his belief: "Silly and inexperienced persons wrongly attempt, by means of dialectics and far-

fetched proofs, either to upset or to establish which things should be confirmed by anatomical dissection and credited through actual inspection. Whoever wishes to know what is in question (whether it is perceptible and visible, or not) must either see for himself or be credited with belief in the experts, and he will be unable to learn or be taught with greater ertainty by any other means." Ibid., 166)] For a discussion of Harvey's scientific style, see Andrew Wear, "William Harvey and 'The Way of the Anatomists,'" *History of Science* 21 (1983): 223–49.

8. Birch, *Works of Thomas Boyle*, 3:427, 1:347; Christiaan Huygens, *Oeuvres complètes*, 22 vols. (The Hague: M. Nijhoff, 1888–1950), 3:311–13.

9. Allen G. Debus, *Science and Education in the Seventeenth Century: The Webster-Ward Debate* (London and New York: Macdonald and American Elsevier, 1970), 25; Isaac Barrow, *The Usefulness of Mathematical Learning Explained and Demonstrated: Being Mathematical Lectures Read in the Publick Schools at the University of Cambridge*, trans. John Kirkby (London: S. Austen, 1734; reprint, London: Frank Cass, 1970), 22, 26, 116; Alan E. Shapiro, *Fits, Passions, and Paroxysms: Physics, Method, and Chemistry and Newton's Theories of Colored Bodies and Fits of Easy Reflection* (Cambridge: Cambridge University Press, 1993), 36 and *passim*.

10. A. R. Hall and Mary B. Hall, eds., *The Correspondence of Henry Oldenburg*, 13 vols. (Madison, Wis., and London: University of Wisconsin Press, Mansell, and Taylor and Francis, 1965–1986), 5:221.

11. Ibid., 5:518, 344–45.

12. Hunter, *Establishing the New Science*, 189–90, 224; William Petty, *The Discourse Made Before the Royal Society the 26. of November 1674 Concerning the Use of Duplicate Proportion in Sundry Important Particulars* (London, 1674), 5.

13. H. W. Turnbull, J. F. Scott, A. R. Hall, and Laura Tilling, eds., *The Correspondence of Isaac Newton*, 7 vols. (Cambridge: Cambridge University Press, 1959–1977), 1:96–97

14. "I doe not therefore see any absolute necessity to believe his theory demonstrated, since I can assure Mr Newton I cannot only salve all the phaenomena of Light and colours by the Hypothesis, I have formerly printed and now explicated yt by, but by two or three other, very different from it, and from this, which he hath described in his Ingenious Discourse. Nor should I be understood to have said all this against his theory as it is an hypothesis, for I doe most Readily agree wth him in every part thereof, and esteem it very subtill and ingenious, and capable of salving all the phaenomena of colours; but I cannot think it to be the only hypothesis; not soe certain as mathematicall Demonstrations." Ibid., 1:113.

15. Similar considerations may have been responsible for Barrow's noninvolvement in the Society's affairs.

16. Turnbull et al., *Correspondence of Isaac Newton*, 1:136, 390.

17. Bodl. MS, Lister, 3 f. 119.

18. Birch, *The History of the Royal Society of London for improving of natural knowledge*, 4 vols. (London: printed for A. Millar, 1756–1757), 4:476; J. P. Bernard, T. Birch, and J. Lockman, eds., *A General Dictionary [. . .] of the Celebrated Mr Bayle*, 10 vols. (London, 1734–1741), 6:605. In public Molyneux was a bit more circumspect. Having extolled the experimental philosophy he nonetheless continued: "There is no part of Philosophy wherein the Mathematicks are not deeply ingredient." William Molyneux, "Epistle Dedicatory" to *Sciothericum Telescopicum; or, A New Contrivance of Adapting a Telescope to an Horizontal Dial for Observing the Moment of Time by Day or Night* (Dublin: printed by Andrew Crook and Samuel Helsham, 1686).

19. Birch, *History of the Royal Society*, 4:479–80, 484, 491.

20. R. T. Gunther, *Early Science in Oxford*, 14 vols. (Oxford, 1921–1945), 12:371–72.

21. Birch, *History of the Royal Society*, 4:489, 505, 516; Richard Waller, ed., *The Posthumous Works of Robert Hooke* (London: S. Smith and B. Walford, 1705; reprint, London: Johnson Reprint Corp., 1971), 329; A. J. Turner, "Hooke's Theory of the Earth's Axial Displacement: Some Contemporary Opinion," *British Journal for the History of Science* 7 (1974): 166–90.

22. Isaac Newton, *Opticks* (repr. New York: 1704; reprint, Doves, 1960), 404–405 (emphasis added).

23. George A. Aitken, *The Life and Works of John Arbuthnot* (Oxford: Clarendon Press, 1892; reprint, New York: Russell and Russell, 1968), 418; John Harris, *Lexicon Technicum*, 2 vols. (London: Daniel Brown et al., 1704–1710), Preface; John Keill, *An Introduction to Natural Philosophy* (London: J. Senex, 1720), viii.

24. Archibald Pitcairne, *The Philosophical and Mathematical Elements of Physick*, trans. John Quincy, 2nd ed. (London: W. Innys, 1745), xviii, xix, xxii; Anita Guerrini, "Archibald Pitcairne and Newtonian Medicine," *Medical History* 31 (1987): 70–83. See also Andrew Cunningham, "Sydenham vs. Newton: The Edinburgh Fever Dispute of the 1690s between Andrew Browne and Archibald Pitcairne," in W. F. Bynum and V. Nutton, eds., *Theories of Fever from Antiquity to the Enlightenment* (*Medical History*, suppl. no. 1, London: Wellcome Institute for the History of Medicine, 1981), 71–98; Stephen M. Stigler, "Apollo Mathematicus: A Story of Resistance to Quantification in the Seventeenth Century," *Proceedings of the American Philosophical Society* 136 (1992): 93–126; Theodore M. Browne, "Medicine in the Shadow of the *Principia*," *Journal of the History of Ideas* 48 (1987): 629–48.

25. Richard Mead, *A Treatise Concerning the Influence of the Sun and Moon Upon Human Bodies and the Diseases Thereby Produced*, trans. Thomas Stack (London: J. Brindley, 1748, published in Latin, 1704), v–vii; John Quincy, *Lexicon Physico-Medicum: or, A New Physical Dictionary*, 2nd ed. (London: E. Bell, 1719), ix–x, cited in Gretchen Finney, "Fear of Exercising the Lungs Related to Iatro-Mechanics 1675–1750," *Bulletin of the History of Medicine* 45 (1971): 345.

26. Nicholas Robinson, *A New Theory of Physick and Diseases, Founded on the Principles of the Newtonian Philosophy* (London: printed for C. Rivington, 1725), ix–x; Peter

Shaw, *A Treatise of Incurable Diseases* (London, 1723), 6, cited in R. W. Gibbs, "Peter Shaw and the Revival of Chemistry," *Annals of Science* 7 (1951): 213–14.

27. Robinson referred to the recently published *An Account of Animal Secretion, the Quantity of Blood in the Humane Body, and Muscular Motion* (London: printed for George Strahan, 1708), in which James Keill relied heavily on theories of attraction and mathematics, but few anatomical observations, to produce some rather odd results.

28. Bodl. MS, Lister, 37, ff. 129, 132, 151; Edward Eizat, *Apollo Mathematicus: or the Art of Curing Diseases by the Mathematicks, According to the Principles of Dr Pitcairn* (1695), 22–23; John Ray, *The Wisdom of God Manifested in the Works of Creation* (London: printed for Samuel Smith, 1691), 27–28; Bodl. MS, Smith, 37, f. 141.

29. S[amuel] P[arker], *Six Philosophical Essays Upon Several Subjects* (London: printed by J. H. for Tho. Newborough, 1700), A3–A3v; F. J. M. Korsten, *Roger North (1651–1734)* (Amsterdam: Holland University Press, 1981), 41–45, 183.

30. *The Works of Walter Moyle*, 2 vols. (London: printed for J. Knapton, 1726), 1:422; Bodl. MS, Lister, 51 ff. 61–62; C. E. Doble, ed., *Remarks and Collections of Thomas Hearne*; 11 vols. (Oxford: Oxford Historical Society, 1885–1921), 9:294.

31. Bodl. MS, Radcl. Trust, C. 10, f. 82v, undated letter to Thomas Petre.

32. *Proceedings of the Massachusetts Historical Society* (1873–1875), 110–11. This was a long-standing concern. Already in 1696 Lister wrote Edward Lhwyd that "some of Leeuwenhoek's letters, though of great importance, had been delathed in publication by the Royal Society over twelve thears: such mean and invidious spirits reign amongst even Societies founded purposely for the promoting of learning in all its parts." Bodl. MS, Ashmole, 1816, f. 124.

33. Bodl. MS, Smith, 51, ff. 71–73. Thomas Smith concurred: "I fully agree with you, that we owe all the *new hypotheses both in physic and Philosophy to the mathematicks*, as also the great discoverthes made in the Heavens and earth and sea to its improvements; but notwithstanding, when these great *Speculativi* have done all wch they can do, it is experiment and practical observations in reducing all things to their proper and immediate natural causes, established by almighty power and infinite wisedome, that must set up the top-stone on this mighty building, as wel as lay the foundation of it." Bodl. MS, Lister, 37, ff. 139–40.

34. Bodl. MS, Lister, 3, f. 246; Peter Shaw, ed., *The Philosophical Works of the Honourable Robert Boyle* 3 vols. (London: W. and J. Innys; 1725), 3:3.

35. Bodl. MS, Smith, 51, ff. 57–58; Bodl. MS, Lister, 37, ff. 139–40, 141. As late as 1776 Oliver Goldsmith, partly paraphrasing the *Encyclopédie*, commented along similar lines that Newton "knew precisely those parts of Nature, which admitted of geometrical applications.... His followers, much less judicious, have expected more from geometry than the art could grant, and, by erroneous imitation, have applied it to parts of Nature, which are incapable of admitting mathematical calculations. Thus we have seen the force of muscular motion determined by numbers, the velocity with

which the blood circulates calculated with geometrical precision, the fermentation of liquors has undergone the same scrutiny, and the most inconstant appearances of Nature have been determined with inflexible demonstration." Oliver Goldsmith, introduction to *A Survey of Experimental Philosophy* (London: printed for T. Carnan and F. Newbery Jun, 1776), in Arthur Friedman, ed., *The Collected Works of Oliver Goldsmith* 5 vols. (Oxford: Clarendon Press, 1966), 5:346–47.

36. Bodl. MS, Eng. Hist., C 11, ff. 85, 94; *Letters of Eminenet Men, Addressed to Ralph Thoresby*, 2 vols. (London, 1832), 1:363; Bodl. MS, Lister, 37, f. 26.

37. See the comments by Hunter, *Establishing the New Science*, 206–207, and Miller, " 'Into the Valley of Darkness,' " 161.

38. Frank E. Manuel, *A Portrait of Isaac Newton* (Cambridge, Mass.: Harvard University Press, 1968), 302–303; Westfall, *Never at Rest*, 541. In a similar vein, Colin Maclaurin, having constructed a sound Baconian genealogy for Newton, admitted that "experiments and observations, 'tis true, could not alone have carried him far in tracing the causes from their effects, and explaining the effects from their causes: a sublime geometry was his guide in this nice and difficult enquiry. This is the instrument, by which alone the machinery of a work, made with so much art, could be unfolded; and therefore he sought to carry it to the greatest height." Colin Maclaurin, *An Account of Sir Isaac Newton's Philosophical Discoveries, in Four Books* (London, 1748), 8.

39. Mordechai Feingold, "Huygens and the Royal Society," in *De Zeventiende Eeuw* 12 (1996): 22–36; "Astronomy and Controversy: John Flamsteed and the Royal Society," in Frances Willmoth, ed., *Flamsteed's Stars: New Perspectives on the Life and Work of the First Astronomer Royal, 1646–1719* (Woodbridge: Boydell, 1997), 31–48.

40. Bodl. MS, Lister, 3, ff. 7, 171. It took the challenge of Folkes and the Newtonians to restore the friendship between Sloane and Sherard. Richardson, *Correspondence*, 286.

41. William Stukeley, *Memoirs of Sir Isaac Newton's Life*, ed. A. Hastings White (London: Taylon and Francis, 1936), 17. Flamsteed, not an unbiased witness, commented in 1709 on Newton's iron rule: "Our Society is ruined by his close politick and cunning forecast." Turnbull et al., *Correspondence of Isaac Newton*, 5:9.

42. Frank N. Egerton, "Richard Bradley's Relationship with Sir Hans Sloane," *Notes and Records of the Royal Society* 25 (1970): 70–71; Huntington Library, Stow MS, 57/31, ff. 29–30.

43. Charles R. Weld, *A History of the Royal Society*, 2 vols. (London: J. W. Parker, 1848; reprint, New York: Arno Press, 1975), 1:486, 483.

44. John Nichols, *Literary Anecdotes of the Eighteenth Century*, 9 vols. (New York: AMS Press, 1966), 3:320–23.

45. Harold B. Carter, *Sir Joseph Banks, 1743–1820* (London: British Museum [Natural History], 1988), 199.

II

Celestial Dynamics and Rational Mechanics

5

Newton's Mature Dynamics: A Crooked Path Made Straight

J. Bruce Brackenridge

Following Newton's death in 1727, John Conduitt acquired his papers and manuscripts and in 1740 the collection passed into the possession of the Portsmouth family, where it remained inaccessible to scholars for over a century. In 1872, the fifth earl of Portsmouth, Isaac Newton Wallop, transferred to Cambridge University the portion of the collection judged to be concerned with scientific and mathematical topics. Rouse Ball partially explored these papers and manuscripts toward the end of the nineteenth century, but not until the second half of the twentieth century did scholars such as Cohen, Hall, Herivel, Westfall, and Whiteside make a systematic study of the mathematical and dynamical papers. In the preface to his *Background to Newton's Principia*, John Herivel notes that upon inspection of the collection he was struck by a wealth of documentary evidence far beyond anything hinted at in Ball's earlier *Essay*. In particular, he felt that the dynamical papers in the *Waste Book* and other manuscripts would provide the basis for some sort of connected account of the growth of Newton's dynamical thought prior to the publication of *Principia*. To that end, Herivel published in 1965 the dynamical writings of the *Waste Book* of 1664, material that had been out of the public domain for three centuries.

The first dynamical writing that appears in the *Waste Book* is an analysis of uniform circular motion, a demonstration often reproduced and discussed since Herivel published it. Newton's cryptic statement that follows that analysis, however, has received much less attention. Herivel transcribes it as follows:

> If the body b moved in an Ellipsis that its force in each point (if its motion in that point bee given) [will?] bee found by a tangent circle of Equall crookedness with that point of the Ellipsis.[1]

In a note to this section, Herivel reflects that here in 1664 "Newton is already pondering the more difficult problem of motion in an ellipse." We have no explicit record of a solution to that problem, however, for some fifteen or twenty years. As a first-time reader I was disturbed, therefore, by the confident tone to Newton's assertion in 1664 that he had a technique for analyzing elliptical motion such "that its force in each point . . . will be found by a tangent circle of Equall crookedness." Eventually, I found an answer that partly satisfied me in Newton's alternate method of solving direct problems in the revised editions of the *Principia*, a method that I called Newton's mature dynamics.[2] This method employs the concept of curvature that Newton developed to measure the rate of bending of curves: a technique in which the tangent circle to a curve at a point measures what Newton called the crookedness (i.e., curvature) of the curve. Because the method employed in the first edition of the *Principia* did not explicitly use curvature, I assumed that it was previous to the curvature method that appeared in the revised editions.[3] That assumption still left unanswered, however, Newton's confident tone in his cryptic statement made some thirty years earlier that he had a curvature method as early as 1665. The recent work of Michael Nauenberg[4] has provided a link between that cryptic statement of 1665 and the *Principia*, and I am now convinced that curvature was central to all of Newton's dynamics, both the early and later analysis. But I get ahead of my story. First I want to review the three basic techniques that Newton employed to solve direct problems and then relate them to the statement on curvature Newton wrote in 1664.

Newton's Three Dynamic Techniques: The Polygonal, Parabolic, and Curvature Methods

Newton's basic methods of analysis can be grouped under three general headings: the polygonal method, the parabolic method, and the curvature method. In the first, an element of a general curve is approximated by a segment of a straight line, in the second by a segment of a parabolic arc, and in the third by a segment of a circular arc.[5] All three techniques appear in his very early writings on dynamics, all appear in some form in the first edition of the *Principia*, and all are important in the final revised editions of the *Principia*.[6]

The Polygonal Measure

Newton's early investigations into dynamics concerned collisions and appear in his bound notebook, the *Waste Book*. The only date among the dynamical entries in the *Waste Book* is the marginalia, "Jan. 20th 1664" (1665 new style) that appears in a section devoted to problems of collisions between two perfectly inelastic bodies.[7] Newton extended the

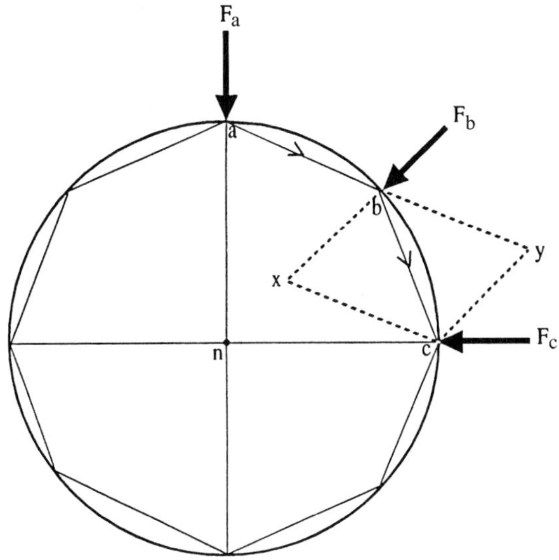

Figure 5.1
Based on the drawing in the *Waste Book* (1665). An octagonal path *abc* approximates the circular path *abc*. An impulsive force *F* acts at each intersection of the circular and octagonal paths.

dynamics of intermittent colliding bodies to his analysis of bodies subject to a continuous force, in particular, to uniform circular motion. In this analysis the circular path is approximated by a polygonal path of *n* sides (shown in figure 5.1 as an octagon), where the number of sides of the polygon will ultimately increase without limit. The ball is envisioned to collide elastically (i.e., the speed or magnitude of the velocity does not change) with the circle or with the circumscribed polygon. At each collision, an impulsive force **F**, directed normal to the tangent and thus toward the center of the circle, acts on the ball; between collisions no force acts on the ball, and it moves with uniform rectilinear motion. Consider the point *b*. In the absence of the impulsive force $\mathbf{F_b}$, the ball would have continued on to the point *y*. If the ball had been at rest at *b* and subject only to the impulsive force, then it would have gone to *x*. The actual displacement *bc* is the diagonal of the parallelogram *bxcy*.

The impulsive force acts for an extremely short time and produces a change in what Newton called the "quantity of motion in a body." He calculated the sum of the magnitudes of these changes in the quantity of motion, which he called "the force of all the reflections," and compared that sum to the quantity of motion of the body's constant orbital motion, which he called "the force of the bodys motion." As the number of sides of the polygon increased, the polygonal path approached a circular path, and the ratio of "the force of all the reflections" and "the force of the body's motion" approached 2π.[8]

The polygonal measure of the very early work in the *Waste Book* is continued in the *Principia*. Figure 5.2 is based upon the drawing that appears in the first and revised editions of the *Principia*. Newton used the polygon *ABCDEF* to approximate the continuous motion of the planet in its path. In this polygonal approximation, the disjointed polygonal path "collides" with the smooth planetary path at a discrete number of points. As in the early analysis of a circular path, an impulsive force, directed normal to the tangent and toward the center of force *S*, acts on the orbiting object at each collision; between collisions no force acts on the object, and it moves with uniform rectilinear motion. Again, consider the point *B*. In the absence of the impulsive force at point *B*, the planet would have continued on to the point *c*. If the object had been at rest at *B* and subject only to the impulsive force, then it would have gone to *V*. The actual displacement *BC* is, as it was in the *Waste Book*, the diagonal of the parallelogram *BcCV*.

In contrast to uniform circular motion, however, the impulsive forces differ in magnitude and they are no longer normal to the tangent, but are always directed radially toward the same fixed point *S*. Newton demonstrated that for a given time between collisions, all the triangular areas *SAB*, *SBC*, *SCD*, etc., are equal. Thus, equal areas are described in equal times. The demonstration of the law is independent of the functional form of the impulsive force. The only restriction is that the force be always directed toward the fixed center of force *S*. Newton ultimately reduced the discontinuous motion along the sides of the polygon to the continuous motion along the smooth orbital path by letting the size of the triangles, such as *SAB*, become infinitely small.

The Parabolic Method

In another manuscript, which Herivel calls *On Circular Motion* and dates about 1665 or 1666, Newton employed the parabolic method rather than the polygonal method to analyze the same uniform circular motion.[9] Figure 5.3 compares the diagram that accompanied this tract on uniform circular motion from *On Circular Motion* with the diagram that accompanied the tract from the *Waste Book*. The obvious difference is that a polygon no longer approximates the circular path. In the previous tract, the distance *bx* is related to the impulse produced by the collision at point

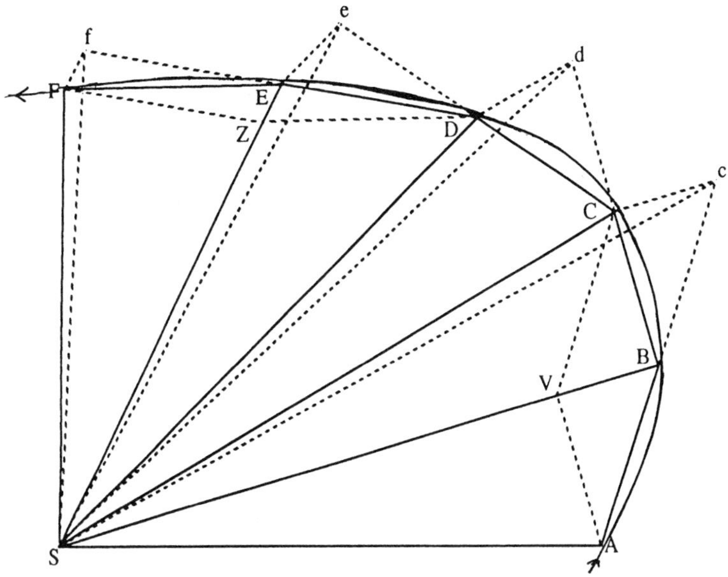

Figure 5.2
Based on the drawing in the *Principia* (1687). The polygonal path *ABCDEF* approximates the continuous motion of the planet. An impulsive force acts at each intersection of the continuous and polygonal paths.

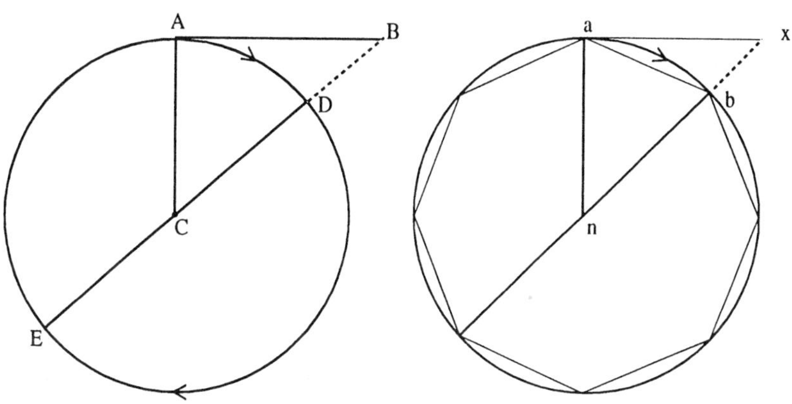

Figure 5.3
A comparison of the diagram from the *Vellum Manuscript* (left) with the diagram from the *Waste Book* (right).

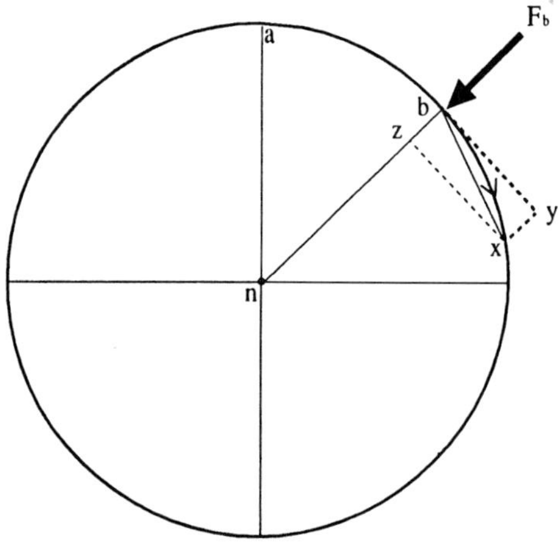

Figure 5.4
Details of the parabolic approximation. The displacement *by* due to the initial tangential velocity is combined with the displacement *xy* due to the constant central force to produce the actual displacement *bx*.

a, and the effects are summed as the number of sides of the polygon increased. Finally, the limit is considered as the number of sides tends to infinity and the polygon tends to the circle.

Figure 5.4 displays the details of the parabolic measure in a figure similar to that employed in the polygonal method. In the parabolic method, the analysis begins with the consideration of the limit as the point *x* shrinks back to the point *b*. In that limit, the force $\mathbf{F_b}$ that produces the displacement *bx* is assumed to be constant in both magnitude and direction. Under the action of a constant force alone (i.e., no initial velocity) the displacement would be *bz*, or its equivalent, *xy*. Under the action of the tangential velocity alone (i.e., no force acting) the displacement would be *by*. The observed motion along the arc *bx* is a combination of these two displacements. Galileo has demonstrated that an initial velocity combined with a constant acceleration gives rise to a parabolic curve as in ideal terrestrial projectile motion. Here the initial projectile velocity is the tangential velocity along the line *by* and the constant acceleration is supplied by the approximate constant force and acts along the radial line *bn*. Thus Newton approximates the element of the circular curve in the vicinity of the initial point *b* by an element of a parabolic curve.

The displacement of any future point *x* on the parabola can be found by using the parallelogram rule to combine the displacement *by* due

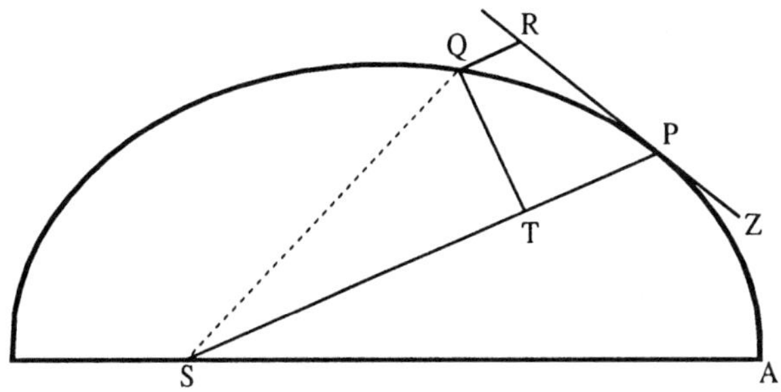

Figure 5.5
Based on the drawing for Proposition 6 in the 1687 *Principia.*

to the initial tangential velocity with the deviation xy due to the constant force. Galileo demonstrated that such a constant force is directly proportional to the displacement and inversely to the square of the time. The challenge therefore is to express that time in terms of meaningful parameters of the uniform circular motion. In this pre-1669 tract, Newton employs (1) the property that the radius of the circle sweeps out equal angles in equal times plus (2) a single proposition from Euclid to demonstrate that (3) the magnitude of the force is proportional to the square of the period of uniform motion and inversely proportional to the diameter of the circle.

The parabolic method used in the very early work in *On Circular Motion* is continued in the *Principia.* Newton employs the parabolic method in Proposition 6, which sets forth the combination of geometric elements that provide the basic paradigm for his primary solution of direct problems. Figure 5.5 renders these basic geometric elements visually. If no force had acted, then the object would have moved along the tangent from P to R. Because the force did act, the object moved along the curve from P to Q. Thus, the displacement QR is the deviation from the linear path PR produced by the centripetal force directed toward the center of force S. In the limit as the point Q approaches the point P, the force is assumed to be constant and its magnitude is proportional to the displacement QR divided by the square of the time (i.e., the parabolic approximation). From Proposition 1, the time is proportional to the triangular area PSQ and is given by one half of the product $QT \times SP$. Thus, combining the parabolic approximation and the area law, the force is shown to be proportional to the linear dynamics ratio $QR/(QT^2 \times SP^2)$, or as Newton preferred, "the centripetal force would be reciprocally as the solid $(SP^2 \times QT^2)/QR$." Newton does not speak to the parabolic approximation explicitly in the 1687 *Principia*. In the revised editions, however, he adds the following corollary to Proposition 1 that makes explicit what is only implicit in the 1687 edition. The approximate constant force **F** is compared to the force of gravity and the deviation *xy* is called the sagitta (literally the arrow in the bow of the arc).

> Corollary 5. And therefore these forces [the forces **F** above] are to the force of gravity as these sagittas [the deviations $Cc = VB$ in figure 5.2] are to the sagittas, perpendicular to the horizon, of the parabolic arcs that projectiles describe in the same time.

It is important to note that the analytical technique of the parabolic approximation Newton employs in Proposition 6 of the 1687 *Principia* for the analysis of general orbital motion is the same as he employs in *On Circular Motion* in the late 1660s for uniform circular motion, except that the force is no longer always normal to the curve but rather acts along the radius directed to a given center of force.

Jo Dobbs has argued that the Cartesian conjecture of celestial vortices was destroyed when Newton demonstrated that the area law was exact.[10] If the mechanical collisions with the celestial aether sweep the planets, then that aether must provide some resistance to motion as well. In the analysis of terrestrial projectile motion, the actual path in the atmosphere differs from the ideal parabolic path calculated under the assumption of no resistance. In the analysis of celestial planetary motion, the actual path in the vortex does not differ from the ideal celestial motion in which equal areas are swept out in equal time. Newton derived the area law under the assumption of no resistance, and since planetary observations are consistent with that law, then the aether offers no measurable resistance to movement, and thus the vortex cannot be the source of celestial motion.

In the tract *On Motion* that Newton sent to Edmond Halley preliminary to the writing of the *Principia*, he presented four theorems and the solutions to four problems, all relating to motion of bodies in a medium devoid of resistance. Then he added three more problems all relating to parabolic motion, the first under the assumption of no resistance and the others under the assumption of resistance. I had been puzzled as to why he included an example of terrestrial motion in a tract motivated by considerations of celestial motion. Seen in the light of Jo Dobbs's conjecture, that inclusion at last made sense. Ideal terrestrial motion differs from the observed motion because the atmosphere provides resistance. Ideal celestial motion does not differ from the observed motion because the aether does not provide measurable resistance.

Whatever the reason, Newton clearly had rejected Cartesian vortex theory of planetary motion by 1685. I am struck, however, by how little Newton's basic kinematic and dynamic analysis changed from its beginning in the *Waste Book* to its fruition in the *Principia*. His earliest writings on dynamics employ a consistent set of elements of orbital analysis that survive into his more mature writings. Domenico Bertoloni Meli makes the following observation: "In his maturity Newton interpreted centrifugal force in orbital motion as a reaction to centripetal force: as such, they were considered to be equal and opposite—whilst in the past they appeared to be only opposite but not necessarily equal—*and then centrifugal force was ignored in the calculations*."[11] Julian Barbour, writing of Newton's evolution of interpretation of the cause of uniform rectilinear motion, comes to a similar conclusion concerning the consistency of the analytic method: "The young Newton appears to have conceived this [uniform rectilinear] motion very much in the manner of medieval impetus theory, whereas the mature Newton moved more to the Cartesian standpoint.... However, this [change in metaphysics] did not alter any of the mathematical consequences of Newton's theory. *The mathematics—and hence the objective content—of an existing theory is indifferent to the metaphysics through which it is interpreted*."[12] Writing of Newton's *Vellum Manuscript* (1665 or 1666), Curtis Wilson sums up this point as follows: "[Newton's] thought was still at this time [before 1679] blinkered by the vortex theory of planetary motion. The centrifugal forces had to be counteracted by some equal and opposite central force, in order that circular motion should be maintained: and of this counteracting force, Newton in the ... [*Vellum Manuscript*] said nothing. Still, the mathematics of the derivation is unexceptionable; it was easy enough, once Newton took up Hooke's idea for the analysis of orbital motion, to adapt the derivation to the new context."[13]

The Curvature Measure

The two techniques for solving direct problems discussed above, the polygonal measure and the parabolic measure, do not exhaust the list of possible attacks to be found in Newton's early papers on dynamics. Herivel and D. T. Whiteside both have recorded the brief statement in the *Waste Book* concerning elliptical motion and curvature. In it, Newton states that the force required to maintain elliptical motion can be found from the circle of curvature. In his early work on mathematics, Newton had developed the circle of curvature as a measure of the bending or "crookedness" of a curve, and in this statement in the *Waste Book* he suggested that curvature could be employed in the analysis of elliptical motion. The circle that represents the best approximation to the curvature has the same first and second derivatives as the curve at the given point. A circle has constant uniform curvature, an equiangular spiral has curvature that changes uniformly, and conics have curvature that change systematically but not uniformly. As early as December 1664 Newton had roughed out a method for finding the center of curvature in an ellipse.[14]

When I first read the statement as given by Herivel, I was bothered that it said, "If the body moved in an ellipse that its force . . . will be found by a tangent circle. . . ."[15] The word "that" bothered me. I wondered if it meant an ellipse *such that* its force *will* be found by a tangent circle. I found Whiteside's transcription of the note in a section of Book 1 of Newton's mathematical papers in a section entitled, "Early Scraps in Newton's *Waste Book*." Whiteside recorded the word y^n (then) in place of Herivel's y^t (that) but still recorded the other key word as "will."[16] I was a bit happier with "then the force . . . will be found," but I thought from context that it should be, "then the force . . . may be found . . . ," that is, one of the multiple ways of solving the problem. Fortunately, Herivel had reproduced the part of folio 1 of the *Waste Book* that contained the quotation.[17] Inspection of that page clearly shows that it was "then" and not "that" as Whiteside had recorded. Unfortunately, the location of the word "will" was on the dark and crumbling edge of a page and I could not read it. I called Tom Whiteside and asked him if he would check the word for me. Fortunately, he had an early copy of the page, and he now reported the word as "may" and not "will." Thus, the statement should read as follows:

> If the body b moved in an Elipsis, then its force in each point (if its motion in that point bee given) may bee found by a tangent circle of Equall crookedness with that point of the Ellipsis.[18]

I was now satisfied with the statement of 1665, but I still did not have an explicit example of the application of the curvature measure to the solution of direct problems before the discovery of the area law in 1679. Michael Nauenberg, however, has demonstrated that before the discovery of the area law Newton could have employed a numerical method based on curvature to evaluate orbits.[19] Nauenberg demonstrates an iterative computational method for reproducing the figure for the orbit of a body subject to a constant central force that Newton sent to Robert Hooke in 1679. In this method, one uses finite portions of successive circles of curvature to approximate the curve generated by a given force law. Thus Nauenberg provides a link between Newton's cryptic curvature statement of 1665 and the *Principia*.[20]

Nauenberg suggests, moreover, that Newton could have used the material in his 1671 *Methods of Series and Fluxions* to produce an analytical measure of the force, independent of the area law, to solve direct problems. There is no record of an analysis in which Newton actually carried out such calculations, but Nauenberg argues that there is considerable circumstantial evidence that he did do so. For example, in the same letter of 13 December 1679 in which Newton sent Hooke the drawing of an orbit obtained by a numerical method for a constant force, Newton also noted that one could suppose a force that increased as the distance decreased such that the body may "by an infinite number of spiral revolutions descend continually till it cross the center."[21] Nauenberg suggests that Newton knew that the reciprocal cube force generated the constant angle spiral (logarithmic spiral). He notes that Newton's observation of "an infinite number of spiral revolutions" into the center cannot be deduced from a numerical solution of orbital motion, because that technique can provide only a finite number of revolutions in approaching the center. Nauenberg also points to Newton's choice of the constant angle spiral produced by a reciprocal cube force directed to the pole of the spiral as an example in the *Principia* for both the spiral/pole direct problem in Proposition 9 and the reciprocal cube inverse problem in Proposition 31.[22] Nauenberg argues that Newton could have solved various direct problems before 1679, and he challenges the received opinion that Newton could not have done so until after his discovery of the area law in 1679.

The first preserved record of such an analysis occurs in a solution that Newton sent to British philosopher John Locke in 1690, but Newton may have produced a first draft of that solution as early as 1684. In the tract *On Motion* that Newton sent to Halley in 1685 and that is recorded in the *Register Book* of the Royal Society, there is no trace of his use of curvature in his solution to the three direct problems: a circular orbit with the center of force on the circumference, an elliptical orbit with the center of force at the center of the ellipse, and an elliptical orbit with the center of force at the focus of the ellipse. In the 1687 *Principia*, however, Newton also presents a solution to the direct problem of motion on an equiangular spiral with the center of force at the pole of the spiral. In this solution he calls upon Lemma 11, which does employ the circle of curvature. The revised editions of the *Principia*, however, include alternate solutions to all the direct problems that explicitly call upon curvature and provide examples of the application of the curvature measure given in the cryptic statement in the *Waste Book* some forty years earlier.

Following the publication of the first edition of the *Principia* in 1687, Newton began to make corrections in his working copy of the text and to propose revisions and additions for a possible second edition. When, twenty-six years later, the second edition was published in 1713, it incorporated many of these revisions. Several revisions, however, never appeared in printed form. Of particular interest are the unpublished revisions of the fundamental dynamics of Sections 2 and 3 of Book I. These unpublished revisions of Newton's dynamics provide insight into what would have been a dramatically different format for these fundamental sections.

Newton was not the only person to note corrections and suggest revisions to the *Principia* following its publication. A select group of scholars, both in Britain and on the continent, struggled with the work and were eager to note its failures as well as its successes. Scottish mathematician David Gregory had aspirations (unfulfilled) of having his notes on the work published in a revised edition or as a separate companion volume. During Gregory's visit to Cambridge in May 1694, Newton showed him the manuscript papers that contained the drafts of the proposed revisions. In a memorandum written in July 1694, Gregory summarized the changes that Newton had revealed to him during his visit. Of particular interest to Newton's dynamics are the opening lines of Gregory's summary: "Many corrections are made near the beginning: some corollaries are added; the order of the propositions is changed and some of them are omitted and deleted. He [Newton] deduces the computation of the centripetal force of a body tending to the focus of a conic section from that of a centripetal force tending to the center [of the conic section], and this again from that of a constant centripetal force tending to the center of a circle; moreover the proofs given in propositions 7 to 13 inclusive now follow from it just like corollaries."[23] Gregory went on to list other revisions proposed by Newton for Books II and III. The fundamental revisions to the opening sections of Book I, however, command our interest. Gregory's statement echoes the cryptic statement on curvature in the *Waste Book* made some thirty-five years earlier. From the circle of curvature, the force to the center of curvature is obtained and then the force to the focus of the conic.

Whiteside has called these proposed revisions of the 1690s "radical restructurings." In the 1687 edition of the *Principia,* Newton employed only the linear dynamics ratio as a measure of the force in producing solutions to the direct problems set in Sections 2 and 3 of Book I. In dramatic contrast, however, the proposed radical restructurings of the 1690s contain major changes in the statements of the propositions. In this proposed radical revision, Newton introduced two other related but distinct methods for generating such solutions: the curvature measure of force and the comparison theorem.[24] In the unpublished revisions, each method appears as an independent proposition and the role of curvature is apparent. The published revisions, however, do not display the role of curvature as clearly: the comparison theorem appears as a corollary to a revised Proposition 7 and the curvature measure as a corollary to a revised Proposition 6.

The Curvature Measure of Force The curvature measure of force gives the force as inversely proportional to the chord of curvature through the center of force and the square of the normal to the tangent through the same center of force. The published revised Proposition 6 in the 1713 *Principia* offers a sense in which the circular dynamics ratio is derived from the parabolic measure. In the preliminary revisions of the early 1690s, however, Newton reversed the procedure and obtained the curvature measure directly without any reference to the parabolic measure. In a draft version of the revised Proposition 6, Newton crossed out the heading "Corol. 6" and inserted above it the title "Prop VI." He then proceeded to derive the curvature measure from first principles, without any reference to the parabolic measure.[25]

Figure 5.6 reconstructs the small sketch that appears in the revised draft version of Proposition 6. The symbols are as in the published edition: *P* for the place of the body, *Y* for a point beyond *P* on the orbit, and *v* and *V* for the terminal of the chords drawn from *P* through *S* of the two circles that touch the orbit at *P*. After much revision and deletion, Newton's draft of the revised, but rejected, new version of Proposition 6 is as follows:

> If circles touch orbits concave to the bodies and if they are of the same curvature with the orbits on the points of contacts, [then] the forces will be reciprocally as the solids comprised of the chords of the arcs of circles from bodies through the centers of forces [*PV* and *Pv*] and by the squares of the perpendiculars descending from the same centers on to the rectilinear tangents [*SY*].[26]

Newton does not give details of the demonstration, but it is not difficult to construct a version. The basic assumption is that the force $\mathbf{F_O}$ (directed

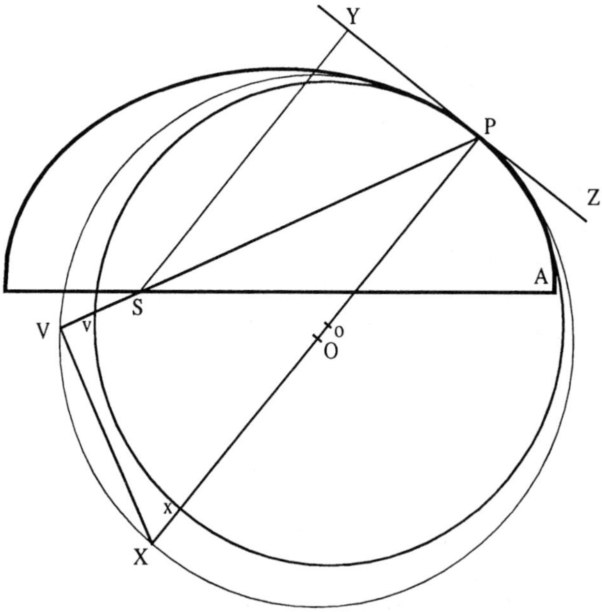

Figure 5.6
A reconstruction of the small sketch that appears in the revised draft version of Proposition 6.

toward the center O) is in proportion to the force $\mathbf{F_S}$ (directed toward the center S) as the chord of the circle of curvature PV drawn through the center S is in proportion to the diameter of the circle of curvature $PX = 2\rho$, where ρ is the radius of curvature. In modern vector notation, $\mathbf{F_O}$ is the component of the force $\mathbf{F_S}$ perpendicular to the tangent.[27] The relationship can be expressed in terms of the angle of the slope α, where $\alpha =$ angle $SPY =$ angle PXV because the angles YPX and PVX are right angles (PX is the normal to the tangent PY and angle PVX is between two chords of a circle to the diameter). Thus, $\mathbf{F_O} = \mathbf{F_S} \sin\alpha$. From Proposition 4, the force $\mathbf{F_O}$ required for uniform circular motion is directly proportional to the square of the speed v at point P and inversely proportional to the radius of curvature ρ, or $\mathbf{F_O} \propto (\text{speed})^2/(\text{radius})$. Thus, one has that $\mathbf{F_S} = \mathbf{F_O}/\sin\alpha \propto v^2/\rho\sin\alpha$, where v is inversely as the perpendicular SY as demonstrated in the revised Proposition 1 and from the diagram, the chord of curvature PV is equal to $2\rho\sin\alpha$. Thus, $\mathbf{F_S} \propto v^2/\rho\sin\alpha \propto [1/(SY^2 \times PV)]$ or, as Newton puts it above, "the centripetal force of the revolving body is reciprocally as the solid $SY^2 \times PV$." Whiteside notes that this fundamental measure of force

Newton developed in the 1690s "was independently discovered some dozen years afterwards by Abraham de Moivre who ... announced the findings to Johann Bernoulli on 27 July 1705 (N.S.)," who in turn developed his own version and published it "without mention of having received its enunciation from de Moivre in the first place or of Newton's earlier equivalent discovery."[28]

The Revised *Principia*

The unpublished radical revisions make no attempt to conform to the general outline of the 1687 *Principia.* The published revised editions of the *Principia* of 1713 and 1726, however, retain the order and wording of the statements of the propositions of the 1687 edition with only minimal changes. It would have been a Herculean task to seek out all the references to earlier lemmas and propositions, revise them, and then renumber them. Rather, Newton contented himself with tucking pieces of the new theorems and solutions into whatever existing nooks and crannies he could find. Thus although the curvature plays a central and fundamental role in the unpublished revised dynamics, it is hidden away in the revised editions. Nevertheless, it is there for the interested observer.

The Revised Proposition 6

The most obvious way to see the inclusion of curvature in the revised *Principia* is to compare the figures for the 1687 edition with those of the revised editions. The revised Proposition 6, which contains the basic paradigm for solutions to the direct problems that follow it, clearly displays the chord of curvature through the center of force, and Corollary 3 explicitly calls upon properties of the circle of curvature. Figure 5.7 compares the diagrams for Proposition 6. In the figure for the first edition, the line from the point *P* terminates at the center of force *S*. In the figure for the revised editions, the line continues through the center of force and terminates at a point *V*. The enhanced diagram in figure 5.8 makes the meaning of the point *V* clear where the circle *PV* represents the circle of curvature at the point *P* to the general curve *APQ*; the line *PSV* is a chord of the circle of curvature drawn through the center of force *S*; and the line *YS* is the perpendicular to the tangent to the curve at point *P* also drawn through the center of force *S*. The arc of the general curve in the vicinity of the point *P* is approximated by the arc of the circle of curvature. In the published revised Proposition 6, Newton demonstrated that the force $\mathbf{F_S}$ is inversely proportional to the square of the normal *SY* and

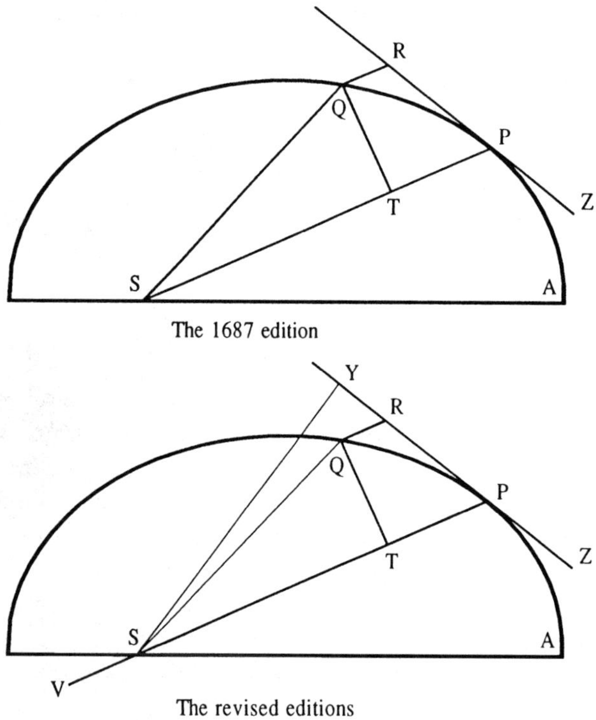

Figure 5.7
A comparison of the diagrams for Proposition 6 from the 1687 *Principia* and the revised (1713 and 1726) *Principia*.

the chord of curvature PV,

$$\mathbf{F_S} \propto 1/(SY^2 \times PV),$$

where $1/(SY^2 \times PV)$ is the curvature measure of force in contrast to $QR/(QT^2 \times SP^2)$, the parabolic measure of force.

The Revised Proposition 9

Proposition 9 presents the direct problem of an equiangular spiral with the center of force at the pole. Figure 5.9 compares the diagrams for the spiral from the first and revised editions. The revised figure also displays the chord of curvature PV through the force center S. Again, the enhanced figure 5.10 shows the circle of curvature. It can be demonstrated, although Newton does not bother to do so, that the chord of curvature PV is equal to twice the radius SP.[29] Further, the tangent YS is equal to $SP \sin \alpha$,

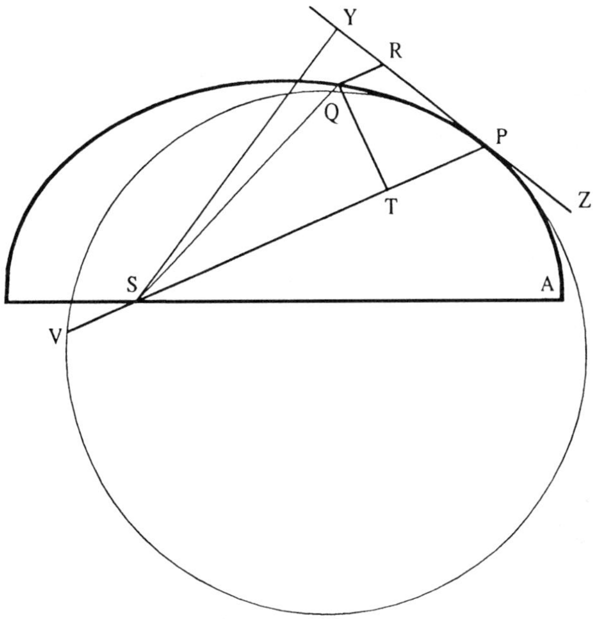

Figure 5.8
The diagram for Proposition 6 for the revised *Principia* with the circle of curvature *PQV* at the point *P* added.

where α is the constant angle *SPY* of the spiral. Thus, from the curvature ratio $1/(SY^2 \times PV)$, the force is proportional to the reciprocal cube of the radius *SP*.

The Revised Proposition 10

Proposition 10 is the direct problem of an elliptical orbit with the center of force at the ellipse's center. Figure 5.11 compares the diagrams appropriate to the first and revised editions for the ellipse, where again Newton has added the point *V* to delineate the chord of curvature through the center of the ellipse. Figure 5.12 is the diagram for Proposition 10 with the circle of curvature at point *P* added. The line *PG* through the center of the ellipse is called the transverse diameter and the line *DK* is constructed parallel to the tangent at the point *P* and is called the conjugate diameter. In the unpublished radical revision, Newton generated an elegant lemma in which he demonstrated that the chord of curvature from a general point *P* through the center of the ellipse is equal to the latus rectum pertaining to point *P*, and then he calculated the chord of curvature

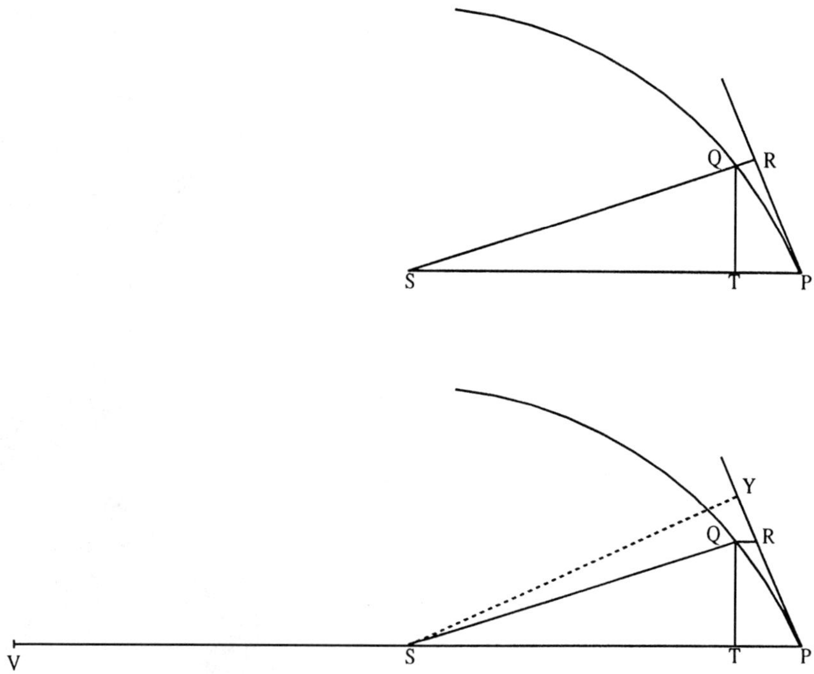

Figure 5.9
A comparison of the diagrams for Proposition 9 from the 1687 *Principia* (top) and the revised (1713 and 1726) *Principia* (bottom).

through the focus, just as Gregory noted in 1690. The conjugate latus rectum L_C pertaining to that conjugate set of diameters is, by definition (see Apollonius, Book I, Proposition 15), in proportion to the conjugate diameter as the conjugate diameter is to the transverse diameter. That is,

$$L_C : DK :: DK : PG,$$

or the conjugate latus rectum is equal to the conjugate diameter squared divided by the transverse diameter,

$$L_C = \frac{DK^2}{PG}.$$

Thus, the chord of curvature $PV = L_C = DK^2/PG$, or what is the same $2DC^2/PC$. From Lemma 12 (which is Apollonius, Book VII, Proposition 31) the area given by the product of the conjugate diameter $2DC$ and the normal PF to it drawn through it to the point P is a constant of the ellipse. In this example the normal PF is equal to the normal YS in the curvature

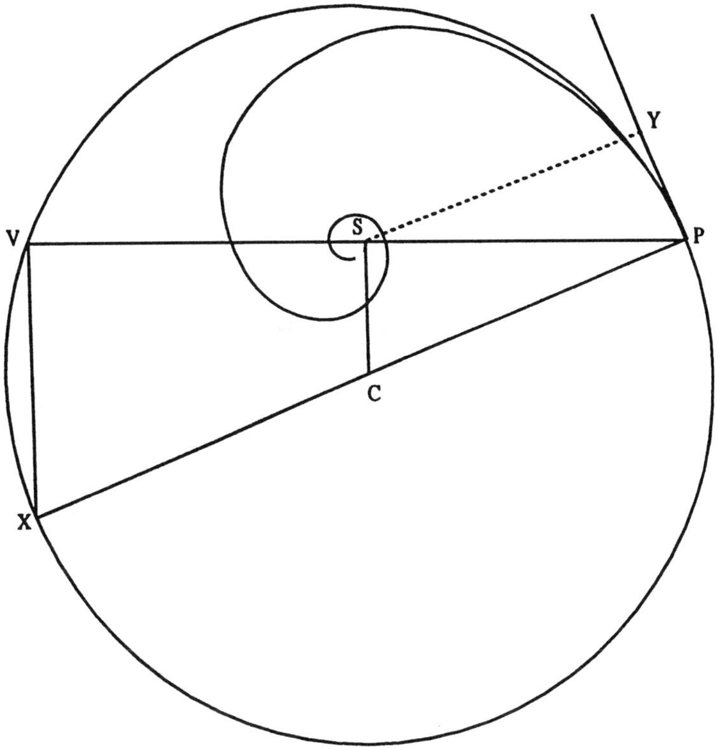

Figure 5.10
The diagram for Proposition 9 for the revised *Principia* with the circle of curvature *PVX* at the point *P* added. The path is a spiral whose pole is the center of force *S*.

measure of force. Therefore the reciprocal of the force is given by curvature measure of force $SY^2 \times PV$ or

$$1/F \propto SY^2 \times PV = \frac{PF^2 \times 2DC^2}{PC}$$
$$= 2(DC^2 \times PF^2)/PC = 1/PC(\text{constant}).$$

Thus, the force is directly proportional to *PC* because the area $DC \times PF$ is a constant of the ellipse.

An Unpublished Revision of Proposition 11

Proposition 11 is the direct problem of an elliptical orbit with the center of force at a focus of the ellipse. It is possible to solve that direct problem in the same manner as the direct problem of Proposition 10. Figure 5.13

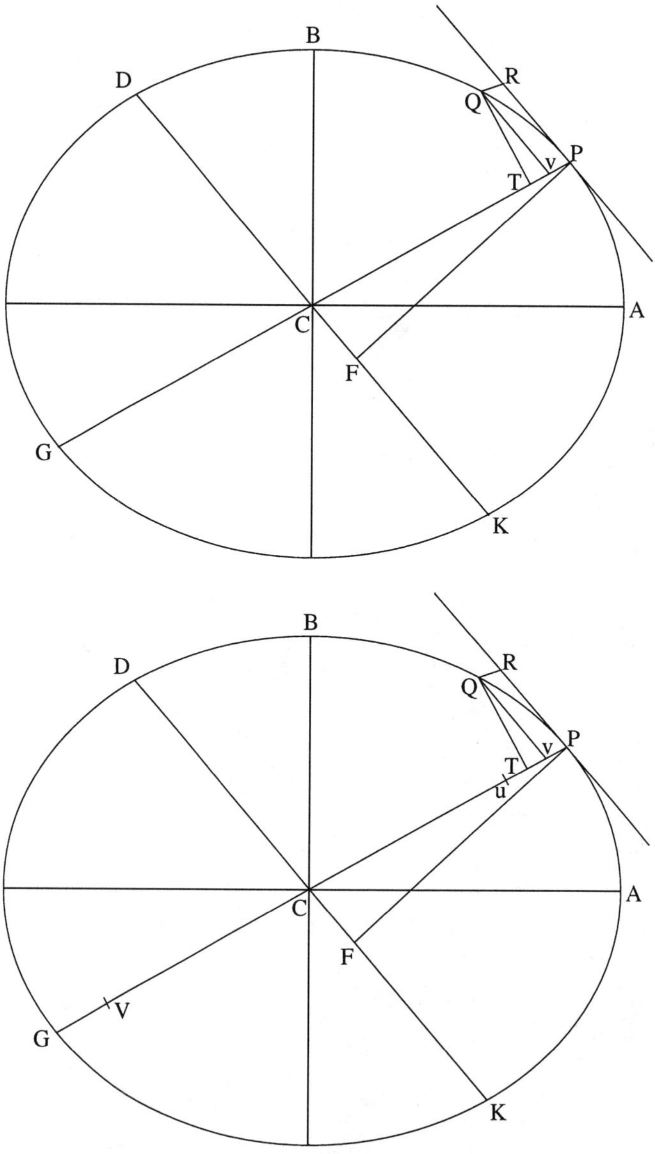

Figure 5.11
A comparison of the diagrams for Proposition 10 from the 1687 *Principia* (top) and the revised (1713 and 1726) *Principia* (bottom).

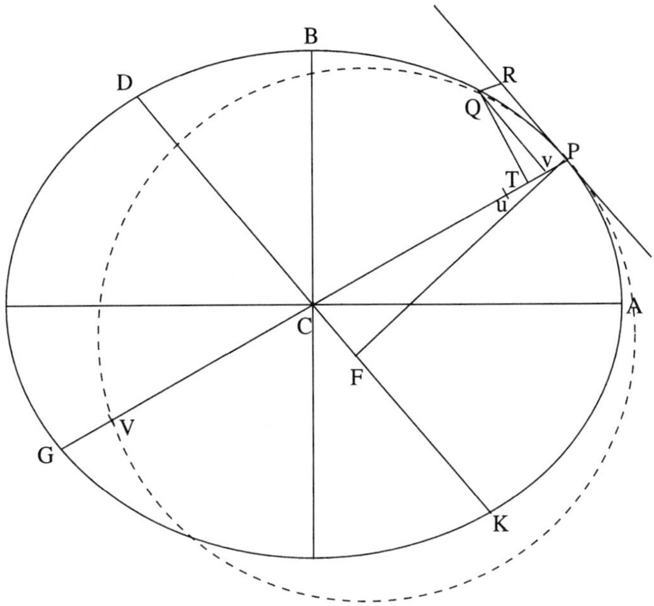

Figure 5.12
The diagram for Proposition 10 for the revised *Principia* with the circle of curvature *PQV* at the point *P* added. The path is an ellipse with the center of force at the center *C* of the ellipse.

shows the chords of curvature through both the center *C* of the ellipse and the focus *S*. It is not difficult to demonstrate that the triangles *PEC* and *PvV* are similar. (From Euclid Book III, Proposition 32, the angle $ZPv =$ the angle PVv. Also, the angle $ZPv =$ the angle *KCG* because *ZP* and *KC* are parallel and angle $KCG =$ angle *PCE*. Thus, triangles *PEC* and *PvV* are equiangular.) Thus the chord of curvature *PV* through the focus of the ellipse is related to the chord of curvature *Pv* through the center of the ellipse by $PV/Pv = PC/PE$ or

$$PV = Pv(PC/PE) = 2(DC^2 \times PF^2/PC)(PC/PE) = 2(DC^2 \times PF^2/PE).$$

The other element in the circular dynamics ratio, the normal *SY*, can be obtained from the similarity of the triangle *SPY* and the triangle *PEF* (see figure 5.14, where the angle $SPY =$ the angle *PEF* because *PF* is normal to *DK* and *DK* is parallel to *YP*).

Thus, the normal *SY* to the tangent through the center of force is equal to $SP(PF/PE)$. The reciprocal of the force is proportional to the curvature measure of force $SY^2 \times PV$ or

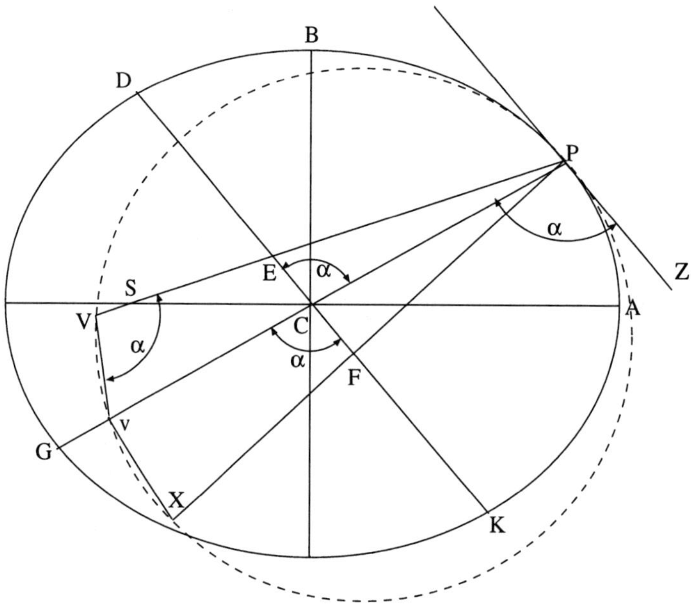

Figure 5.13
The figure shows diameter of curvature *PX* and the chords of curvature *PV* and *Pv* through the focus and center, respectively, of the ellipse *APB*.

$$1/F \propto SY^2 \times PV = SP^2(PF/PE)^2 \times (DC^2 \times PF^2/PE)$$
$$= SP^2 \times (DC^2 \times PF^2/PE^3).$$

Newton demonstrated that the line $PE =$ the line AC (the constant major semi-axis) and since by Lemma 12 the constant area $DC \times PF =$ the area $AC \times BC$, then the force is given as follows:

$$1/F \propto SP^2 \times (AC^2 \times BC^2/AC^3) = SP^2(BC^2/AC)$$
$$= SP^2(L_P/2) = SP^2(\text{constant}),$$

where L_P is the principal latus rectum, $2BC^2/AC$. Thus, the force is proportional to the inverse square of the radius *SP*.

The Published Revision of Proposition 11 Newton did not, however, elect to solve Proposition 11 in that way in the revised *Principia*. In contrast to the diagrams for Propositions 6, 9, and 10, the diagram for Proposition 11 in the revised *Principia* does not contain a representation for the chord of curvature. Figure 5.15 is the plate from the 1729 Motte translation of the 1726 *Principia*. There is much to criticize about the

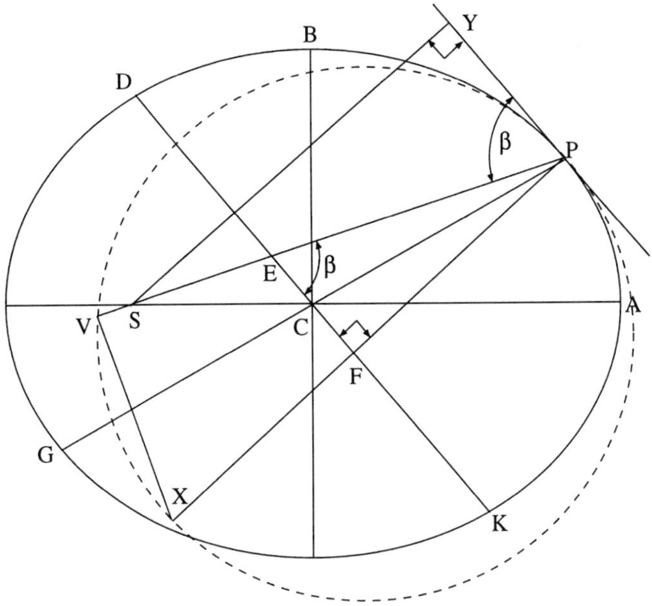

Figure 5.14
The figure shows diameter of curvature *PX*, the chord of curvature *PV*, and the normal to the tangent through the center of force *YS*.

figures (note that in the upper figure the conjugate diameter *DK* should be parallel to the tangent *PR* and the line *PF* should be normal to *DK*), but they are faithful to the *Principia* in that the upper figure, for Proposition 10, contains the point *V* for the chord of curvature but the lower figure, for Proposition 11, does not display that point. That does not mean, however, that Newton did not employ curvature in the alternate solution for Proposition 11. In place of the direct extension of the alternate method of Proposition 10 to solve Proposition 11 discussed above (and employed by Newton in the unpublished radical revision), he used a version of the comparison theorem that he had made central to the unpublished radical revisions. To maintain the general format of the 1687 edition, Newton revised Proposition 7 to accommodate the new theorem rather than to present it as a new and independent proposition. Figure 5.16 compares the diagrams for Proposition 7 from the 1687 *Principia* and the revised *Principia*. In the 1687 edition, Proposition 7 is a simple example of a direct problem, that is, given a circular orbit, find the force directed toward a force center on the circumference of the circle. In the revised editions the orbit remains a circle but the center of force is any

Plate IV. Vol. I. P. 80.

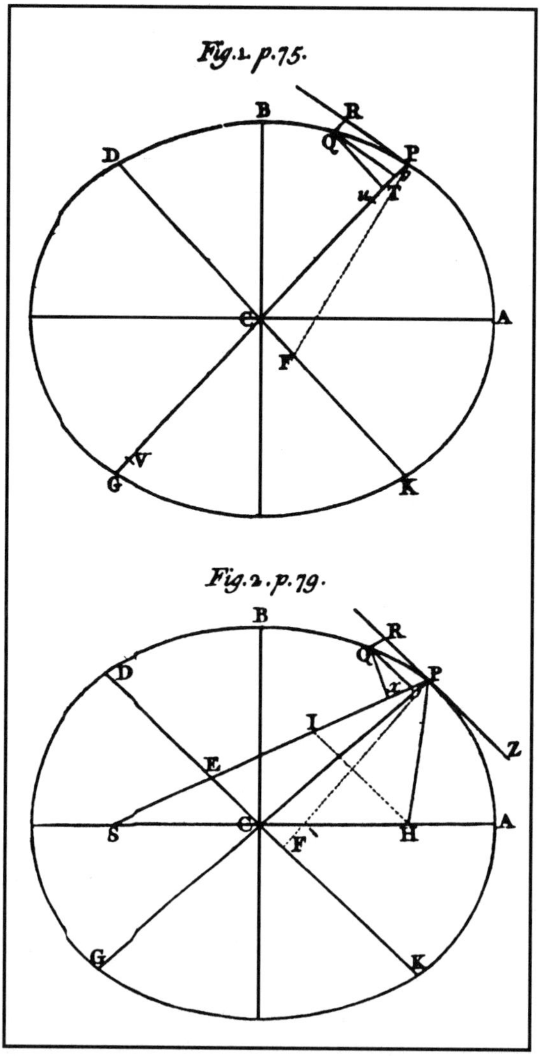

Figure 5.15
The plate containing the diagrams for Propositions 10 and 11 from the 1729 Motte translation of the 1726 *Principia*.

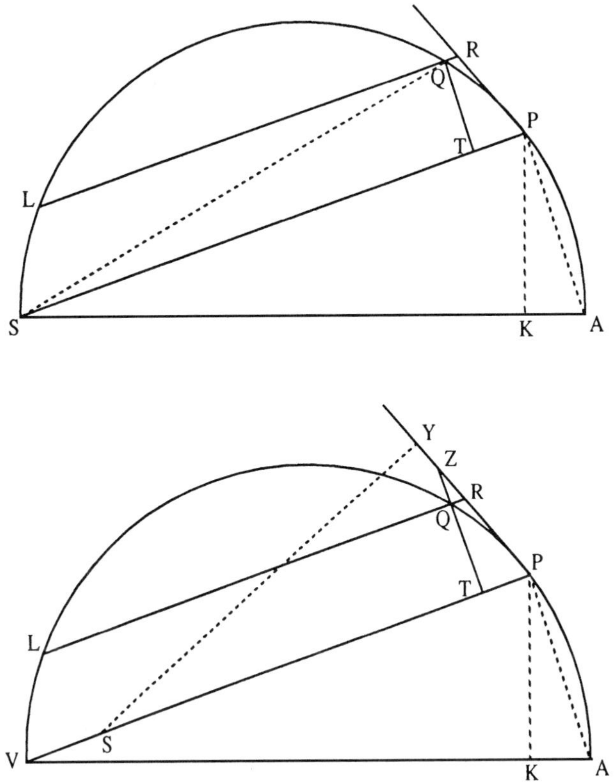

Figure 5.16
A comparison of the diagrams for Proposition 7 from the 1687 *Principia* (top) and the revised (1713 and 1726) *Principia* (bottom).

point rather than a point on the circumference, where *PSV* is the chord of curvature through the force center *S*. In the first corollary the general point is take to lie on the circumference; in the second corollary a relationship is derived for the force at two different points for a given circular orbit; and in the third and final corollary, the relationship is extended to any orbit with the following defense: For the force in this orbit at any point P is the same as in a circle of the same curvature.

This hidden theorem is the one that plays a major role in the unpublished radical revisions, but here it is tucked away at the end of a problem on circular motion and the reference to curvature given by a single sentence. The first-time reader must surely wonder why Newton has wandered down this rather obscure path. Not until the alternate

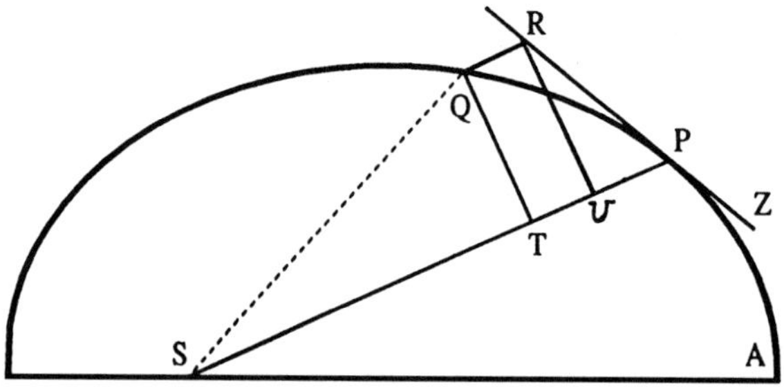

Figure 5.17
Based on the diagram for Proposition 6 for the 1687 *Principia,* with the line *RU* added.

solution to Proposition 11 is one referred back to this hidden comparison theorem.[30]

CURVATURE IN MODERN DRESS

In the 1687 edition of the *Principia,* it is difficult to find traces of Newton's use of curvature in the first three sections of Book I. (It appears twice in Lemma 6 and once in the scholium of Lemma 11.) In the revised editions, however, it is distributed throughout the work, but it is not featured in a central role as it is in the unpublished radical revisions. It is most obvious in Proposition 6 and Lemma 11, less so in Propositions 7, 9, and 10, and quite hidden in Proposition 11 unless one traces it back to the hidden theorem in Corollary 3 of Proposition 7. It is possible to cast the analysis in modern terminology, however, and to display quite clearly the central role of curvature that appears hidden in the *Principia.*

The first relationship is an expression of Newton's Proposition 1, Kepler's law of equal areas in equal times. Figure 5.17 is based upon the diagram for Proposition 6 in the 1687 edition of the *Principia.* The general curve is APQ, the tangent to the curve is ZPR, and the center of force is S. The area A of the triangle SPQ is equal to $(1/2)SP \times QT$, or what is equivalent, $(1/2)SP \times PR\sin(\alpha)$, where $QT = PU = PR\sin(\alpha)$ and α is the angle SPR between the radius SP and the tangent ZPR. If the radius SP is written as r and the tangential velocity at P as v, then in a given time Δt, the chord PR is equal to $v\Delta t$, and the area ΔA is given by $rv\Delta t\sin(\alpha)$. Thus, the rate at which twice the area ΔA is swept out is given as follows:

$$2\Delta A/\Delta t = (rv\Delta t\sin(\alpha))/\Delta t = rv\sin(\alpha) = K,$$

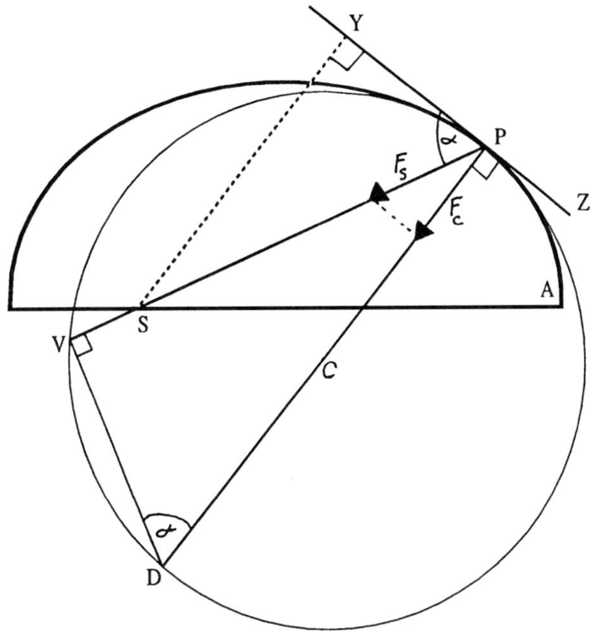

Figure 5.18
Based on the diagram for Proposition 6 for the revised *Principia*, with the points *Q*, *R*, and *T* removed and the circle of curvature at point *P* explicitly displayed.

where *K* is a constant proportional to the area swept out per unit time. This relationship is mathematically equivalent to the modern law of conservation of angular momentum, where *K* is the angular momentum per unit mass.

The second relationship is an expression of Newton's curvature measure: the replacement of motion along an incremental arc of the general curve at point *P* with uniform circular motion along an incremental arc of the circle of curvature at the same point. Figure 5.18 is based upon the diagram for Proposition 6 in the 1713 edition of the *Principia*. As you may recall, in his revised diagram, Newton added the normal to the tangent through the center of force *YS* and extended the line *PS* to the point *V* to display the chord of curvature through the center of force *S*. I have removed the points *Q*, *R*, and *T*, and I have explicitly displayed the circle of curvature *PD* at the point *P*, where the center of the circle of curvature is at *C* and the diameter of curvature is *PD* (i.e., 2ρ). The angle *PVD* between the chord and diameter of the circle is a right angle and α is the angle *SPY* (equal to the angle *SPR* in figure 5.17). The component of force $\mathbf{F_C}$ directed toward the center of curvature C is given by

$$\mathbf{F_C} = \mathbf{F_S}\sin(\alpha) = v^2/\rho,$$

where $\mathbf{F_S}$ is the force per unit mass directed toward the center of force S, and ρ is the radius of curvature. The force $\mathbf{F_C}$ directed toward the center of curvature C provides the centripetal circular acceleration v^2/ρ as required in Proposition 4 of the *Principia*. Newton has explicitly employed this uniform circular replacement in Lemma 11, Corollary 3, and in Proposition 7, Corollary 5, and he has implicitly employed it in other discussions.

The third relationship, the curvature orbital measure of force, is obtained by combining the two relationships. The force $\mathbf{F_S}$ directed toward the center of force S can be written as follows:

$$\mathbf{F_S} = K^2/(r^2\rho\sin^3(\alpha)).$$

This curvature orbital measure is identical with the circular dynamics ratio $1/(SY^2 \times PV)$, which Newton introduced into the revised edition. From figure 5.18, the angle SPY is equal to the angle PDV or α (both YS and PD are normal to the tangent YPZ and the angles PVD and YPD are right angles). The chord of the circle of curvature through the center of force PV is equal to $PD\sin(\alpha)$ or $2\rho\sin(\alpha)$, and the normal to the tangent SY is equal to $SP\sin(\alpha)$ or $r\sin(\alpha)$. Thus, the circular dynamics ratio $1/(SY^2 \times PV)$ also is equal to twice the curvature orbital measure.[31]

If the force is expressed in terms of the curvature orbital measure, $1/[r^2\rho\sin^3(\alpha)]$, then the solutions to the direct problems that Newton selected for the 1687 *Principia* fall into an interesting pattern. In Proposition 7, Newton's first choice of an example of a direct problem, the radius of curvature ρ is a constant, and in the second example, Proposition 9, the angle α is a constant. The solutions to the curvature orbital measure are therefore quite simple.

PROPOSITION 7: *Circle/Circumference*
$\rho =$ constant, $\sin(\alpha) = r/2\rho$.
Thus, the force $\mathbf{F} \propto 1/[r^2\rho\sin^3(\alpha)] = 1/[r^2\rho(r/2\rho)^3] = (\text{constant})/r^5$.

PROPOSITION 9: *Spiral/Pole*
angle $\alpha =$ constant, $\rho = r/\cos(\alpha)$.
Thus, the force $\mathbf{F} \propto 1/(r^2\rho\sin^3(\alpha)) = 1/[r^2(r/\cos(\alpha))\sin^3(\alpha)] = (\text{constant})/r^3$.

The solution to the next two direct problems, the elliptical orbits of Propositions 10 and 11, require a bit more analysis, but not much. The radius of curvature of the ellipse ρ is equal to DC^2/PF and from inspection of the diagrams one can determine $\sin(\alpha)$.

PROPOSITION 10: *Ellipse/Center*
$\sin(\alpha) = (PF/r)$, $\rho = DC^2/PF$.
Thus, the force $\mathbf{F} \propto 1/[r^2\rho\sin^3(\alpha)] = 1/[r^2(DC^2/PF)(PF/r)^3] = r/(DC^2 \times PF^2) = r/(\text{constant})$,
where $(DC \times PF)$ is a constant for the ellipse.

PROPOSITION 11: *Ellipse/Focus*
$\sin(\alpha) = (PF/PE)$, $\rho = DC^2/PF$.
Thus, the force $\mathbf{F} \propto 1/(r^2\rho\sin^3(\alpha)) = 1/[r^2(DC^2/PF)(PF/PE)^3] = 1/[r^2(DC^2 \times PF^2/PE^3)] = (\text{constant})r^2$,
where $PE = AC$, the constant semi-major axis of the ellipse.

The physical problem of the planets dictated the choice of the orbit and focal center for Proposition 11 in 1684, but the choice of the other examples was arbitrary. They may have been suggested by the work on curvature, which Newton began as early as 1664.[32] In his *Methods of Series and Fluxions* of 1671, Newton calculated the radius of curvature for a number of examples, including conic sections and spirals.[33] Newton's statement of 1664 indicated that he intended to use his work on curvature to solve direct problems of orbital motion. Whatever method he had in mind when he made the statement in 1664, however, it could not have been in the form of the curvature orbital measure $1/(r^2\rho\sin^3(\alpha))$, because that measure entails the area law, which Newton did not discover before 1679. Nevertheless, curvature in some form was clearly important for Newton well before 1679, and I think that it is reflected in his choice of examples.

CONCLUSION

The title of this chapter is "Newton's Mature Dynamics: A Crooked Path Made Straight." It is, as you may now have observed, a crude attempt at a pun. The "crooked path" is a reference to the curvature method and the "straight path" is a reference to the parabolic measure. The question is in the tradition of the chicken and the egg: which solution to the elliptical problem came first, the curvature or parabolic measure? It may be that they came into being simultaneously. Certainly the statements in the *Waste Book* and the *On Circular Motion* all occur some twenty years before Newton produced the tract *On Motion* and the first edition of the *Principia*. Arguing for the primacy of the parabolic measure is its role in *On*

Motion, where it appears in Theorem 3 as the sole paradigm for the solution of direct problems and then is applied to the problems that appear as Propositions 7, 10, and 11 in the first edition: the circle/circumference, the ellipse/center, and the ellipse/focus. There is no direct reference to curvature in *On Motion* and only minimal reference to it in the 1687 *Principia*. Not until the proposed radical revisions of the 1690s does Newton employ curvature in a clear and fundamental manner.

The only argument in favor of the primacy of the curvature measure is the cryptic statement of the *Waste Book* and the possibility that the Locke solution of 1690 was perhaps the alternate solution of 1679, a point I have defended elsewhere.[34] But those were the only two items that supported the primacy of the curvature measure, and when I wrote of Newton's "mature dynamics," I referred to the curvature measure and thought of it as coming after the parabolic measure: mature in the sense of coming later. Michael Nauenberg has provided what are for me persuasive arguments that Newton employed curvature in a numerical solution to direct problems, and he also argues that Newton may have had an analytical solution independent of the area law of 1679. You may judge for yourself in his paper "Newton's Early Computational Method for Dynamics."[35] If Newton did indeed employ curvature in such a fashion before 1679, then clearly the void that bothered me has been filled. Moreover, given Newton's very early development of the details of curvature, then perhaps it is the curvature measure that precedes the parabolic measure. In that case, I can still call curvature the "mature dynamics," except now I intend "mature" to mean not that which comes last, as in a mature opinion, but rather "mature" in the sense of that which has been around the longest, as in a mature cheddar (and in my dairy state of Wisconsin, comparison to a good piece of cheese is taken as a compliment).

Notes

1. John Herivel, *The Background to Newton's Principia: A Study of Newton's Dynamical Researches in the Years 1664–84* (Oxford: Clarendon Press, 1965), 130.

2. J. Bruce Brackenridge, "Newton's Mature Dynamics: Revolutionary or Reactionary?" *Annals of Science* 45 (1988): 451–76; and "Newton's Unpublished Mature Dynamics: A Study in Simplicity," *Annals of Science* 47 (1990): 3–31.

3. J. Bruce Brackenridge, "The Critical Role of Curvature in Newton's Developing Dynamics," in P. M. Harman and Alan E. Shapiro, eds., *An Investigation of Difficult Things: Essays on Newton and the History of the Exact Sciences* (Cambridge: Cambridge University Press, 1992), 231–60.

4. Michael Nauenberg, "Newton's Early Computational Method for Dynamics," *Archive for History of Exact Sciences* 46 (1994): 221–52.

5. In contrast to the polygonal and parabolic measures, which require an eventual limiting process, the circle of curvature already matches the first and second derivatives of the general orbit at that point.

6. J. Bruce Brackenridge, *The Key to Newton's Dynamics: The Kepler Problem and the Principia* (Berkeley and Los Angeles: University of California Press, 1995). See chapter 3 for a discussion of these three techniques in the context of Newton's early dynamics before 1669, chapters 4 and 5 for the application of the polygonal and parabolic measures after 1679, chapter 8 for the demonstration of the curvature measure after 1690, and chapter 9 for these techniques in the final revised editions of the *Principia.*

7. Herivel, *Background to Newton's Principia*, 133–35.

8. Ibid., 129–30.

9. Ibid., 192–97.

10. Betty Jo Teeter Dobbs, *The Janus Faces of Genius: The Role of Alchemy in Newton's Thought* (Cambridge: Cambridge University Press, 1991), 185–86.

11. Domenico Bertoloni Meli, *Equivalence and Priority: Newton versus Leibniz* (Oxford: Claredon Press, 1993), 182 (emphasis added).

12. Julian B. Barbour, *Absolute or Relative Motion? A Study from a Machian Point of View of the Discovery and the Structure of Dynamical Theories*, vol. 1, *The Discovery of Dynamics* (Cambridge: Cambridge University Press, 1989), 610 (emphasis added).

13. Curtis Wilson, "Newton's Orbit Problem: A Historian's Response," *College Mathematics Journal* 25 (1994): 195.

14. D. T. Whiteside, trans. and ed., *The Mathematical Papers of Isaac Newton*, 8 vols. (Cambridge: Cambridge University Press, 1967–1981), 1:252–55.

15. Herivel, *Background to Newton's Principia*, 135.

16. Whiteside, *Mathematical Papers of Isaac Newton*, 1:456.

17. Herivel, *Background to Newton's Principia*, plate 2 (between p. 12 and p. 13).

18. Whiteside, *Mathematical Papers of Isaac Newton*, 1:456, n. 3. In a footnote to this statement, Whiteside notes that "Newton equates the (centripetal) force at a general point on an ellipse with that towards the same centre instantaneously at rest in the plane of the circle of curvature at the point. This insight, in its immediate extension to a point on an arbitrary curve, allowed Newton to reduce all problems of centripetal forces in arbitrary orbits to the equivalent ones which treat of centripetal attractions in circular paths. (The concept is fundamental in Newton's account of centripetal force in his *Principia*: Book One.)"

19. Nauenberg, "Newton's Early Computational Method for Dynamics," 221–52.

20. Ibid., 235–39. On 13 December 1679, Newton sent Hooke a letter on orbital dynamics containing a diagram for the orbit of a body subject to a constant central force. In the letter Newton did not reveal any details of his method of construction beyond a cryptic reference to "the method of indivisibles." This figure has been the subject of considerable scholarly concern, with most finding it to be wanting in many respects. Nauenberg, however, demonstrates that Newton's method was correct, and that the error was in Newton's drawing of the figure. He demonstrates that Newton used discrete arcs of the circle of curvature to approximate the curve and uses the method to reproduce the correct form of Newton's figure.

21. H. W. Turnbull, J. F. Scott, A. R. Hall, and Laura Tilling, eds. and trans., *The Correspondence of Isaac Newton*, 7 vols. (Cambridge: Cambridge University Press, 1959–1977), 2:308.

22. Nauenberg, "Newton's Early Computational Method for Dynamics," 234.

23. Whiteside, *Mathematical Papers of Isaac Newton*, 3:384.

24. Ibid., 6:569–89. The statement of the proposed Proposition 1, Kepler's area law, remains as it is in the 1687 edition, although six new corollaries are added. The statements of the next four propositions also remain unchanged. The statement of the proposed Proposition 6, however, undergoes a dramatic revision. In the 1687 edition, Proposition 6 introduces the parabolic measure of force, and several propositions follow in which it is used to generate solutions for a series of direct problems. The proposed revised Proposition 6 is the first of three new propositions that introduce a new technique for solving direct problems, the comparison theorem, that bears no resemblance to the original Proposition 6. The parabolic measure of force of the original Proposition 6 is transferred to the proposed Proposition 9, to which is added yet another measure of force, the curvature measure of force. The proposed Proposition 10 provides a measure of force for motion in a conic directed to any point, and the proposed Proposition 11 produces yet another method of attack.

25. For a reproduction of the original manuscript page, see Brackenridge, "Newton's Mature Dynamics," 474.

26. Cambridge University Library, Manuscript Add.3965.6:37r. The following text then supports the statement of the proposition:

> Let S be the centre of forces, the body P revolving on the orbit PR, the circle[s] PVX [and Pvx] touch the orbit at P, at the concave parts of it; whichever [circle] is of the same curvature with the orbit at the point of contact, the chord PV of this circle constructed from the body P through the centre S; the straight line PY touching the orbit at P and SY perpendicular from the center S falling on this tangent: I say that the centripetal force of the revolving body is reciprocally as the solid $SY^2 \times PV$.

27. For a discussion of this point, see "Newton's Unpublished Mature Dynamics," 14.

28. Whiteside, *Mathematical Papers of Isaac Newton*, 6:548, n. 25. See this note for an extended discussion of the claims and counterclaims of British and continental scholars concerning the demonstration of this measure of force.

29. S. Chandrasekhar, *Newton's Principia for the Common Reader* (Oxford: Clarendon Press, 1995), 82–86. This recent work by a distinguished physicist drastically distorts this figure for Proposition 9 of Book I. The figure has an added second point and the text attributes to Newton a solution not presented by him, but by Clarke (published in 1730). Even more disturbing is the alternate solution in Proposition 9. The body of the text correctly demonstrates that the chord of curvature *PV* through the pole *S* of the spiral is equal to twice the radial distance *SP*. In the diagram, however, *PV* is more than three times *SP* and any attempt to visualize the circle of curvature is destroyed.

30. Dana Densmore, *Newton's Principia: The Central Argument*, with translations and illustrations by William H. Donahue (Santa Fe, NM: Green Lion Press, 1995). As an example of the confusion that Newton's revision of Proposition 7 can cause for even veteran readers, consider the choices made by the author of a scholarly guided study of the *Principia*. The foreword notes that the translation "is based upon the third edition ... [but] all the alternative proofs are omitted" (*xv*). Thus, the author omits the major display of curvature, the curvature measure of force, $1/SY^2 \times PV$, from her presentation of Proposition 6 and "the central argument" is to be based on the parabolic measure of force, $QR/(QT^2 \times SP^2)$. It is very difficult, however, to escape from Newton's explicit inclusion of curvature in the revised editions. Recall that in the 1687 edition, Proposition 7 was the relatively simple example of a circular orbit with the center of force on the circumference. To derive the "comparison theorem" used in the alternative proof of Proposition 11, Newton revised and extended Proposition 7 from a simple example of a direct problem to a given force center (the circumference) to a more complex problem with the force center at any point. The solution to the simple example is given as a special case in the first corollary, which the author elects to omit. Retained, however, are the next two corollaries that provide the comparison theorem that is designed to be used in the alternate solution of Proposition 11, which the author has omitted. Thus, the simple teaching example of the 1687 edition is lost and a more complex example is retained, one that has immediate application only to a solution that is itself omitted (150–60).

31. For a discussion of this point, see Brackenridge, "Critical Role of Curvature," 254.

32. Whiteside, *Mathematical Papers of Isaac Newton*, 1:245–97.

33. Ibid., 3:169.

34. J. Bruce Brackenridge, "The Locke/Newton Manuscript: Conjugates, Curvatures, and Conjectures," *Archives internationales d'histoire des sciences* 43 (1993): 280–92.

35. Nauenberg, "Newton's Early Computational Method for Dynamics."

6

Newton on the Moon's Variation and Apsidal Motion: The Need for a Newer "New Analysis"

Curtis Wilson

Newton in his assault on the lunar problem during the 1680s relied crucially on techniques he had developed in his period of intense algebraic exploration, from the mid-1660s to the early 1670s. These techniques included use of the binomial theorem for approximations, determination of the curvature of algebraic curves at given points, and integration of sinusoidal functions. How far could they take him: at what point would he be brought to a standstill and why?

The Continental mathematicians took up the lunar problem only in the 1740s: why this long delay? Their methods, it is generally acknowledged, were "more powerful" than Newton's, but can we specify precisely in what this greater power consisted?

I propose here a moderately detailed comparison of Newton's methods with those of Euler, Clairaut, and d'Alembert, in the calculation of two quantities involved in the determination from theory of the moon's longitude: the inequality known as the variation (or more strictly, that part of it that is independent of the eccentricity), and the apsidal motion. In the case of the variation, the values Newton obtained in the *Principia*, Book III, Propositions 26, 28, and 29, for the relevant numerical parameters, namely, the maximum value of the inequality in longitude, and the ratio of the major to the minor axis of the variational orbit, do not differ significantly from those obtained by L. Euler in 1753, by J. A. Euler in 1766, or by G. W. Hill in 1878. But Newton's methods contrast sharply with those of the Eulers and Hill. They were approximative from the start. Their success, it turns out, rested on an assumption that Newton did not justify. And Newton's procedure, unlike those of his successors, provided no check on the accuracy of his results.

In the case of the apsidal motion the tale is different. The manuscript embodying Newton's attempted derivation of it, probably dating from 1686, was not included in the *Principia*. It was published in full for the first time by Whiteside in 1974, along with detailed technical commentary.[1] Newton, Whiteside shows, changed the premises of his calculation more than once. His final calculation of the annual motion of the apse, in Whiteside's phrase, was "adeptly fudged,"[2] though it differs from Flamsteed's empirical value by nearly 1°50′. In the Scholium to Book III, Proposition 35, in the first edition of the *Principia*, Newton described this calculation as "too complicated and cluttered with approximations, and insufficiently accurate" to be included. The attempted derivation, I shall argue, begins with lemmas from which the first-order term of the apsidal motion can be derived but contains fatal flaws in the two main propositions. In the later editions Newton deleted all mention of it, presumably having concluded (correctly, I shall argue) that it was unsalvageable.

The first successful derivation of the Moon's apsidal motion (or rather, of most of it) was announced some sixty-three years later, by Alexis-Claude Clairaut, in May 1749. Euler obtained a derivation in good agreement with Clairaut's by mid-1751. These derivations left unelucidated the question of the solution's completeness. Jean le Rond d'Alembert published a more perspicuous derivation, with the degree of approximation made explicit, in 1754. Success came for Newton's successors only with a new approach, different from any he had envisaged: algorithmic and global.

The Continental mathematicians began with the differential equation, the bequest of Leibniz. It made possible, at the start, the framing of dynamical problems in an exact way, essentially free of approximations. Whether or how the differential equations could be solved was, to be sure, another question. But up to the 1740s, the Continental mathematicians failed even to formulate the lunar problem, probably because they lacked a requisite procedure: the integration of sinusoidal functions. Newton had applied this procedure in both the *Principia* and in his manuscript on the apsidal motion, though in a geometrical mode that makes it almost unrecognizable today. Euler from 1739 onward systematized the calculus of trigonometric functions as a symbolic discipline and applied it in the solution of differential equations. It was he who first attempted the derivation of a lunar theory by the (approximative) solution of differential equations. In this way perturbational problems entered the purview of the Continental mathematicians.

Yet the problem of the motion of the moon's apse did not yield at once to the new techniques. An unexpected further hurdle had to be surmounted: it was necessary to abandon a residual trust in geometrical conceptualization. *Successive* approximations were required, and these needed to be ordered according to degree of approximation. The algorithmic machine had to be prompted to churn out, at successive levels

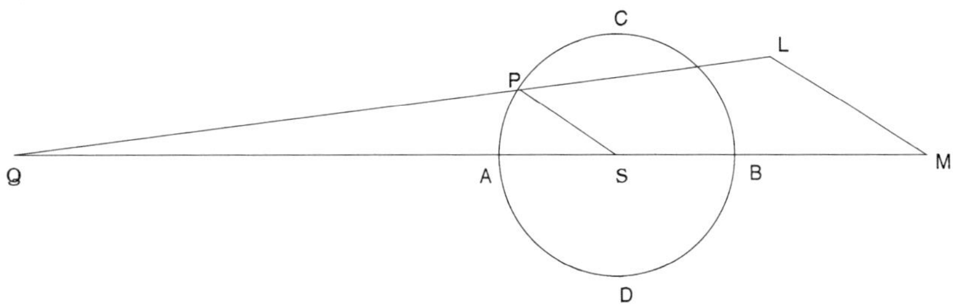

Figure 6.1
Newton's diagram for Book III, Proposition 25.

of approximation, the successive contributions to the apsidal motion of algebraic terms whose relevance to that motion was indiscernible to direct insight.

The Variation

Variation, discovered by Tycho in the 1590s, is an inequality that involves a sinusoidal variance in the moon's mean motion. The effect is that, other things being equal, the moon maximally lags behind its mean position in the octants before the syzygies and maximally exceeds its mean position in the octants after the syzygies. Tycho assigned $40'30''$ as the amplitude of this inequality; a more accurate value is $39'31''$. In Book III, Propositions 26, 28, and 29, which remain essentially unaltered in the three editions of the *Principia*, Newton gives his quantitative account of the part of the variation that is independent of the eccentricity.[3]

Proposition 26 presupposes the result of Proposition 25, which determines a quantitative relation between the solar forces that perturb the moon's motion about the earth and the far greater force by which the earth attracts the moon. The diagram for Proposition 25 (here figure 6.1) shows a circular orbit *CADB* for the moon; the moon is at *P*, the earth at *S*, and the sun at *Q*. (These are the designations used in the first edition; in the later editions *S* is replaced by *T*, and *Q* by *S*.) The length *SQ* in the diagram is used to represent the acceleration **SQ** of the earth toward the sun. (In Newton's text a designation like "*SQ*" signifies either the length of a line or the magnitude of an acceleration, depending on context; for clarity's sake I shall use bold type for a length representing the magnitude of an acceleration, and indicate the direction of the acceleration by the order of the letters.) The acceleration **LQ** of the moon toward the sun is then given, in accordance with the inverse-square law, by

$$\frac{\mathbf{LQ}}{\mathbf{SQ}} = \frac{SQ^2}{PQ^2}. \tag{6.1}$$

Newton resolves **LQ** into two components, **LM** parallel to *PS*, and **MQ** parallel to *SQ*. By subtracting **SQ** from **MQ**, he obtains **MS**, the net acceleration parallel to *QS* with respect to the earth considered at rest.

Because of the great distance of the sun, *LMSP* is very nearly a parallelogram, and **LM** varies but slightly in magnitude as the moon circles the earth, so that its mean value may be taken to be **PS**. Thus

$$\frac{\text{Mean value of } \mathbf{LM}}{\mathbf{SQ}} = \frac{\mathbf{PS}}{\mathbf{SQ}} = \frac{PS}{SQ}. \tag{6.2}$$

But since, by Corollary 2 of Book I, Proposition 4, the centripetal forces of bodies moving in circles are as the radii divided by the squares of the periodic times,

$$\frac{\mathbf{SQ}}{\text{Earth's attraction of moon}} = \frac{\text{SQ}/(365.256)^2}{\text{PS}/(27.3215)^2}. \tag{6.3}$$

Compounding (6.2) and (6.3) yields

$$\frac{\text{Mean value of } \mathbf{LM}(=\mathbf{PS})}{\text{Earth's attraction of moon}} = \frac{1}{178.725}. \tag{6.4}$$

In contrast to **LM**, the component **MS** varies round the orbit from a value of zero near the quadratures to a maximum value at the syzygies. To obtain a quantitative expression for it, Newton assumes the sun to be sufficiently distant so that *LP* can be regarded as parallel to *MS*.[4] See figure 6.2. Under this assumption, not only is **LM** reduced to its mean quantity **PS**, but Newton says that **MS** is reduced to its "mean quantity" 3**KP**, a varying quantity.

Today we would write, in notation deriving from Euler, $\mathbf{KP} = \mathbf{SP}\sin\alpha$, where α is the angle *CSP* between **SP** and the line of quadratures *CD*. Newton in his "Epitome of Trigonometry," written in the 1670s, designated sines by the letter "s", but these were lines rather than ratios: "s *A*" meant what we would write as "$R\sin A$," *R* being the radius of a circle.[5] Yet Newton's grasp of the quantitative relations cannot have been essentially different from ours.

Newton undoubtedly established that "the mean quantity" of **MS** is 3**KP** by applying the binomial theorem. Equation (6.1) implies that

$$\frac{\mathbf{MQ}}{\mathbf{SQ}} = \frac{\mathbf{LQ}}{\mathbf{SQ}} \cdot \frac{\mathbf{MQ}}{\mathbf{LQ}} = \frac{SQ^3}{PQ^3}; \tag{6.5}$$

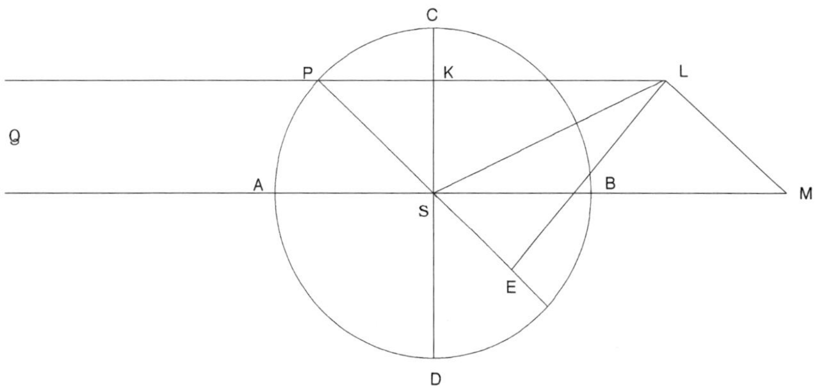

Figure 6.2
Diagram for Newton's derivation of **MS** = 3**KP**.

separando we then have

$$\frac{\mathbf{MQ}-\mathbf{SQ}}{\mathbf{SQ}}=\frac{\mathbf{MS}}{\mathbf{SQ}}=\frac{SQ^3}{PQ^3}-1. \tag{6.6}$$

But $PQ = SQ - PK$, very nearly, and therefore

$$\frac{SQ^3}{PQ^3}=\left(1-\frac{PK}{SQ}\right)^{-3}. \tag{6.7}$$

The first two terms of the expansion of the right-hand member of (6.7) are $1+3(PK/SQ) = 1+3(PS/SQ)\sin\alpha$, and since $PS/SQ = 1/178.725$, the further terms, involving as they do the square and higher powers of this fraction, are relatively negligible. Hence from (6.6)

$$\frac{\mathbf{MS}}{\mathbf{SQ}}=\frac{3PK}{SQ}=\frac{3PS\sin\alpha}{SQ}. \tag{6.8}$$

Compounding (6.8) with (6.3), we obtain

$$\frac{\mathbf{MS}}{\text{Earth's attraction of moon}}=\frac{3\sin\alpha}{178.725}. \tag{6.9}$$

Combining (6.4) with (6.9) and writing **PS** for the mean value of **LM**,

$$\mathbf{MS} = 3\mathbf{SP}\sin\alpha = 3\mathbf{KP}. \tag{6.10}$$

In Proposition 26 Newton adds the two components **LM** and **MS** to obtain the total acceleration **LS** produced by the perturbing force, then resolves **LS** into transverse and radial components, **LE** and **ES**. Proposition 26 requires that we know the magnitude of **LE**. In modern notation, since MS is assumed equal to LP,

$$\mathbf{LE} = \mathbf{MS}\cos\alpha = 3\mathbf{SP}\sin\alpha\cos\alpha = \tfrac{3}{2}\mathbf{SP}\sin 2\alpha. \qquad (6.11)$$

Newton, by contrast, uses the similar triangles PLE and PSK to derive

$$LE = \frac{3PK \cdot SK}{SP}; \qquad (6.12)$$

since $PK = PS\sin\alpha$ and $SK = PS\cos\alpha$, (6.12) is quantitatively equivalent to (6.11).

The task of Proposition 26 is "to find the hourly increment of the area which the Moon, by a radius drawn to the Earth, describes in a circular orbit." In III.28 the orbit with which we are concerned turns out not to be circular: perturbing forces acting on a pristinely circular orbit flatten it in the direction of the sun. Newton does not construct this orbit starting from the forces, but rather assumes a circular orbit, then shows how it must be modified. The assumed circular orbit offers the special advantage that the angle and area swept out by the radius are proportional to each other.

To find the increment in the moon's speed as it moves from C to any point P between C and A, it is necessary to integrate the acceleration **LE** over time. Newton presents the integration geometrically. He takes the angle α to represent the moon's angle of mean motion, increasing proportionally to the time. In (6.8) and (6.11) it represented the moon's actual motion. Since in the sequence Propositions 26. 28, 29 Newton's aim is to find the maximum difference between the angle of mean motion and the angle of actual motion, these two angles will later have to be distinguished. The difference is luckily small enough not to influence the result of the integration significantly. Constructing CG at right angles to SC and equal to it, and connecting GS (see figure 6.3), Newton shows that the area $FKkf$ is proportional to $\mathbf{LE} \cdot d\alpha = 3PK \cdot SK \cdot d\alpha/SP$ (see equation (6.12); $FK = SK = SP\cos\alpha$ and $Kk = Pp\sin\alpha = SP\sin\alpha\, d\alpha = PKd\alpha$, while SP is assumed constant). The integral from 0 to α, namely the area $FKCG$, then represents the total increment in the moon's speed (let us call it Δv) as it moves from C to P. From equation (6.11) it is apparent that this integral is

$$\Delta v = \int_0^\alpha \mathbf{LE} \cdot d\alpha = \frac{3\mathbf{SP}}{4}(1 - \cos 2\alpha). \qquad (6.13)$$

From Newton's geometrical formulation, (6.13) is scarcely obvious. But later in the proposition Newton states that the increment in rate of

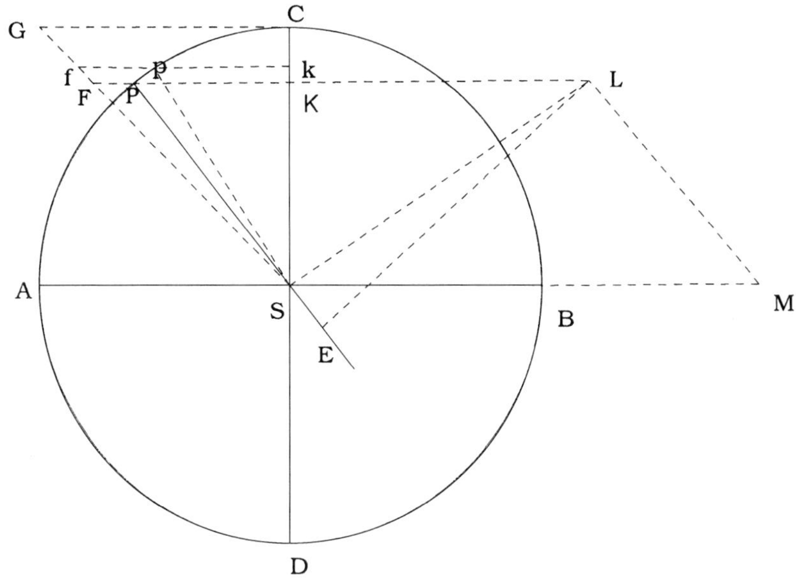

Figure 6.3
Newton's diagram for Book III, Proposition 26.

areal description is proportional to "the versed sine of the double distance of the Moon from the nearest quadrature," that is, to $(1 - \cos 2\alpha)$. In Proposition 29 he says that this increment is proportional to $\sin^2 \alpha$. Newton evidently knew his way among the relevant trigonometrical identities and knew that he was integrating the sine of twice α.

Equation (6.13) gives the increment in speed Δv, as the moon passes from quadrature at C to point P through an angle α about the earth S. When $\alpha = \pi/2$, this increment becomes $(\Delta v)_{CA} = 3\mathbf{SP}/2$. On the left we have an increment in speed, and on the right an acceleration **SP**. The equation is correct numerically, but to have the units right, we must understand the acceleration to be multiplied by unit time so as to yield a speed. Since we are measuring time in terms of α, the unit of time will be the time required for the moon to move through one radian at its mean speed v_{av}; that is, it will be r/v_{av}, where r is the mean earth-moon distance. Thus,

$$(\Delta v)_{CA} = \frac{3SP}{2} \cdot \frac{r}{v_{av}}. \tag{6.14}$$

Now by (6.4), $\mathbf{SP} =$ (Earth's mean attraction of the moon)/178.725. In accordance with Corollary 1 of Book I, Proposition 4, of the *Principia*, the earth's mean attraction of the moon may be expressed by v_{av}^2/r, and if this accelerative force were to act over unit time, accelerating a body from rest, the body's final velocity would be $(v_{\mathrm{av}}^2/r) \cdot (r/v_{\mathrm{av}}) = v_{\mathrm{av}}$. Hence

$$(\Delta v)_{CA} = \frac{3v_{\mathrm{av}}}{2(178.725)} = \frac{v_{\mathrm{av}}}{119.15}. \tag{6.15}$$

Thus in the moon's passage from C to A, its linear speed, and therefore (because the orbit is at present assumed to remain circular) its areal velocity, is increased by 1/119.15 of its mean value. (Newton's route to this result is a little different but equivalent.)

In the subsequent passage of the moon from syzygy to quadrature, this increase in speed is reversed, so that v_{av} retains the same value in each quadrant. Newton represents this mean rate by the number 11915, so that the increment is 100. The moon will be moving at this mean rate when $\alpha = 45°$, and therefore Newton puts for the least rate in the quadratures $11915 - 50 = 11865$, and for the greatest rate in the syzygies $11915 + 50 = 11965$.

"But these things take place only in the hypothesis that the Sun and the Earth are at rest, and that the synodical revolution of the Moon is completed in $27^{\mathrm{d}}.7^{\mathrm{h}}.43^{\mathrm{m}}$" (Book III, Proposition 26, second paragraph). The numbers need therefore to be increased in the ratio of the moon's synodical to its sidereal period, or 1.080853 to 1. The whole increment then becomes 100/11024 parts of the mean rate, and the least rate in the quadratures comes to be to the greatest rate in the syzygies as 10973 to 11073.

This result is used in Proposition 28, "to find the diameters of the orbit in which, without eccentricity, the Moon would move." The assumption is once again that, in the absence of the perturbing accelerations **LE** and **ES**, the moon would move uniformly in a circle about the earth's center. Since the perturbing forces as previously approximated are symmetrically disposed with respect to the lines of quadratures and syzygies, Newton assumes the resulting orbit will be symmetrical with respect to these same two lines. His general strategy is to determine, in two different ways, the ratio of the curvature of the orbit at syzygy to its curvature at quadrature. The two expressions of the ratio are set equal, and from the resulting equation he obtains algebraically a numerical value for the ratio of the orbital radius at quadrature to the orbital radius at syzygy.

The operations involved in the solution of the final algebraic equation are straightforward and need not occupy us. I focus rather on the two determinations of curvature. At the start of the proposition Newton

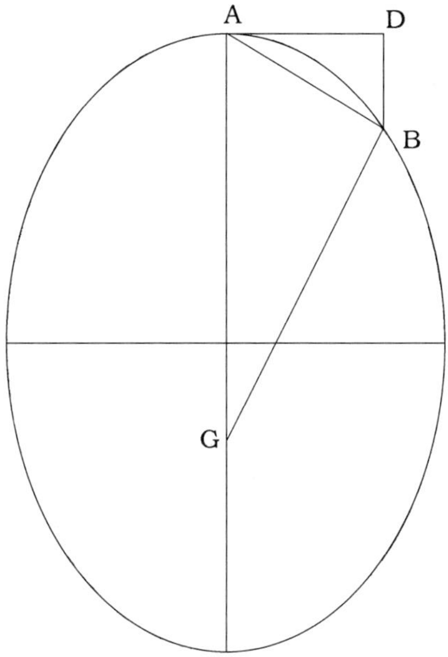

Figure 6.4
Diagram for Newton's Lemma 11.

announces: "The curvature of the orbit which a body describes, if attracted in lines perpendicular to the orbit, is directly as the force of attraction, and inversely as the square of the velocity. I estimate the curvature of lines compared one with another according to the evanescent ratio of the sines or tangents of their angles of contact to equal radii, supposing those radii to be infinitely diminished." In Lemma 11 of the *Principia* Newton had implicitly defined the diameter of curvature of a curve at a point: in the figure for that lemma (see figure 6.4, in which ABG is a right angle), it was given by

$$\lim_{B\to A}\left(\frac{AB^2}{BD}\right). \tag{6.15}$$

But years before, from 1664 onward, Newton's manuscripts show an intense concern with the fluxional determination of the curvature of algebraic curves.[6] In his *De Methodis Serierum et Fluxionum*, completed around 1671, he derived an analytic procedure for determining the limiting position of the intersection of the normals to a curve at P and Q as Q approaches P, hence for determining the curvature. He was apparently the first to do so, though not the first to publish the result. His procedure is essentially equivalent to the following Leibnizian-style formula for the radius of curvature:

$$\rho = \frac{\left[1 + \left(\frac{dy}{dx}\right)^2\right]^{3/2}}{\frac{d^2y}{dx^2}} = \frac{[1 + z^2]^{3/2}}{\frac{dz}{dt} \cdot \frac{dt}{dx}}, \tag{6.16}$$

where $z = (dy/dt)/(dx/dt)$.[7] (Newton's final formula lacks the factor dx/dt, which he sets equal to unity.) The parametric definitions of x and y in terms of an imaginary time t in which the curve is traced lend themselves easily to dynamical applications, the second-order derivative being interpreted as an acceleration. If the acceleration is entirely in the y direction, and the velocity at the point where curvature is to be determined is entirely in the x direction, (6.16) can be shown to reduce to

$$\rho = \frac{(dx/dt)^2}{d^2y/dt^2}, \tag{6.17}$$

so that the curvature, which Newton identifies with the reciprocal of the radius of curvature, is precisely the force (that is, the "accelerative force" or acceleration) divided by the square of the velocity, as implied in the above quotation from Proposition 28.

To calculate the curvature of the lunar orbit at quadrature and syzygy, it is thus necessary to know the total radial accelerative forces in those places, consisting of the earth's attraction and the added or subtracted radial perturbing force due to the sun. If the latter acceleration is derived in a manner parallel to the derivation of **LE** in equation (6.11), we obtain

$$\mathbf{ES} = \mathbf{MS}\sin\alpha - \mathbf{SP} = 3\mathbf{SP}\sin^2\alpha - \mathbf{SP} = \frac{\mathbf{SP}}{2}(1 - 3\cos 2\alpha). \tag{6.18}$$

By this formula, **ES** reduces at quadrature, where $\alpha = 0°$, to $-\mathbf{SP} = \mathbf{PS}$, a radial acceleration inward, equal by (6.4) to the Earth's attraction/178.725. At syzygy, where $\alpha = 90°$, it reduces to $2\mathbf{SP}$, a radial acceleration outward. These conclusions can be derived from (6.4) and (6.10), without recourse to an explicit formula for **ES** such as (6.18).[8]

Newton now compounds the ratio of the accelerative forces with the square of the inverse of the ratio of the speeds (as found in III.26) to obtain the ratio of the curvatures. If (as in figure 6.5) AS is the orbital radius at syzygy, CS the orbital radius at quadrature, and N their mean, or

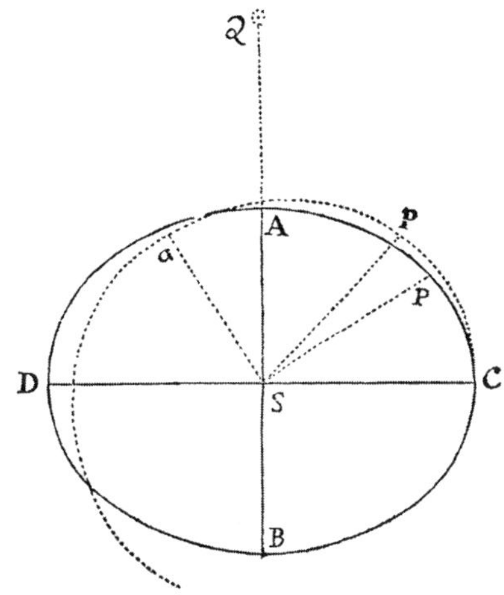

Figure 6.5
Newton's diagram for Book III, Proposition 28 (as in first edition).

$(AS + CS)/2$, then the curvatures at syzygy and quadrature prove to be as $(2151969AS \cdot CS \cdot N - 24081AS^3)$ to $(2191371AS \cdot CS \cdot N + 12261CS^3)$.

To obtain a second expression for the ratio of the curvatures at A and C, Newton introduces an assumption: "Because the figure of the Moon's orbit is unknown, let us, in its stead, assume the ellipse DBCA, in the center of which we suppose the Earth to be situated, and the greater axis CD to lie between the quadratures as the lesser AB between the syzygies" (Book III, Proposition 28, second paragraph). Thus the ellipse is rotating with the sun; the curvatures we are concerned with belong, not to points in the stationary ellipse, but to points in the curve produced by the rotating ellipse: the curve indicated in figure 6.5 by the dotted line *Cpa*.

How good is the approximation that substitutes a rotating ellipse for the variational orbit? G. W. Hill investigated this question by numerical integration in his "Researches in the Lunar Theory" of 1878.[9] The answer, as he found, depends on the number of lunations per year. For our moon, with 12.36875 months per year, the approximation is very close. For imaginable moons with three or fewer lunations per year—moons much farther from the earth than ours—the approximation becomes unsatisfactory; and for "the moon of maximum lunation"—a moon with the longest possible month (= 0.56 year), and placed as far from the earth as can be while still moving round the earth—the variational orbit has sharp cusps at the points of quadrature.

To obtain the ratio of the curvature of the curve *Cpa* at *a* to its curvature at *C*, Newton sets up a series of proportions which, by the Euclidean operations of compounding, *separando* and *componendo*, yield the ratio sought. In finding these proportions, Newton says he proceeded "by computation": "all these relations are easily derived from the sines of the angles of contact, and of the differences of those angles" (Book III, Proposition 28, second paragraph). Newton is here using a consequence of formula (6.16). Also, he must be applying the binomial theorem. One of the proportions, for instance, states that the curvature of the stationary ellipse at A is to curvature of a circle of radius SA as SA^2 is to SC^2. The sine of the angle of contact, in each case, is the vertical distance between the curve and the tangent to the curve at A, for a given abscissa or horizontal distance from S along SC. Taking SC as the x-axis, SA as the y-axis, and setting $SC = \mathbf{a}$, $SA = \mathbf{b}$, we find that, for a given value of x, the quotient of the sine of the angle of contact for the ellipse by the sine of the angle of contact for the circle is

$$\frac{\mathbf{b} - (\mathbf{b}/\mathbf{a})\sqrt{\mathbf{a}^2 - x^2}}{\mathbf{b} - \sqrt{\mathbf{b}^2 - x^2}}. \tag{6.19}$$

For $x = 0$, (6.19) reduces to the indeterminate quotient 0/0. To find its value, Newton presumably expanded the radicals by the binomial theorem; this done, the result is found to be $\mathbf{b}^2/\mathbf{a}^2$, as Newton states.

One of Newton's five proportions (the first) involves a passage from the moving ellipse to the stationary ellipse, and another (the last) involves the reverse passage. In the passage from stationary to moving ellipse, the radius *SP* for a given point *P* in the stationary ellipse is moved forward to the position *Sp* in such a way that the angles from last quadrature, *CSP* and *CSp* bear to one another the ratio of the sidereal to the synodic period of the moon, namely $27^{\mathrm{d}}7^{\mathrm{h}}43^{\mathrm{m}}$ to $29^{\mathrm{d}}12^{\mathrm{h}}44^{\mathrm{m}}$, or 1 to 1.080853. The curve is thus expanded angularly about the center *S* in this same ratio, and the curvature is thereby reduced. Newton's overall result is that the curvature at *a* on the curve *Cpa* is to the curvature at *C* on the same curve as $(\mathbf{b}^3 + 0.16824\mathbf{a}^2\mathbf{b})$ to $(\mathbf{a}^3 + 0.16824\mathbf{b}^2\mathbf{a})$.

The final step in III.28 is to set the two ratios equal to one another; and putting $N = 1$, $SA = \mathbf{b} = 1 - x$, $\mathrm{SC} = \mathbf{a} = 1 + x$, to solve for x. Newton obtains $x = 0.00719$, whence $\mathbf{a}$ is to $\mathbf{b}$ as $70^{1/24}$ to $69^{1/24}$, or, rounded, as 70 to 69.[10] This derivation is the first to be published showing that the variation, when dynamically considered, implies a noncircular orbit.

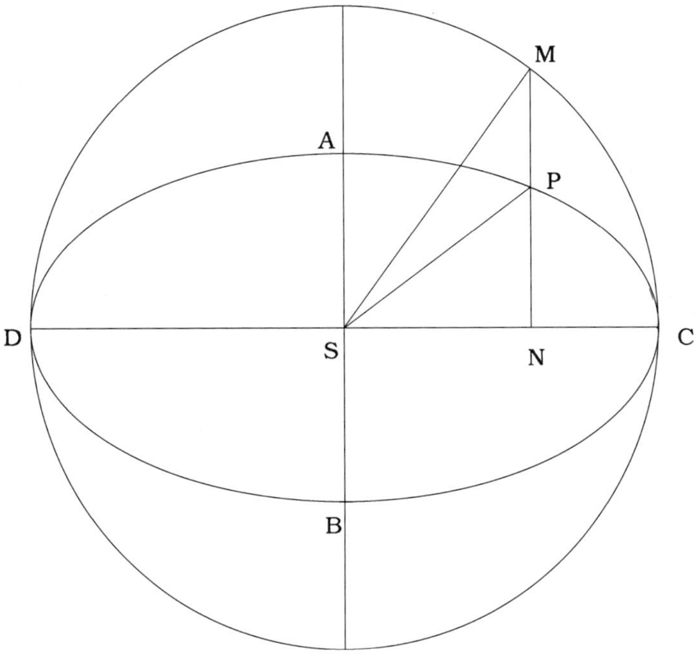

Figure 6.6
Diagram for deriving the part of the *variation* implied by the area law.

The outcomes of Propositions 26 and 28 are applied in Proposition 29 to obtain the part of the moon's variation that is independent of the orbital eccentricity. If the moon revolved in the orbit *DBCA* about the earth quiescent in *S*, and by the radius *SP* described areas proportional to the times, then, says Newton, $\tan\alpha$, where α is the angle *CSP* measured from quadrature, would be to $\tan M$, where M is the angle of mean motion, as $\mathbf{b} = SA$ to $\mathbf{a} = SC$. (I here explicitly introduce the symbol M as distinct from α.) This is apparent from figure 6.6, in which a circle has been circumscribed about the ellipse *DBCA*; through *P* a perpendicular has been dropped to the major axis *SC*, and then produced backward to meet the circle at *M*, and the radius *SM* has been drawn. If the radius-arm *SP* in the ellipse sweeps out equal areas in equal times, the same will be true of the corresponding radius-arm *SM* in the circle, since for all points *P* the area *CSP* in the ellipse bears to the area *CSM* in the circle the constant ratio of **b** to **a** or 69 to 70.

The relation $\tan\alpha : \tan M = 69 : 70$ allows us to derive the part of the variation implied by the area law in the now flattened orbit *DBCA*.

When $M = 45°$, so that $\tan M = 1$, α must be $44°35'16''$, whence $M - \alpha = 24'44''$.

But the acceleration established in III.26 modifies the uniformity of areal description. The increment in velocity, it was there shown, is proportional to the versed sine of $2M$ or $(1 - \cos 2M)$. Thus if the velocity at quadrature is represented by 10973, the velocity at any angle α from quadrature will be $10973 + 50(1 - \cos 2M) = 11023 - 50 \cos 2M$; or if we take the mean velocity to be represented by unity, the velocity when the angle of mean motion from quadrature is M will be $1 - 0.004536 \cos 2M$. The integral of this expression, yielding angle traversed (assuming the orbit circular) is $M - 0.002268 \sin 2M$. If $M = 45°$, this integral gives $44°52'12''$, differing from $45°$ by $7'48''$. The sum of $7'48''$ and the $24'44''$ found in the preceding paragraph is $32'32''$. Newton obtained the same result by a different route, as follows.

The rate of description of area in the quadrature is to that in the syzygy as 10973 to 11073, and in any intermediate place P is as the square of the sine of the angle CSP; which, Newton says, "we may effect with accuracy enough, if we diminish the tangent of the angle CSP in the ratio obtained from the square root of the ratio of the number 10973 to the number 11073, that is, in the ratio of the number 68.6877 to the number 69." Thus Newton sets $\tan M/\tan \alpha = 70/68.6877$, and for $M = 45°$ finds $\alpha = 44°27'28''$, so that $M - \alpha = 32'32''$. This procedure is strictly accurate only for $M = 45°$, but its errors elsewhere do not exceed about 4 arcseconds.

The value $32'32''$ that Newton has found for the maximum coefficient of the variation presupposes that the moon in passing from quadrature to syzygy describes an angle of $90°$ only. The angle is in fact larger, because the sun is meanwhile moving with respect to the earth. To obtain the true maximum of the variation the number $32'32''$ must be multiplied by the synodical divided by the sidereal period of the moon, or 1.080853; the result is $35'10''$. Newton takes this to be the mean value of the maximum coefficient. In the second and third editions he claims that this maximum coefficient varies annually with changes in the sun's distance from the earth, being $33'14''$ for the sun in apogee and $37'11''$ for the sun in perigee.

How accurate is Newton's result for the mean value? L. Euler in his *Theoria Motus Lunae*[11] of 1753 determined the parameters of the variation by solving his differential equations (by the method of undetermined coefficients) with all terms involving the orbital eccentricities of the moon and the sun deleted; he found $35'15''$ for the maximum coefficient and a ratio of major to minor axis of the variational orbit essentially identical with Newton's. In April 1766 his son, J. A. Euler, read to the Berlin Academy his "Réflexions sur la variation de la lune,"[12] in which he calculated the variational parameters in the same way but carried the determination of the coefficients to a higher degree of approximation. For the ratio of the earth-moon distances in quadrature and syzygy he obtained 1.00725/0.99285, which is equal to 70/68.99926; and for the maximum value of the variation in the octants he found 35′5″.6.

Treatment of the variation in isolation from the other inequalities was again a feature of Leonhard Euler's third and final lunar theory of 1772.[13] G. W. Hill studied this work, and in his "Researches in the Lunar Theory" of 1878 took the variational orbit as the "intermediate orbit" or starting point for a new lunar theory.[14] Earlier lunar theorists had generally taken as the pristine lunar orbit, to be modified by perturbations, an ellipse with the earth in one focus. This supposition, although in agreement with the solution of the two-body problem, brought with it the difficulty that the ellipse's eccentricity was difficult to determine, being always implicated with other inequalities. Hill's new starting point had the advantage that the variation was dependent on only one empirical constant, the ratio of the moon's period, sidereal or synodic, to the sun's, and this ratio was more accurately known than any other constant of the lunar theory, having been refined and verified over centuries. Using 1/12.36875 as the ratio of the moon's synodical period to the year, Hill determined by numerical integration a starting velocity for the moon on the line of quadratures such as to cause the orbit to cross the line of syzygies perpendicularly. He obtained 1.01446/1 (but with an accuracy of fifteen significant figures) as the ratio of the orbital radius in the quadratures to the orbital radius in the syzygies, and a maximum variation of 35′6″. From these results Newton's corresponding numbers differ by 0.002% and 0.2%.

Newton's determination of the variation is thus quantitatively successful. On the other hand, the differences in method between Newton's calculation and the later calculations of the Eulers and Hill are profound. Both the Eulers and Hill started from differential equations that stated exactly the conditions of the problem. Reference to the differential equations stated initially controls the successive approximations of J. A. Euler's derivation and the successive steps of Hill's numerical integration. Newton's calculative procedure, which depends on the close approximation of the variational orbit to a circle concentric to the earth and introduces approximative assumptions from the start, provides no internal check on the accuracy of these assumptions or of his results.

Nor could an empirical check be conclusive. The total variation included a part dependent on the eccentricity as well as the part Newton calculated. Newton followed Jeremiah Horrocks in treating the equation of center and evection in combination, by means of an elliptical orbit with varying eccentricity and an apsidal line oscillating to either side of the advancing mean apsidal line. In Corollaries 7–9 of Book I, Proposition 66, all editions, Newton showed that this Horrocksian theory followed qualitatively from a consideration of the perturbing forces in the quadratures and syzygies. Of all the known lunar theories to date Flamsteed had found the Horrocksian to be in best agreement with observations. Thus Newton undoubtedly regarded it as strong support for his gravitational theory. But this theory's numerical parameters could only be determined empirically. In the 1690s Newton undertook to determine them and found the mean eccentricity to be 0.05505, with the range of variation ± 0.011732. The product of these two numbers multiplied by (5/2) and changed from radian into degree measure yields the new contribution to the variation, namely $5'33''$. This result, added to the $35'10''$ found previously, gives $40'43''$, a little more than the value determined by Tycho, which was about $1'$ too large.

The problem that Newton set himself and solved in Propositions 26, 28, and 29 of Book III is in a sense artificial: the variational orbit is a construction diverging from the moon's actual path. The variable ellipse of the Horrocksian theory is also an artifice and also diverges from the moon's actual path. The geometrical eccentricity of the variational orbit is larger than that of the mean Horrocksian ellipse, 0.16838 as compared with 0.05505, so that the former differs more in shape from a circle than the latter; but astronomically the variational orbit is not eccentric, since the earth is at its center, whereas in the Horrocksian ellipse the earth is displaced from the center by about 1/20 of the radius. Each of these theories implies something true about the moon's longitudinal motion, namely the larger of the inequalities to which that longitudinal motion has been found to be subject: an equation of center with a first-order term cresting at $6°17'.32$, the evection with a maximum of $1°16'.45$, and the variation with a maximum of $39'.52$ (when the part dependent on eccentricity is included). In orbital shape, they differ, so both cannot be true.

An accurate theory for the moon's longitude requires many further longitudinal terms. The largest of these is the annual inequality, at maximum $11'.17$; the next eight in order of descending magnitude have maxima of $3'.54$, $3'.45$, $3'.20$, $2'.77$, $2'.47$, $1'.8$, $1'42$, $0'.98$. Many additional terms have to be taken into account when the observations become precise to arcseconds. Are all these terms to be connected with modifications in orbital shape? Dynamically, it must be presumed so. For the Newtonian dynamicist, the motions must be conceived as derivable from the forces, which depend on, and also cause, alterations in moon-earth distance.

Newton on the Motion of the Moon's Apse

The moon's orbital apse moves some 3° per month or 40° per year. What is this "apse"? According to the *Oxford English Dictionary*, the apses of a celestial body's orbit are the two points "furthest and nearest to the body round which it moves." By this definition, finding the position of either apse would be a matter of measuring distances. But measurements of the distance of the moon (up to the 1950s when laser and radar ranging were applied to the task) were always much less precise than determinations of longitude. The filar micrometer, invented by William Gascoigne in the late 1630s, came into general use in the second half of the seventeenth century for measuring astronomical angles, including angular diameters of the moon, from which the moon's relative distances from the earth could be deduced. The precision of such measurements, previously limited to 1 or 2 arcminutes, was thus increased by about an order of magnitude; Flamsteed's angular measures, for instance, were good to about 5 arcseconds. Micrometer measurements of lunar diameters were among the chief data that satisfied Flamsteed as to the superiority of Horrocks's lunar theory over the theories of Boulliau, Wing, and Streete, which had the moon approaching the earth when it was actually receding, and vice versa. But such measurements, although supplying a check on the moon's distances, were poorly suited for the precise determination of the longitude at which the moon's distance was greatest or least. In either of these regions, the Moon's apparent diameter changes with exceeding slowness, diminishing only about 1.5 arcseconds in 10° of motion from the lower apse. Even at 90° from either apse, the moon's diameter changes only about 2 arcseconds per degree of longitudinal motion.

For the astronomer, the apse was more importantly the position in the orbit from which the planet's or satellite's "mean anomaly" and "true anomaly" were measured. Before 1800 the higher apse (apogee in the case of the moon or sun, aphelion in the case of a planet) was used for this purpose; thereafter the lower apse (perigee or perihelion). In locating the apsidal line, the astronomer sought two longitudes exactly 180° apart, such that mean motion and true motion could there be assumed to coincide, while symmetrically to either side of the line connecting these points, mean and true motion diverged, reaching a maximum difference (the maximum equation of center) at a little more than 90° from the upper apse. The orbital eccentricity was calculated from the maximum equation of center. The determination of both the apse's longitude and the eccentricity was primarily a matter of longitude measurements rather than distance determinations. For a satellite subject to perturbation, greatest or least distance from the primary need not coincide with zero anomaly.

But in the Newtonian theory, forces and distances are interdependent. The variational orbit and the Horrocksian ellipse implied two disparate pictures of the lunar orbit. As Newton put it, "the eccentric orbit in which the Moon really revolves is not an ellipse but an oval of another kind."[15] How does one derive, by dynamical theory, the motion of the apse of the *actual* orbit?

Newton's manuscript on the apsidal motions begins with two lemmas, which Whiteside labels [α] and [β]. They deal with the apsidal motion induced by radial and transverse forces, when the moon is assumed to move in an eccentric elliptical orbit of small (indeed vanishing) eccentricity. The effect of a radial perturbing force is determined by [α], that of a transverse perturbing force by [β]. The perturbing forces are imagined as acting in pulses; the time is divided into equal, small particles of time, with the forces acting only at the beginning of each time particle.

In figure 6.7, assume the radial perturbing pulse acts at P toward S (the earth) at the beginning of a time particle, so as to change the moon's path from Pp to PG. The moon's path in the preceding time particle, πP, makes with the path Pp an angle that Newton takes to be negligibly different from the central angle PSp;[16] the perturbing force is therefore to the force the earth exerts on the moon as pG is to $Pp\cdot$(angle PSp), or as the angle pPG is to the angle PSp. Newton proceeds to show that the line from P to the second focus must, by the properties of the ellipse, move through an angle FPf equal to twice pPG. Because a radial impulse leaves the transverse axis unchanged in length, $Pf = PF$; hence $Ff = PF$(angle Fpf). Let fh be the perpendicular from f to the former transverse axis QR; the axis will thus be moved through an angle given, very nearly, by fh/SF. With V representing the magnitude of the radial perturbing force, and P the moon's weight toward the earth, Newton expresses his result as

$$\frac{\angle fSF}{\angle PSp} = \frac{2V}{P} \cdot \frac{SE}{SF}, \tag{6.20}$$

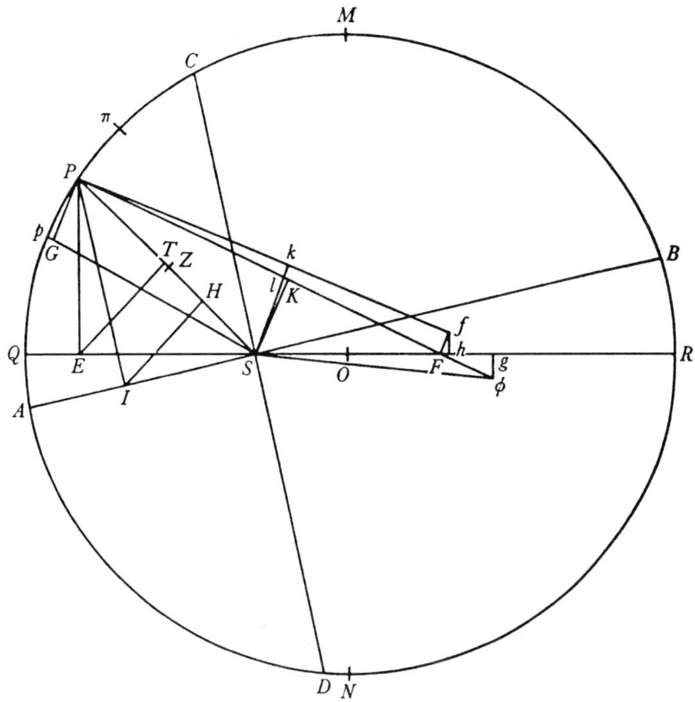

Figure 6.7
Newton's diagram for deriving the forces causing motion of the moon's apse (from *Mathematical Papers of Isaac Newton*, 6: 508).

where angle *fSF* is the angle through which the apsidal line moves, and *PSp* is the angle through which the moon moves about *S*, during the time particle considered. Let us write $\delta\theta$ for angle *PSp*, where θ is the moon's longitude measured from a designated point in the ecliptic (say ♈, the first point of Aries); $\delta\omega$ for angle *fSF*, where ω is the longitude of the lower apse of the ellipse (Q in figure 6.7); $2ae$ for *SF* (where a is the transverse axis and e the eccentricity of the ellipse); and $a\cos\varphi$ for SE (where $\varphi = \text{angle PSQ} = \theta - \omega$), an approximation good enough to give a result accurate to $O(e)$. Newton's equation then takes the form

$$\left(\frac{\delta\omega}{\delta\theta}\right)_r = \frac{V}{Pe}\cos\varphi. \tag{6.21}$$

In [β] Newton considers the effect of an impulse W acting at P perpendicularly to *SP*. Suppose T is the moon's speed before W acts, and $T + t$ its speed afterward. The speed with which area is swept out about S is increased very nearly in this same ratio, and hence by Proposition I.14 the latus rectum is increased in the square of this ratio, or as $(T+t)^2/T^2 \approx 1 + 2t/T$. It follows, Newton shows, that the major axis $2a$ increases by the amount $4at/T$; and, since SP does not change in length, PF increases by $F\phi = 4at/T$. From similar triangles $g\phi = (4at/T)(PE/PF)$. The angle $FS\phi$ through which the apsidal line is moved by the transverse perturbing force W is very nearly to the angle FSf through which it is moved by the radial perturbing force V as $g\phi$ is to fh. In addition, t is to the speed generated by V as W is to V, and the speed generated by V is to the moon's speed as $Ff/2$ is to PF. By combining these relations and ignoring terms proportional to the eccentricity in the numerator, Newton obtains a result expressible as

$$\left(\frac{\delta\omega}{\delta\theta}\right)_t = \frac{2W}{Pe}\sin\varphi. \qquad (6.22)$$

(For P as placed in figure 6.7, $\sin\varphi$ is negative, but the effect of a forward perturbing force W is a backward motion of the apse; hence the sign of the right-hand member of (6.22) is positive.)

Newton's result (6.22) is inaccurate; it lacks a second term proportional to $e\cos\varphi$, as a correct derivation from the diagram would show.[17]

The remainder of the manuscript consists of two propositions, which Whiteside labels [A] and [B]. [A] is an attempt to derive, for a given angle between the lines of apsides and syzygies (angle QSA or BSR in figure 6.7, hereafter designated β), "the hourly motion of the Moon's apogee," that is, the average advance of the apsidal line, per sidereal month, when β has the given value. For a different β, the result would be different, so that later, in [B], Newton undertakes the further task of calculating the total advance of the moon's apse over a year.

Proposition [A] is flawed: into (6.21) and (6.22) Newton substitutes mistaken expressions for V and W. Let us first see what correct substitutions would imply.[18]

In integrating (6.21), we need an expression for V/P. In the case of the variation, Newton assumed a concentric circular orbit, and we found the radial perturbing force to be

$$\mathbf{ES} = \frac{\mathbf{SP}}{2}(1 - 3\cos 2\alpha), \qquad (6.18)$$

where $\mathbf{SP}$ is a constant, equal numerically to (1/178.725) of the mean accelerative force exerted by the earth on the moon. Putting μ/a^2 for the latter force, where a is the radius of the circle, and m for the ratio of the sidereal month to the sidereal year, so that $1/178.725 = m^2$, we can express **SP** in (6.18) as $-m^2\mu/a^2$ (the minus sign indicating that **SP** is outward). But in (6.21) the moon is taken to be moving in an eccentric elliptical orbit, with varying radius vector r and mean radius vector a, and therefore, since the perturbing force varies directly as the radius vector, the varying acceleration **SP** is given by $-m^2\mu r/a^3$. The required revision of (6.18) is then $V = -(m^2\mu r/2a^3)(1 - 3\cos 2\alpha)$. The accelerative force P exerted by the earth on the moon also varies, but with a variation that follows the inverse-square law, so that $P = \mu/r^2$. For the quotient V/P we thus find $-(m^2r^3/2a^3)(1 - 3\cos 2\alpha)$, which to $O(e) \approx -(m^2/2)\cdot(1 - 3e\cos\varphi)(1 - 3\cos 2\alpha)$. Equation (6.21) becomes

$$\frac{(\delta\omega)_r}{\delta\theta} \cong -\frac{m^2}{2e}\cos\varphi\cdot(1 - 3e\cos\varphi)(1 - 3\cos 2\alpha). \qquad (6.23)$$

Equation (6.23) gives a rate of change of ω with respect to a change in θ, supposing the moon to be at a certain distance φ from the lower orbital apse, and at a certain distance α from the last quarter. To find the rate for any position of the moon in its orbit we need to average the rate given by (6.23) over a full cycle of the moon, integrating with respect to θ from 0 radians to 2π, then dividing by 2π. Now $\varphi = \theta - \omega$, and angle $\alpha = \varphi - \beta - 270° = \theta - \omega - \beta - 270°$. We can express β approximately in terms of θ and ω if we set θ_0 = the longitude of the moon when the sun and the apse were last coincident; the sun's longitude at any later time will be approximately $m(\theta - \theta_0)$, and β will be $m(\theta - \theta_0) - \omega$. With these substitutions the factor $(1 - 3\cos 2\alpha)$ becomes $[1 + 3\cos 2(\theta\{1 - m\} + m\theta_0)]$, and the integral of the right-hand side of (6.23) becomes

$$-\frac{m^2}{2e}\cdot\frac{1}{2\pi}\int_0^{2\pi}\cos(\theta - \omega)\cdot[1 - 3e\cos(\theta - \omega)]$$
$$\times[1 + 3\cos 2(\theta\{1 - m\} + m\theta_0)]\,d\theta = \frac{3m^2}{4}. \qquad (6.24)$$

This result was obtained under the assumption that ω and θ_0 had particular values, but it is evidently independent of these values; hence we can conclude, without restriction, that the total secular apsidal motion arising from (6.23) when θ increases by 2π will be $\Delta\omega = (3m^2/4)\cdot 360° = 1°30'39''$, about half the moon's observed apsidal motion in a sidereal month.[19]

Accepting Newton's equation (6.22)—we recall that it lacks a necessary term—we can proceed similarly to obtain an appropriate expression for W, one that allows for the variability of the radius vector, by modifying equation (6.11). The result is $W = -(3m^2/2)(\mu r/a^3)\sin 2\alpha$. As before, P is given by μ/r^2. Equation (6.22) thus takes the form

$$\left(\frac{\delta\omega}{\delta\theta}\right)_t \cong -\frac{3m^2}{e}\sin\varphi\cdot(1-3e\cos\varphi)\sin 2\alpha. \qquad (6.25)$$

Replacing φ by $\theta-\omega$ and $\sin 2\alpha$ by $-\sin 2[\theta(1-m)+m\theta_0]$, integrating with respect to θ from 0 to 2π, we obtain 0 as a result: Newton's form of (6.22) yields no contribution to the apsidal motion. The total apsidal motion extractable from Newton's equations is therefore $(3m^2/2)\Delta\theta$, or $1°30'39''$ per sidereal month. As we shall see, this is the term of lowest order in d'Alembert's derivation of the first four terms in the motion of the lunar apse. Had (6.22) contained the second term that properly belongs to it, a very considerable additional contribution to the apsidal motion would have resulted from it.[20]

We now turn to Newton's proposition [A]. In this proposition, he has a different notion as to the forces to be substituted for V and W.

The perturbing forces, Newton says, "do not, when the Moon is located in the orbit in which by their means it might revolve without eccentricity, contribute at all to the motion of the apogee."[21] The orbit referred to is the variational orbit, as found in Proposition 28. In this orbit the center coincides with the earth's center; hence there is no apse in the sense of apocenter or pericenter, and thus no apsidal motion. Newton proceeds to take the perturbing forces in the variational orbit as the starting point for a surprising computation: "The motion of the apogee arises from the differences between these forces [the perturbing forces acting on the moon in its actual orbit] and forces which in the Moon's recession from that concentric orbit decrease, if they are radial, in the duplicate ratio of the distance between the Moon and the Earth's center, but, if they act in the transverse direction, in the triplicate of the same ratio—as I found once I undertook the calculations."[22]

The implicit assertion is that only a certain fraction of the radial and transverse perturbing forces acting on the moon in its actual orbit produces apsidal motion, the remainder in each case being "used up" in changing the orbit's shape and in producing periodic changes in the moon's speed. In each case this remainder is taken to be derivable from the perturbing forces acting in the variational orbit, by modifying the latter forces in a certain ratio—the inverse square of the distance in the case of the radial component, the inverse cube in the case of the transverse component. For this notion, Newton offers no justification.

The notion is surprising because the instantaneous forces are acting, not on the moon's orbit, but on the moon itself, which at each instant has a position and velocity. Equation (6.21) and a corrected equation (6.22) show that in an elliptical orbit any radial or transverse perturbing force produces apsidal motion. Apsidal motion is a holistic effect resulting from a summation of all the instantaneous actions. Whence Newton's notion that it can be traced to some specifiable fraction of the instantaneous actions?

The only section of the *Principia* devoted thematically to apsidal motion is Book I, Section IX, Propositions 43–45. In Propositions 43–44 Newton shows that a body may be made to move in an ellipse with rotating apse if an inverse-cube force is added to the inverse-square force that produces motion in a stationary ellipse. Newton expresses the total force toward the center by the formula

$$\frac{F^2 A}{A^3} + \frac{R(G^2 - F^2)}{A^3},$$

where A is the variable distance from the center of force to orbiting body, R is the semi–latus rectum of the ellipse, and $G{:}F$ is the ratio of the angular motion about the center to the orbital motion relative to the (moving) upper apse. Newton's result corresponds exactly to the solution of the differential equation

$$\frac{d^2 u}{d\theta^2} = \frac{1}{h^2 u^2}(\mu u^2 + b u^3),$$

where u is the reciprocal of the radius vector, θ the angle the radius vector makes with a stationary line, h the double of the rate of sweeping out of area about the center, and $\mu u^2 + b u^3$ the sum of the inverse-square and inverse-cube forces. The solution of the foregoing differential equation may be written

$$u\left(1 - \frac{b}{h^2}\right) = \frac{\mu}{h^2}(1 + e \cos k\theta),$$

where $k =$ Newton's F/G, e is the eccentricity, $1 - b/h^2 = k^2$, kh is the rate of sweeping out of area in the moving ellipse, $\mu =$ Newton's F^2, $k^2 h^2/\mu$ is the semi–latus rectum of the ellipse, hence equal to Newton's R, and b is therefore identical with Newton's $R(G^2 - F^2)$.

In Proposition 45 Newton uses the result of Proposition 44 to show that, given a centripetal force proportional to A^{n-3}, and provided that the orbital eccentricity is vanishingly small, the angle between higher and lower apse is given by $180°/\sqrt{n}$. This angle is therefore 180° if $n = 1$ (the inverse-square case) and other than 180° if $n \neq 1$. Thus for orbits of vanishingly small eccentricity, the apsidal line is immobile if and only if the centripetal force is inverse square. Where the centripetal force consists of an inverse-square force (given by b/A^2) plus or minus a centripetal force varying as cA^{n-3}, Newton shows that—assuming once more a vanishingly small orbital eccentricity—the angle between higher and lower apse is $180°/\sqrt{(b \pm c)/(b \pm nc)}$. Taking the average radial perturbing force on the moon (it is subtractive; see note 8) as 1/357.45 times the accelerative force due to the earth, he finds the resulting apsidal motion to be $1°31'28''$ per revolution. In this application as in all the other applications made in I.45, the total centripetal force forms a radially symmetrical field; indeed, the proposition is derived on that assumption.

That Newton was imagining his procedure in Proposition [A] to be justifiable on the basis of Proposition 45 is suggested by the fact that his initial choice for the ratio in which the perturbing forces in the variational orbit—both radial and transverse—were to be altered, so as to yield the non-apse-moving forces in the actual orbit, was the inverse square.[23] Only when this choice yielded a disappointing result did he try other ratios. In the end he chose the inverse cube for the transverse force and retained the inverse square for the radial force. The sole reason for this choice appears to have been his desire to obtain an empirically plausible result, whence Whiteside's conclusion that the result was "adeptly fudged."

Proposition 45 had a special role in Newton's thinking. Propositions 11–13 of Book I, together with Corollary 1 of Proposition 13 (assuming the latter proved),[24] show that conic-section orbits with center of force at the focus are caused by inverse-square forces and conversely. These orbits have fixed apsides, so that the planet or satellite returns to its upper apse after precisely 360° of motion. Newton wanted to use the fixity or quiescence of the apsides as a sign that the centripetal force is inverse square, and I.45 establishes this relation for the case where the orbit's eccentricity (taken as the difference between the greatest and least distances to the center divided by the sum of those distances) is negligible. In Proposition 2 of Book III Newton assumes that this relation holds for the planets, and in the General Scholium he assumes that it would apply to a comet.[25] Evidently he believed that the apsides of an orbit would be quiescent if and only if the law of the centripetal force were inverse square, whatever the orbital eccentricity.

In [A] Newton is assuming the eccentricity to be small, but his procedure, unfortunately, is not justifiable on the basis of Proposition 45, even for the radial perturbing forces. Let the inverse-square field of accelerative force due to the earth's attraction of the moon be defined by μ/r^2, and suppose that at each angle α measured about the earth's center from the line of quadratures a further radial force is added or subtracted, and that this force also varies inversely as the square of the distance, so that it can be defined by b/r^2 for an appropriate choice of b. For any given α, the total centripetal force will be inverse square, expressible by $(\mu \pm b)/r^2$. But in Newton's application, the force b/r^2 is determined to be identical with the radial perturbing force due to the sun when r is equal to the radius vector in the variational orbit. If we approximate the variational orbit by means of a concentric circle (which is what Newton does, as we shall see), we find that

$$b = \frac{m^2\mu}{2}(1 - 3\cos 2\alpha).$$

The value of b thus depends on the angle α. Consequently, the accelerative field of force given by $(\mu \pm b)/r^2$ is not radially symmetrical and so is not an inverse-square field of force. Proposition 45 does not apply. The same conclusion applies a fortiori in the case of the transverse perturbing forces, Proposition 45 being applicable only to centripetal (or radial) forces.

Newton's initial error was to imagine that inverse-square variation of the force, even when holding only pointwise for each angle α and not for the field as a whole, would imply quiescence of the apsides. His application of the same kind of pseudoargument to the transverse perturbing force is reckless daring gone amok.

Let me now summarize the further course of Newton's argument in [A]. He begins with a simplifying assumption: "[The variational orbit] has above been shown to be oval [in Book III, Proposition 28]. And thence it results that the eccentric orbit in which the Moon really revolves is not an ellipse but an oval of another kind. But if, to lessen the difficulty of the computation, we should suppose that the concentric orbit is a circle, the eccentric orbit will turn out to be an ellipse whose major axis QR will be equal to the circle's diameter."[26] Newton is thinking of the actual orbit as primarily a conflation of the variational orbit and an eccentric ellipse. He proposes a qualitative proportion:

Actual orbit : variational orbit :: eccentric ellipse : circle.

He proceeds to compute with the circle and eccentric ellipse as stand-ins for the variational orbit and (unknown) actual orbit, in the hope that quantitative results relevant to the actual orbit can thereby be obtained.

Consider first the radial perturbing force. In an initial step, Newton imagines the moon to be transferred from its eccentric elliptical orbit to the concentric circular orbit. Since the perturbing forces are as the moon-earth distance, if the radial perturbing force in the circle is given by $-(m^2\mu/2a^2)(1 - 3\cos 2\alpha)$, the radial perturbing force in the ellipse is then $-(m^2\mu r/2a^3)(1 - 3\cos 2\alpha)$, where r is the radius vector in the ellipse and a the radius vector in the circle. Next, the radial perturbing force in the circle is altered in the inverse square of the distance to obtain the supposedly non-apse-moving radial perturbing force in the ellipse; the result is $-(m^2\mu/2r^2)(1 - 3\cos 2\alpha)$. Finally, we subtract the latter force from the total radial perturbing force in the ellipse, obtaining

$$-(m^2\mu/2r^2)[(r^3/a^3) - 1](1 - 3\cos 2\alpha)$$
$$\approx -(m^2\mu/2r^2)(-3e\cos\varphi)(1 - 3\cos 2\alpha) \text{ to } O(e).$$

This, according to Newton, is the radial perturbing force V in (6.21). But μ/r^2 is the earth's attraction or P; whence $V/P = (3/2)m^2 e\cos\varphi \cdot (1 - 3\cos 2\alpha)$. Substituting this expression into (6.21), we have

$$\left(\frac{\delta\omega}{\delta\theta}\right)_r \cong \frac{3m^2}{4}(1 + \cos 2\varphi)(1 - 3\cos 2\alpha). \tag{6.26}$$

The corresponding calculation for the transverse perturbing force involves transferring the force from the eccentric ellipse to the concentric circle using the direct ratio of the distances, then the reverse transfer, but using this time the inverse cube of the ratio of the distances, to yield the non-apse-moving force in the ellipse. This force, subtracted from the actual transverse force in the ellipse, yields according to Newton the fraction of the transverse force actually producing apsidal motion, or W in equation (6.22). We find

$$W \approx -\frac{3m^2\mu}{2r^2} \cdot 4e\cos\varphi \cdot \sin 2\alpha,$$

and since P is μ/r^2, (6.22) becomes

$$\left(\frac{\delta\omega}{\delta\theta}\right)_t \cong -6m^2 \sin 2\varphi \cdot \sin\alpha. \tag{6.27}$$

Newton's next step is equivalent to substituting $\theta - \omega$ for φ, and $\theta - \omega - \beta - 270°$ for α, then integrating (6.26) and (6.27) with respect to θ from 0 to 2π, so as to obtain the average increase in ω per sidereal month for given β and ω. The integral for (6.26) is

$$\frac{3m^2}{4}\int_0^{2\pi}[1+\cos 2(\theta-\omega)][1+3\cos 2(\theta-\omega-\beta)]\,d\theta$$
$$\cong \frac{3m^2}{4}\left(1+\frac{3}{2}\cos 2\beta\right)\cdot 2\pi, \tag{6.28}$$

and that for (6.27) is

$$6m^2\int_0^{2\pi}\sin 2(\theta-\omega)\cdot\sin 2(\theta-\omega-\beta)\,d\theta = 2m^2\cos 2\beta\cdot 2\pi. \tag{6.29}$$

The sum of these two increments, divided by 2π, gives the average rate of increase of ω per sidereal month for fixed β and can be expressed as

$$\frac{3m^2}{4}\left(1+\frac{11}{2}\cos 2\beta\right).$$

Newton, putting $4/3m^2 = 238.3$, writes it as

$$\frac{\frac{2}{11}\pm\cos 2\beta}{43\frac{18}{55}},$$

where the "$\pm$" sign means that $\cos 2\beta$, being for Newton a line in a diagram rather than a signed ratio, must be taken as positive for 2β in the first and fourth quadrants, otherwise as negative. In modern (Eulerian) notation, we can simply write

$$\left(\frac{\delta\omega}{\delta\theta}\right)_{\beta=\text{const.}} = \frac{3m^2}{4}\left(1+\frac{11}{2}\cos 2\beta\right). \tag{6.30}$$

Equation (6.30), we must insist, has been obtained by illegitimate processes. Had Newton's equation (6.22) been correct, and had the further processes constituted a correct derivation, the coefficient 11/2 would have turned out to be 5.[27] Newton's value approximates the correct value as a result of a fudge.

Because Newton's result for [A] contains $\cos 2\beta$, a further integration is required to obtain the average value of the rate of increase of ω independent of β. The problem thus posed is a delicate one, since β, being the difference between the sun's longitude and that of the lunar apse, does not increase uniformly with time. Newton takes up this problem in Proposition [B]. He expresses the problem as that of finding the motion of the lunar apogee during the course of an entire year. He conceives the moon's apogee to be drawn back at the end of each hour to its position with respect to the stars at the beginning of the hour, while the sun moves out of the line of the moon's apsides at the start of the year and, after

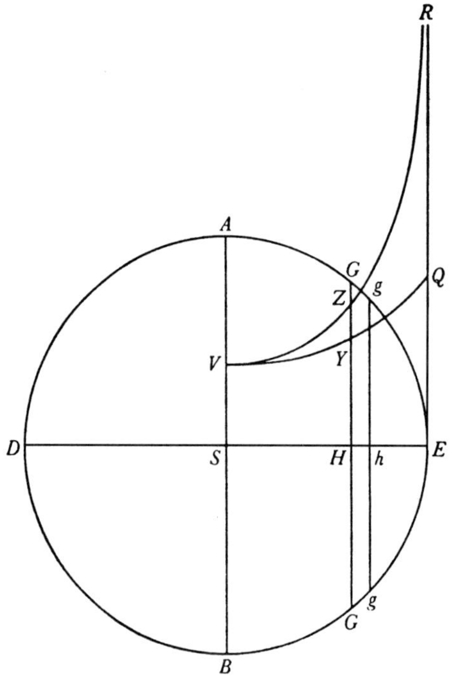

Figure 6.8
Newton's diagram for deriving the annual motion of the moon's apse (from *Mathematical Papers of Isaac Newton*, 6: 524).

traversing 360°, returns to it at year's end. In brief summary, his procedure is as follows.

In figure 6.8 angle ASG is 2β, equal to angle 2(QSA) in figure 6.7. $\mathrm{Gg}/\mathrm{AS} = \mathrm{d}(2\beta)$ is an increment of the angle 2β, and GH, in Newton's terminology, is the cosine of angle ASG, where AS is the circle's radius. According to Newton's result in [A], the motion of the apse when $2\beta = ASG$ is to the moon's mean motion as $(\frac{2}{11}AS \pm GH)$ is to $43\frac{18}{55}AS$, or as $(\frac{2}{11}Gg \pm Hh)$ is to $43\frac{18}{55}Gg$, where $Hh = Gg\cos 2\beta$. If, says Newton, t is the time in which the motion of the moon is $43\frac{18}{55}Gg \cdot t$, then the simultaneous motion of the apogee is expressed by $(\frac{2}{11}Gg \cdot t \pm Hh \cdot t)$. The ratio of $43\frac{18}{55}Gg \cdot t$ to $\frac{2}{11}Gg \cdot t$ is 238.3 to 1. Since the moon moves $4812°45'38''$ in a year, the contribution of all the increments $\frac{2}{11}Gg \cdot t$, when t equals a year, is $20°11'46''.4$ (This result, differently put, is just $3m^2/4$ times the moon's motion in a year.) Newton proposes to evaluate the sum of the increments $\pm Hh \cdot t$ in relation to the sum of the $Gg \cdot t$.

Now because ω does not advance uniformly, 2β does not increase uniformly with time. What must be substituted for t, Newton says, is an expression proportional to the time in which the change $Gg/AS = d(2\beta)$ occurs. This time is reciprocally as the difference between the hourly motion of the sun and the hourly advance of the apogee, proper account being taken of the sign of Hh. Since the mean motion of the sun is to that of the moon as 360° is to 4812°.76056, or as 3.24093 is to $43\frac{18}{55}$, the mean motion of the sun is to that of the moon's apogee as 3.24093 Gg is to $(\frac{2}{11}\,Gg) \pm Hh$. The difference of the motions of the sun and the moon's apogee is thus proportional to 3.0591113 $Gg \mp Hh$. Replacing 3.0591113 by n, Newton writes

$$t \propto \frac{1}{n \cdot Gg \mp Hh},$$

equivalent in modern notation to

$$t \propto \frac{1}{n - \cos 2b}.$$

In effect Newton changes variables from t to β. We moderns would likely formulate the problem that then needs to be solved as the integration

$$\frac{3m^2}{4}\int_0^{2\pi} \frac{1 + \frac{11}{2}\cos 2\beta}{n - \cos 2\beta}\,d\beta.$$

Newton poses the problem somewhat differently: he integrates only from $\beta = 0$ radians to $\beta = \pi$ and pairs contributions from the first quadrant with those from the second. I shall not here follow his process in detail, as Whiteside has commented on it in detailed technical thoroughness.[28] The ratio of apsidal motion to the moon's motion, Newton finds, is the ratio of

$$\int_0^{\pi/2} \frac{\cos^2 2\beta \cdot d(2\beta)}{n^2 - \cos^2 2\beta} \quad \text{to} \quad \int_0^{\pi/2} \frac{n \cdot d(2\beta)}{\cos^2 2\beta(n^2 - \cos^2 2\beta)}.$$

The two integrals are represented in figure 6.8 by the areas *SVYQES* and *SVZRES*, respectively. The integrands can be reduced to series,[29] and Newton finds the value of the ratio to be $(n - \sqrt{n^2 - 1})$ to 1, or 0.1680627 : 1. It follows that all $Hh \cdot t$ are to all $\frac{2}{11}\,Gg \cdot t$ as 0.1680627 is to $\frac{2}{11}$, or as 1.848689 is to 2. *Componendo*, all $(Hh \cdot t + \frac{2}{11}\,Gg \cdot t)$ are to $\frac{2}{11}\,Gg \cdot t$ as 3.848689 is to 2. Since the sum of all the $\frac{2}{11}\,Gg \cdot t$ in a year is $20°11'46''$, the total apsidal motion is by this calculation $38°51'51''$, to be compared with $40°41\frac{1}{2}'$ as obtained empirically by Flamsteed.

This brave conclusion is unfortunately worthless, because of the fatal flaw in [A]. Nevertheless, we should acknowledge that Newton's integrations were, in Leibniz's phrase, "worthy of his genius."[30]

WHY IT IS UNLIKELY THAT NEWTON WOULD EVER HAVE SOLVED THE PROBLEM OF THE MOON'S APSIDAL MOTION CORRECTLY

As a boy, Newton was good with his hands, an accomplished whittler, fashioning model windmills and sundials.[31] A certain skill in drawing is evident in the diagrams of his manuscripts.[32] "A sober, silent, thinking lad,"[33] he focused much of his thought on spatial relations, problems of a geometrical and mechanical character. I would guess that Leibniz's bent was quite different: his mathematical innovations had an arithmetic and algorithmic direction from the start.

How does Newton's bent jibe with his early devotion to algebraic studies? At the start of his tract *De Methodis Serierum et Fluxionum* (completed in 1671 but published only posthumously in 1735), Newton praised analysis: "Observing that the majority of geometers, with an almost complete neglect of the Ancients' synthetical method, now for the most part apply themselves to the cultivation of analysis and with its aid have overcome so many formidable difficulties that they seem to have exhausted virtually everything apart from the squaring of curves and certain topics of like nature not yet fully elucidated: I found it not amiss, for the satisfaction of learners, to draw up the following short tract in which I might at once widen the boundaries of the field of analysis and advance the doctrine of curves."[34] The center of interest is evidently the understanding of curves and curvature.

Toward the mid-1670s Newton's early passion for algebraic investigations cooled. John Collins, writing to James Gregory in October 1675, says that he has not corresponded with Newton for eleven or twelve months, "not troubling him as being intent upon Chimicall Studies and practices, and both he and Dr Barrow &c beginning to thinke math[all] Speculations to grow at least [at last?] nice and dry, if not somewhat barren . . ."[35]

Later in the decade we find Newton criticizing algebraic-style mathematics for the complication to which it can lead. In his defense of the ancients' treatment of solid loci against Descartes, he wrote: "their [the ancients'] method is more elegant by far than the Cartesian one. For he [Descartes] achieved the result by an algebraic calculus which, when transposed into words (following the practice of the Ancients in their writings), would prove to be so tedious and entangled as to provoke nausea, nor might it be understood."[36] In his *Geometria Curvilinea* (written ca. 1680), Newton allows that analysis may be appropriate to some problems, synthesis or geometry to others, but his strictures against excessive algebraic complication remain strong: "Men of recent times, eager to add to the discoveries of the Ancients, have united the arithmetic of variables with geometry. Benefiting from that, progress has been broad and far-reaching if your eye is on the profuseness of output but the advance is less of a blessing if you look at the complexity of its conclusions. For these computations, progressing by means of arithmetical operations alone, very often express in an intolerably roundabout way quantities which in geometry are designated by the drawing of a single line."[37]

During the years 1684–1687, when Newton was writing the *Principia*, there is no sign that his attitude toward algebraic-style mathematics changed. He used algebraic devices on occasion but mostly sought to express his arguments in plain Latin. Algebra, apparently, was to be used in the immediate service of geometrical or mechanical insights but eschewed otherwise. Newton appears to have viewed Cartesian and Leibnizian claims of omnicompetence for a symbolic and algorithmic way of proceeding as so much empty boasting.

Years later, in his anonymous "Account of the *Commercium Epistolicum*," Newton would claim that "by the help of the new *Analysis* Mr. Newton found out most of the Propositions in his *Principia Philosophiae:* but because the Ancients for making things certain admitted nothing into Geometry before it was demonstrated synthetically, he demonstrated the Propositions synthetically, that the Systeme of the Heavens might be founded upon good Geometry."[38] These lines, penned in the heat of the dispute with Leibniz over priority in the invention of the calculus, were long taken (for instance, by Laplace) to mean that Newton had first derived the propositions analytically, then translated the proofs into geometric form. According to Whiteside, the successive revisions preserved in the manuscripts confute any such hypothesis. They show Newton refining a text that was all along basically geometrical in character.[39]

Indeed, the major insights entering into the initial draft (the *De Motu*) from which the *Principia* sprang were geometrical: (1) the equivalence between centripetal force and the equable description of area; (2) the generalization of Galileo's law of free fall to make it applicable to centripetal force fields in which the force varies from point to point.[40] On the other hand, the *Principia* contains telltale signs of Newton's dependence on analytical devices. Thus Book I, Proposition 9, assumes without proof the logarithmic property of the equiangular spiral; years earlier Newton had derived it analytically by an integration.[41] In some cases he brings in an analytically derived result without apology or geometrical disguise. According to Corollary 2 of Book I, Proposition 40, if the centripetal force is as A^{n-1}, where A is the distance from the center, and if P is the greatest distance from the center to which a body may rise in virtue of its velocity, then the velocity at any altitude A will be as $\sqrt{P^n - A^n}$, a result surely arrived at in the first instance analytically.

We, like Laplace, may be confused by an error of perspective: for us "analysis" has come to mean the posing and solving of differential equations as systematized in mid-eighteenth century, principally in the works of Euler. "Analysis" for Newton was something less general and less abstract, usually tied to the study of curves. A somewhat similar limitation held true for Leibniz in the 1680s: his "differential calculus" was originally a study of curves by algebraic means. But it differed from Newton's "analysis" in that it aimed at divorcing itself from a reliance on figures; in the pursuit of that aim, Leibnizian analysis would eventually be generalized so as to apply to functions. Solving the problem of the motion of the lunar apse turned out to depend on that generalization.

Newton, to be sure, distinguished early on between independent and dependent variables: "I shall have no regard to times formally so considered, but from quantities propounded which are of the same kind shall suppose some one to increase with an equable flow: to this all the others may be referred as though it were time, and so by analogy the name of 'time' may not be improperly conferred on it."[42] The notion of function is implicit here—and fated to remain so. Explicit formulation would come about when the central focus was on algorithmic method. Leibniz's differential algorithms were developed in analogy with algebraic algorithms for finite quantities and without special concern for logical rigor. Newton developed his "method of first and last ratios" because of a concern with logical rigor, and this worked against his developing a purely symbolic algorithm. When faced with particular geometrical and mechanical problems, he freely adopted whatever method came to hand, without seeking to subsume all such procedures under a general algorithmic method.

Leibniz introduced into mathematics the term *functio* as referring to a relation between the ordinate and abscissa of a geometrical curve.[43] Leibniz later claimed to be the first to constitute out of the new infinitesimal calculus "an algorithm, whereby the imagination would be freed from the perpetual attention to figures."[44] The correspondence of Leibniz and

Johann Bernoulli in the 1690s shows how the need for a general term representing the dependence of one quantity on another brought about the use of the term "function" in the sense of an analytic expression.[45] In particular, this usage proved helpful in removing ambiguity from the definition of the higher-order differentials.[46] The first published definition of the term "function" was due to Johann Bernoulli in 1718: "A function of a variable quantity is a quantity composed in any way of this variable quantity and of constants."[47] Euler echoes this definition in his *Introductio in Analysin Infinitorum* of 1748: "A function of a variable quantity is an analytic expression composed in any way from this variable quantity and numbers or constant quantities."[48] By 1755, the problem of the vibrating string had led Euler to generalize the notion still further, so that it became virtually identical with the modern definition in terms of ordered pairs.[49] But what is important for our story is the identification of a function as an analytic expression. The stress upon functions meant that problems were henceforth formulated by means of equations between analytic expressions.

Newton on many occasions formulated and solved problems algebraically,[50] but he did not do so always or as a matter of policy. In the *Principia* the initial attack on orbital motion (Propositions 7–13 of Book I) was geometrical rather than algebraic: assume an orbital shape, then find the law of the centripetal force that would make that orbit possible. In the case of the moon's variation, whereas the forces were known, the shape of the variational orbit that these forces would produce was unknown. Newton assumed a fictive shape (concentric circle) and reasoned to the modification in it that the perturbing forces would require; the approach was feasible because of the symmetry and relative smallness of the perturbing forces.

In the case of the apsidal motion, Newton viewed the actual orbit as mainly a conflation of the variational orbit previously found and an eccentric elliptical orbit consonant with the inverse-square law. Unable to deal with this actual orbit directly, he assumed its relation to the variational orbit to be analogous to the relation of an eccentric elliptical orbit to a concentric circular orbit and attempted to derive relevant consequences from the latter relation. In this attempt, he assumed that the perturbing forces could be partitioned into those causing apsidal motion and those not doing so, although such a partition was certainly impossible for an elliptical orbit. The assumptions of the derivation were merely analogical and conjectural. They constituted what Socrates would have called a "wind-egg."

If, as I will urge in what follows, the problem of the apsidal motion admitted only of an algorithmic solution, and that of a blind, iterative character, it seems unlikely that Newton would have found the key to it. With respect to the proper scope of algorithmic methods, he had firmly chosen a restrictive view in the 1670s.

EULER AND THE FIRST APPLICATION OF THE LEIBNIZIAN CALCULUS TO THE PERTURBATIONAL PROBLEM

Euler was the first to employ the Leibnizian calculus in determining the perturbations of the moon. Years later he claimed to have done this already in 1742,[51] but according to a letter of his to Delisle, his efforts were successful only in 1745.[52] He first published lunar tables in the latter year,[53] then revised them for his *Opuscula Varii Argumenti* of 1746. In the remarks prefaced to the tables in the *Opuscula* he stated that they were derived from Newton's theory of attraction except for the coefficients of some of the terms, which he based on observations.[54] He omitted all explanation of the theoretical basis for lack of space.

Why had he not attempted this application earlier? For more than a decade Euler had been engaged in applying the Leibnizian calculus to problems of Newtonian mechanics; he devoted his *Mechanica* of 1736 to such problems. Yet in that work he made no attempt to cope with the perturbational problem or with its outstanding example, the problem of the moon's motions. The handsome prize established by the British Parliament in 1714 for a method of determining longitude at sea—£20000 if it gave the longitude to within $\frac{1}{2}^\circ$, half as much if to within 1°—would presumably have been incentive enough. A lunar theory accurate to 2 arcminutes would have sufficed for the second prize.

Victor Katz, in an article on "The Calculus of the Trigonometric Functions,"[55] has shown that differentiation and integration of trigonometric functions did not become standard procedures in the solution of differential equations before 1739, when Euler made them so. Euler established the practice of treating sines and cosines as ratios and indeed as functions rather than as lines, established the equivalence between trigonometric functions and complex exponential functions, showed that the exponential function together with the trigonometric functions could be used to solve higher-order linear differential equations with constant coefficients, and developed the trigonometrical identities whereby powers of sines and cosines could be systematically replaced by sines and cosines of the multiple angles. Euler discovered many of the relevant rules and systematized all of them.[56] These procedures would play a crucial role in his own as well as in Clairaut's and d'Alembert's treatment of the lunar problem. Their use gradually became known to other Continental mathematicians from Euler's memoirs,[57] and from his *Introductio in Analysin Infinitorum* of 1748.

Clairaut on the Motion of the Lunar Apse

Euler in two treatises on planetary perturbations completed in June 1747 expressed strong doubt as to the exactitude of the inverse-square law of gravitation.[58] Enhancing this doubt was his metaphysical commitment to the transmission of all forces by material contact; the inverse-square law could hardly be valid to all distances on such an assumption. But Euler's doubt was chiefly founded on his conviction that Newton's law led to but half the observed motion of the lunar apse.

The second of the above-mentioned treatises, on the inequalities of Jupiter and Saturn, was destined for submission in the Paris Academy's prize contest of 1748; it arrived in Paris during the summer of 1747. Clairaut, as one of the prize commissioners for the contest, read it in September. He was delighted to discover from it that he and Euler were in agreement on the moon's apsidal motion: by calculation both had found only half of it to be derivable from Newton's law. Clairaut announced this conclusion to the Paris Academy in November 1747, thereby unleashing a storm of controversy.[59]

Clairaut's retractation came in May 1749. In broad outline, this reversal came about as follows.[60]

In his initial derivation, Clairaut obtained two equations for the motion of the Moon in an assumed orbital plane:[61]

$$rd^2v + 2dr\,dv = \Pi\,dt^2, \tag{6.31}$$

$$rdv^2 - d^2r = \Sigma\,dt^2, \tag{6.32}$$

where r is the radius vector, v the longitude, t the time, Π the sum of the transverse forces, and Σ the sum of the radial forces. By two successive integrations of (6.31) he deduced an expression for the differential of the time; substituting this into (6.32) and integrating twice, he arrived at

$$\frac{f^2}{Mr} = 1 - g\sin v - q\cos v + \sin v\int\Omega\,dv\cos v - \cos v\int\Omega\,dv\sin v, \tag{6.33}$$

where f, g, and q are constants of integration, M is the sum of the masses of the earth and the moon, and Ω is a function of r and the perturbing forces—a precursor of Lagrange's perturbing function. How could one solve this equation for r, given that r occurs under the integral signs on the right?

The left member of (6.33) together with the first three terms of the right member define r in terms of a fixed elliptical orbit, expressible by an equation of the form $p/r = 1 - c\cos v$. But it would not do, Clairaut explained, to use the r of this formula in the integrals on the right, because empirically it was known that the lunar apse rotates, and a value of r given by a fixed ellipse would soon, because of the apsidal motion, become fatally inaccurate.[62] Instead, he proposed using the formula for a moving ellipse: $k/r = 1 - e\cos mv$, where k, e, and m are constants to be determined later, and v is the true anomaly, or the longitude measured from apogee. Any difference between m and unity would imply apsidal motion. After substitution of this approximate value for r, Clairaut hoped in the simplification of (6.33) to be able to evaluate k and m in terms of the other constants in the equation in such a way that the resultant motion could be largely accounted for—small oscillations represented by the remaining terms being set aside—as motion on a moving ellipse.

In the initial outcome, this hope appeared to be gratified: Clairaut's modified equation took the form

$$\frac{k}{r} = 1 - e\cos mv + \beta\cos\frac{2v}{n} + \gamma\cos\left(\frac{2}{n} - m\right)v + \delta\cos\left(\frac{2}{n} + m\right)v, \quad (6.34)$$

where n is the moon's mean motion divided by the difference between the moon's and sun's mean motions, or 13.36881/12.36881; and β, γ, δ evaluated in terms of other constants in the theory were 0.007090988, −0.00949705, and 0.00018361, respectively, hence small relative to e (known empirically to be approximately 1/20).

The chief processes whereby Clairaut arrived at an equation of the form of (6.34) were as follows: the formula for r in the hypothesized moving ellipse was substituted into Ω; all angles in Ω were expressed (often by approximation) in terms of the angle v; products of the cosines and sines of two angles were replaced by the equivalent cosines and sines of the angle sum or difference; and the integrations indicated in (6.33) were carried out, whereby the constants multiplying angle v in the arguments of sines and cosines came to appear in the coefficients of the sinusoidal terms. Clairaut thus obtained an equation for $1/r$ in which the multinomial coefficient of $\cos v$ could be set equal to 0, and the coefficient of $\cos mv$, namely $3e\alpha/2p(m^2 - 1)$, could be set equal to $-e/k$. The latter equation, when p was replaced by k, to which it could be assumed to be approximately equal, yielded $m = 1 - 3\alpha/4$, where α is 1/178.725, the constant that we met with in Newton's lunar theory. It followed that $m = 0.9958036$, implying that the moon returns to its apse after $361°30'38''$ of motion, so that the apse was found, by this derivation, to advance only $1°30'11''$ per month, about half its observed advance.

From the beginning, Clairaut had supposed that a second-stage approximation was eventually to be carried out, to refine the coefficients of the several terms of the theory preparatory to constructing tables. In this second calculation, formula (6.34), with the coefficients β, γ, δ left as symbols rather than evaluated numerically, was to be substituted back into Ω and all the previous steps repeated. But Clairaut did not expect this second approximation to lead to a much improved value of m. If Newton's theory was correct, and if the approximation in terms of a rotating ellipse was a reasonably good one, as the results of Clairaut's first calculation seemed to indicate, then the result for the apsidal motion would not have been 50% in error, as in fact it was. Clairaut concluded that Newton's theory was at fault.

The other alternative, of course, was that the first-stage approximation was highly inaccurate with respect to the apsidal motion, although relatively satisfactory in other respects. This would be Clairaut's revised conclusion, after he had carried out the second, more refined calculation. Substitution of (6.34) into Ω led to new contributions to the coefficient of $\cos mv$. In particular, the contributions due to the term with coefficient γ were sizable.

The surprise was that such very small terms in the reciprocal radius vector could make such a large difference in the motion of the lunar apse. The visual representation in terms of a rotating ellipse had proved misleading.

The algebraic expressions of the coefficients γ, δ of the last two terms of (6.34) contain the constant m. Both these terms are proportional to the *transverse* perturbing force, whereas the initially computed contribution to m had been proportional to the *radial* perturbing force. When these terms are substituted into Ω, they lead to terms in $\cos mv$ whose coefficients have then to be taken into account in computing m—the logical circularity is unavoidable. In a first stage of his more refined computation, Clairaut set $m = 1$ in the coefficients γ, δ, then computed the effect of the last two terms of (6.34) on the coefficient of $\cos mv$, and finally set this coefficient equal to $-e/k$, and recomputed m. His result was $m = 0.99164$, implying a monthly apsidal motion of $3°2'6''$. The empirical value accepted by Clairaut was $3°4'11''$, implying for m the value

0.991545. At this point Clairaut felt justified in assuming that the full apsidal motion was deducible from the inverse-square law. For the further development of the theory he proceeded without further ado to substitute the observational value of m into the coefficients γ, δ.

Euler on the Moon's Apsidal Motion

Euler's first publication of a theoretical derivation of the lunar inequalities came in 1753, with his *Theoria Motus Lunae Exhibens Omnes Eius Inaequalitates*. This work had no other aim than to test the truth of Clairaut's claim, announced in May 1749, that Newton's inverse-square law sufficed to account for the full motion of the lunar apse. During the late spring and early summer of 1749 Euler attempted to discover by his own method whether Clairaut could be right, but wrote the latter in mid-July to say he could find no error in his own derivation. Meanwhile the St. Petersburg Academy was establishing a series of prize contests, and at Euler's suggestion, it chose for the first of these (for the year 1751) the question of the accordance of the moon's motions with Newton's inverse-square law.[63] Euler was named as one of the commissioners to examine the memoirs submitted.

In March 1751 Euler wrote Clairaut to say that he had received four essays thus far, one of them recognizably Clairaut's, and that the three others were "abominable not only in relation to your piece but in themselves." After further praise of Clairaut's piece, he went on to say that he was pursuing his own effort to rederive the apsidal motion, but with certain changes of procedure:

> It is with infinite satisfaction that I have read your piece, which I have waited for with such impatience. It is a magnificent piece of legerdemain, by which you have reduced all the angles entering the calculation to multiples of your angle v, which renders all the terms at once integrable. In my opinion this is the principal merit of your solution, seeing that by this means you arrive immediately at the true motion of the apogee; and I must confess that in this respect your method is far preferable to the one I have used. However I see clearly that your method cannot give a different result for the motion of the apogee than mine; in which I have recently made some change, for having previously reduced all angles to the eccentric anomaly of the Moon, I have now found a way to introduce the true anomaly in its place. Thus while your final equation has as its two principal variables the distance of the Moon from the Earth and the true longitude, I have directed my

> analysis to the derivation of an equation between the longitude of the Moon and its true anomaly, which seems to me more suitable for the usage of astronomy.[64]

Euler's lunar theory of 1753, like Clairaut's theory, assumes a rotating elliptical orbit as the pristine orbit to which perturbations are to be applied, but Euler is concerned with expressing the results in a more phenomenological form. At the end of chapter 7 he arrives at the equation

$$\varphi = \text{Const.} + Or + \text{periodic terms}, \tag{6.35}$$

where φ is the longitude, r the true anomaly, and O is a constant, the reciprocal of Clairaut's m, taken as determined by observation.[65] The equation eliminates all reference to the moon's distance from the earth and gives the astronomers what they actually determine by observation.

To arrive at (6.35), Euler started from equations of the same form as Clairaut's equations (6.31) and (6.32) but represented the moon's motion as projected onto the ecliptic rather than in an assumed orbital plane. Having developed detailed expressions for the radial and transverse forces, Euler integrated the first equation, obtaining a result of the form

$$z^2\, d\varphi = dp\left(C - \frac{S}{n^2}\right), \tag{6.36}$$

where z is the factor (always close to 1) by which the moon's mean distance from the earth must be multiplied to yield its actual distance at any moment, φ is the moon's longitude, p is the Moon's mean motion, C is a constant of integration, n is the ratio of the moon's mean motion to the sun's mean motion, and S consists of periodic terms that perturb the equable description of areas by the moon's radius vector. Euler then used (6.36) to eliminate $d\varphi$ from the second equation, a differential equation giving d^2z in terms of dp, z, and angular variables.

Euler's next step (in chapter 3) was to introduce the true anomaly r in place of the mean anomaly p, assuming the orbit to be an ellipse with the mean eccentricity of the moon's orbit (call it k), and with apse rotating at the observed mean rate of the moon's apse (namely $1 - [1/O]$). (A decade later, in an essay presented to the Berlin Academy, Euler would say that the choice of this orbit was arbitrary but ought to be such as to make the difference between the approximate theoretical and observed motions small;[66] I would hypothesize that distrust of the representation of the lunar orbit in terms of a rotating ellipse came, for Euler as for Clairaut, only with the realization that the apsidal problem could be solved on a

Newtonian basis.) It was assumed that the standard Keplerian rules gave the relation of r to p.

At this point Euler may seem to have assumed what is in question, but a symbol μ also appears in his equation, standing for a possible deviation from the inverse-square law. If this proves to be zero, the observed apsidal motion is in accordance with Newton's law, and if not, not.

While replacing p by r, Euler replaced z by tu, where t is the radius vector in the theoretical ellipse, given by $(1-k^2)/(1-k\cos r)$, and u is the factor, always close to 1, by which the actual radius vector differs from the radius vector in this ellipse. The resulting equation contained the constant C of integration introduced in (6.36), and an additional constant $m=(T+L)a^3/Sb^3$, where T, L, and S are the masses of the earth, moon, and sun, respectively, and a is the mean moon-earth distance and b the mean earth-sun distance. Euler obtained approximate values for these constants by setting $u=1$ in his equation and collecting homologous terms, those containing and those free of the factor $\cos r$. The results were

$$C^2 = 1 + \frac{3+4\mu}{2n^2}, \tag{6.37}$$

$$\frac{m}{n^2} = 1 + \frac{2+3\mu}{n^2}. \tag{6.38}$$

As these equations gave only approximate values, Euler introduced unknowns γ, δ to make them exact, and at the same time replaced C by $\kappa\sqrt{1-k^2}$, so that κ differed slightly from C:

$$\kappa^2 = 1 + \frac{3+4\mu+\delta}{2n^2}, \tag{6.39}$$

$$\frac{m}{n^2} = 1 + \frac{2+3\mu+\gamma}{n^2}, \tag{6.40}$$

where the difference between C^2 and κ^2 is understood to be included in $\delta/2n^2$.

In chapters 4–6 Euler, using the method of undetermined coefficients, derived successively the inequalities independent of the eccentricity k, those dependent on k, and those dependent on k^2, while ignoring the complications due to the eccentricity of the sun's path and the latitudinal deviations of the moon. This process produced certain constants, leading to progressive refinements in the values of γ, δ, and κ. Thus on completing the derivation of the inequalities independent of eccentricity (the terms of the variation), Euler had an equation of the form

$$\varphi = (\kappa - 0.000080)r + \text{terms in} \sin 2\eta, \sin 4\eta, \text{ etc.},$$

where η is the difference in longitude of the moon and the sun; but his final equation was to have the form $\varphi = Or +$ periodic terms, and thus, to this point, $\kappa = O + 0.000080$. With the introduction of the eccentricity k, in chapter 5, it became possible to determine a value of γ, namely 1.40673, and, as depending on it, a value of δ, namely 2.84830. The latter value was refined when terms in k^2 were introduced (in chapter 6) to 2.80226. Euler then corrected the coefficients of all the inequalities in chapter 7 to take account of these values of γ and δ, and these corrections led to a new value of δ, namely 3.20892. Meanwhile, κ became equal to $O + 0.000429$ at the end of chapter 6, and finally $O + 0.000364$ in chapter 8. The postulated deviation from the Newtonian law, namely μ, at this point became 0.010095, so that its contribution to κ, namely $\mu/2n^2$, was only 0.000029. Euler had already decided, at the end of chapter 6, that it could be set equal to zero.

Thus Euler satisfied himself that the inverse-square law was sufficient to account for the motion of the lunar apse. His final equation was free of all references to orbital shape. His process of derivation, though very different from Clairaut's, was similar in that it required successive circlings around a logical circle, whereby the values of certain constants were successively refined. Both for Euler and for Clairaut, dynamical insight was required for the formation of the initial differential equations, but not thereafter. A complex dynamical system was represented by an analytic function, and its characteristics derived by standard algorithms and judiciously chosen steps of successive approximation applied to the function.

Neither Clairaut nor Euler conducted the successive approximations in such a way as to be able to lay out their sequence in ordered stages, with a measure of the degree of approximation at each stage. That task was left for d'Alembert, who in addition went Euler one better in the phenomenological direction by avoiding any appeal to an intermediary ellipse.

D'Alembert on the Motion of the Moon's Apse

D'Alembert did not publish his lunar theory till 1754,[67] although he had developed much of it[68] before he embarked, in late 1748, on a task he deemed more urgent because he imagined himself to be in competition with the English, namely, the derivation of the precession of the equinoxes and nutation of the earth's axis from Newton's law.[69] When he got around to completing his lunar theory, his discussion included critical comment on Clairaut's and Euler's treatments of the lunar problem, and in particular on Clairaut's claim to have demonstrated agreement between Newton's law and the observed apsidal motion.[70]

D'Alembert's first step in the lunar theory was to derive a differential equation for the lunar orbit as projected onto the ecliptic. The independent variable was the longitude z as measured from some initial position, and the dependent variable was the reciprocal radius vector u. Taking the moon's orbit as approximately a circle concentric to the earth, he substituted $K + t$ for u, where K is a constant and t is always relatively small. By this means he obtained an equation of the form

$$d^2t + N^2t\,dz^2 + M\,dz^2 = 0, \tag{6.41}$$

where N is a constant to be determined, and M is a function of z. He expressed the form of the solution to this equation in terms of complex exponentials:

$$t = \frac{\delta}{2}\left(e^{Nzi} + e^{-Nzi}\right) + \frac{\varepsilon}{2Ni}\left(e^{Nzi} - e^{-Nzi}\right) - \frac{e^{-Nzi}}{2N}\int Mi \cdot dz \cdot e^{Nzi} + \frac{e^{Nzi}}{2N}\int Mi \cdot dz \cdot e^{-Nzi}. \tag{6.42}$$

Here δ and ε are constants to be determined by observation; I have substituted i for $\sqrt{-1}$ and e for d'Alembert's c, the base of natural logarithms. The complex exponential terms and factors are ultimately to be replaced by cosines and sines of Nz.

The detailed solution proceeded by clearly defined stages of approximation, each stage consisting of a solution of (6.42) that expressed the function $M(z)$ with a specified degree of precision. For the sake of this specification, d'Alembert classed the small quantities involved in the calculation according to order of magnitude. Thus among the quantities that he assigned to the first order of smallness were $n = 1/13.3688$, the ratio of the moon's period to the year, and $m \approx 1/11.4$, the tangent of the inclination of the moon's orbit to the ecliptic. The eccentricity of the sun's orbit, λ, was less, about 1/59. D'Alembert defined "infinitely small quantities of the second order" as those in which the quantities n^2, m^2, $n\lambda$, etc., occurred; "infinitely small quantities of the third order" as those in which the quantities n, m, λ, δ, etc., formed a product of three dimensions, or λ^2 occurred; and so on.[71]

In the first approximation d'Alembert carried out the determination of t to the order of $n^2 = 1/178.725$.[72] In this approximation N proved to be

$$\sqrt{1 - \frac{3n^2}{2}} \cong 1 - \frac{3n^2}{4},$$

yielding just about half the motion of the lunar apse, $1°30'39''$. D'Alembert considered Clairaut's proposal of 15 November 1747, to modify the inverse-square law by the addition of an inverse–fourth power term in order to make up the difference, to be wrongheaded. Even if the full apsidal motion was not derivable from Newton's law, the law was not invalidated: the extra apsidal motion could be due to some nongravitational force.[73]

In his second, more refined calculation of t, d'Alembert carried the precision to the third order of small quantities.[74] For this, some quantities had to be calculated to the fifth order, namely, factors in terms that were greatly augmented in the double integration that the solution entailed, by the emergence of small divisors.

In this second approximation N proved to be, very nearly,

$$1 - \frac{3n^2}{4} - \frac{225n^3}{32},$$

implying a monthly apsidal motion of $2°34'13''$, still $30'$ shy of the observed value.[75] It was therefore necessary, D'Alembert concluded, to carry the precision of the calculation farther than to the fourth order of small quantities, because (a) if the third term proved to be very different from $30'$, the calculation would have to be pushed at least to the fifth order to give assurance that the observations agreed with the theory; and (b) if the third term was close to $30'$, it would still be necessary to know that the fourth term was very small, before one could regard the sum of the first three terms as representing approximately the sum of the series.[76]

In the end, d'Alembert carried the computation of the apsidal motion to the precision of terms of the sixth order of smallness. His third term added $23'30''$, and his fourth term $5'5''$, to the calculated apsidal motion, leading to a total apsidal motion per sidereal month of $3°2'33''$, just $1'4''$ short of the value he accepted as given by observation.[77] He provided an algebraic formula for the motion of the apse,[78] thus carrying the algebraic formulation of the theory farther than Clairaut or Euler. At the same time, he was acutely aware of the impossibility of demonstrating rigorously the accuracy of his formulas, above all because of the augmentation of terms the integrations produced.[79]

D'Alembert's work on the lunar theory presents a theoretical rather than a practical triumph. Having put the Newtonian lunar theory, as set forth in Le Monnier's *Institutions astronomiques* (1746), into a form permitting this theory to be compared with his own theory as well as the theories of Euler and Clairaut,[80] d'Alembert concluded that Le Monnier's tables were never in error by more than 5′; but he did not dare affirm that the terms of his own theory were accurate to 1′.[81] Further tests would show that the errors in his own as well as in Euler's and Clairaut's tables went as high as 2′ or 3′. Tobias Mayer achieved tables accurate to 1′ by 1753, using a statistical fitting of the theory to observations to determine not only empirical constants but the coefficients of numerous inequalities.[82] The later Laplacian tables of Bürg (1806) and Burckhardt (1812) would likewise be dependent on statistical fitting for determining many coefficients. It is not inconceivable that a statistical comparison of observations with the Newtonian theory, such as the one Halley attempted, could have led to a fairly reliable solution of the longitude problem without benefit of differential equations.[83]

From Newton to d'Alembert, the essential theoretical advance in the lunar theory consisted in the decision to invest all dynamical insight in an initial set of differential equations, while relinquishing the demand for direct insight into the particularities of the lunar motions, and to entrust the derivation of consequences to algorithmic processes and successive approximations, carefully controlled.

Acknowledgments

To Professor George Smith of Tufts and his former student Michael Herstine I am indebted for a close reading of an earlier version of this chapter and the detection of a number of errors. To Professor Michael Nauenberg I am indebted for a crucial insight, as indicated in note 17.

Notes

1. *The Mathematical Papers of Isaac Newton*, ed. D. T. Whiteside, 8 vols. (Cambridge: Cambridge University Press, 1967–1981), 6:508ff.

2. Ibid., 6:518, n. 26.

3. Newton did not understand the limiting clause, and later writers have not always understood it. S. B. Gaythorpe in "On Horrocks's Treatment of the Evection and the

Equation of the Centre [...]," *Monthly Notices of the Royal Astronomical Society* 85 (1925): 858–65, and in "Jeremiah Horrocks and his 'New Theory of the Moon,'" *Journal of the British Astronomical Association* 67 (1957): 134–44, failed to understand it, and was corrected by N. T. Jørgensen, "On the Moon's Elliptic Inequality, Evection, and Variation and Horrox's 'New Theory of the Moon,'" *Centaurus* 18 (1973–1974): 316–18.

4. The two lines diverge from parallelism by at most about sixty times the horizontal solar parallax. In the first edition Newton put the latter figure at $20''$, but reduced it in the second edition to $10''$, and in the third to $10''.55$.

5. *Mathematical Papers of Isaac Newton*, 4:116ff.

6. See, for instance, *Mathematical Papers of Isaac Newton*, 1:146, 215, 248ff, 420.

7. Ibid., 3:157. In his English translation (facing the Latin text), Whiteside uses the dot notation for a fluxion; Newton introduced this in 1691, and it does not appear in the actual text of 1671. Our use of Leibnizian notation is equally or more unfaithful.

8. The following fact suggests that Newton had in his possession a formula equivalent to (6.18). To obtain the mean radial accelerative force round the orbit, it is necessary to integrate (6.18) or its equivalent from $\alpha = 0°$ to $\alpha = 360°$, then divide by 2π; the result is $-\mathbf{PS}/2 = \mathbf{SP}/2$. In the first edition of the *Principia*, in Corollary 2 of I.45, which has to do with apsidal motion due to a radial perturbing force, Newton introduces without explanation as an example an outward radial perturbing force equal to 1/357.45 times the central force and finds that it produces an apsidal motion of $1°31'14''$ per revolution. But $357.45 = 2 \cdot 178.725$, and thus the extraneous outward force is precisely $-\mathbf{PS}/2$. Clairaut later conjectured that Newton did not know how to integrate the term of (6.18) containing $\cos 2\alpha$ [see Craig B. Waff, "Universal Gravitation and the Motion of the Moon's Apogee ..." (Ph.D. diss., The Johns Hopkins University, 1976), 110]. Newton's integration of (6.11) would suggest the contrary.

The issue is complicated by Newton's insertion, in the second edition of the *Principia*, of the following statement in Book III, Proposition 4: "The action of the Sun, attracting the Moon from the Earth, is nearly as the Moon's distance from the Earth; and therefore ... is to the centripetal force of the Moon as ... 1 to $178^{29/40}$." The implication would seem to be that the radial perturbing force was such as to yield the full motion of the lunar apse. Newton had certainly known better in the 1680s, as we shall see. Was his insertion forgetful, or an inadvertent or dishonest fudge?

9. *American Journal of Mathematics* 1 (1878): 257–60; *Collected Mathematical Works of George William Hill*, 4 vols. (Carnegie Institution of Washington, 1905), 1:333–335. For bringing the problem here mentioned to my attention, I am indebted to Professor George Smith of Tufts University and to his erstwhile student Michael Herstine.

10. In the first edition Newton obtains for this ratio $69\frac{11}{12}$ to $68\frac{11}{12}$.

11. Euler, Leonhard, *Theoras Motus Lunae* (St. Petersburg, 1753), in *Opera Omnia*, Series secunda, eds. L. Courvoisier and J. O. Fleckenstein (Basel: Orell Füssli Turici, 1969) 23:64ff.

12. *Histoire de l'Académie des Sciences et Belles-Lettres* (Berlin, 1766), 334–53.

13. Leonhard Euler, *Theoria Motuum Lunae Nova Methodo Pertractata* (St. Petersburg, 1772), in *Opera Omnia* II, ed. L. Courvoisier, 22 (Lausanne: Orell Füssli Turici, 1958).

14. *American Journal of Mathematics* 1 (1878): 5–26, 129–47, 245–60; *Collected Mathematical Works of George William Hill*, 1:284–335.

15. *Mathematical Papers of Isaac Newton*, 6:519.

16. The cosine of the maximum angle between the tangent to the curve and the perpendicular to the radius vector is given, very nearly, by $1 - (e^2/2)$, and at the apses this angle is 0. If the eccentricity is 0.01, for instance, the maximum value of the angle is $0^\circ.57$. From one time particle to the next, the angle between radius vector and tangent changes by only a small fraction of the maximum angle.

17. See equation (7.39) (in chapter 7).

18. I am indebted for the analysis that follows to the prompting and tutoring of Michael Nauenberg during the course of the Dibner Institute's Symposium on Isaac Newton's Natural Philosophy, held at the Dibner Institute at MIT, Cambridge, Mass., 10–11 November 1995. The impetus supplied by Nauenberg led to a total revision of this second section of this chapter.

19. In Corollary 2 of Book I, Proposition 45, Newton derived a formula for the apsidal motion due to a radial perturbing force, and applying it to the case of the moon, found an apsidal motion of $1^\circ 31' 28''$. See note 8. The formula of Proposition 45, however, is accurate only in the case of vanishing eccentricity, whereas the result just found from equation (6.21) is the exact first-order term due to the radial perturbing force.

20. See equation (7.72) (in chapter 7).

21. *Mathematical Papers of Isaac Newton*, 6:516, 517.

22. Ibid., 6:516–19.

23. See ibid., 6:518–19, n. 26.

24. For an unpacking of Newton's sketch of a proof in the second and third editions of the *Principia*, see Bruce Pourciau, "On Newton's Proof That Inverse-Square Orbits Must Be Conics," *Annals of Science* 48 (1991): 159–72, and "Newton's Solution of the One-Body Problem," *Archive for History of Exact Sciences* 44 (1992): 125–46.

25. See Alexander Koyré and I. Bernard Cohen, eds., *Isaac Newton's Philosophiae Naturalis Principia Mathematica*, 3d ed. with variant readings (Cambridge: Harvard University Press, 1972), 2:565 (lines 19–23 of p. 395 of the third edition text) and 2:764 (lines 11–12 of p. 530 of the third edition text).

26. *Mathematical Papers of Isaac Newton*, 6:518–19. The major axis *QR* must be equal to the circle's diameter in order that the period of the motions on the two orbits should be the same.

27. See equations (7.68)–(7.72) (in chapter 7).

28. *Mathematical Papers of Isaac Newton*, 6:528, note 53 and following notes.

29. Newton actually has more than one way of evaluating these integrals, but his preferred way is the reduction to series. See Whiteside's detailed account in *Mathematical Papers of Isaac Newton*, 6:528–30, nn. 54, 56.

30. See *The Correspondence of Isaac Newton*, 7 vols., ed. H. W. Turnbull (Cambridge: published for the Royal Society at the University Press, 1960), 2:57 (letter of Leibniz to Henry Oldenburg, 17 August 1676).

31. See Richard S. Westfall, *Never at Rest* (Cambridge: Cambridge University Press, 1980), 60–62.

32. For instance, the diagrams for his *De Motu Corporum in Gyrum*, published in *The Preliminary Manuscripts for Isaac Newton's 1687 Principia 1684–1685* (Cambridge: Cambridge University Press, 1989).

33. For the source of this phrase see Richard S. Westfall, *Never at Rest* (Cambridge: Cambridge University Press, 1980), 59.

34. *Mathematical Papers of Isaac Newton*, 3:33.

35. *Correspondence of Isaac Newton*, 1:356.

36. *Mathematical Papers of Isaac Newton* 4:277. The intensity of Newton's attack on Descartes's algebra is reminiscent of Hobbes's earlier attack on John Wallis's algebra. "Symbols," said Hobbes, "are poor, unhandsome, though necessary, scaffolds of demonstration; and ought no more to appear in public, than the most deformed necessary business which you do in your chambers" ["Six Lessons to the Savilian Professors of the Mathematics," in *The English Works of Thomas Hobbes of Malmesbury*, 11 vols., ed. Sir William Molesworth (Darmstadt, Germany, 1966; a reprint of the 1839 ed.) 7:248].

37. *Mathematical Papers of Isaac Newton*, 4:421.

38. *Philosophical Transactions* 29 (1714–1716): 206.

39. See D. T. Whiteside, "The Mathematical Principles Underlying Newton's *Principia Mathematica*," *Journal for the History of Astronomy* 1 (1970): 116–38, esp. 119.

40. See François De Gandt, *Force and Geometry in Newton's Principia*, trans. Curtis Wilson (Princeton, N. J.: Princeton University Press, 1995), chap. 1.

41. See Herman Erlichson, "Newton's Solution to the Equiangular Spiral Problem and a New Solution Using Only the Equiangular Property," *Historia Mathematica* 19 (1992): 402–13, and Curtis Wilson, "Newton on the Equiangular Spiral: An Addendum to Erlichson's Account," *Historia Mathematica* 21 (1994): 196–203.

42. *De Methodis Serierum et Fluxionum*, in *Mathematical Papers of Isaac Newton*, 3:72–73.

43. A. P. Youschkevitch, "The Concept of Function up to the Middle of the 19th Century," *Archive for History of Exact Sciences* 16 (1976/77): 37–85; 56.

44. G. W. Leibniz, *Mathematische Schriften* ed. C. I. Gerhardt (Hildesheim: Olms, 1971), 5:393.

45. Youschkevitch, "Concept of Function," 57, n. 41.

46. See J. M. Bos, "Differentials, Higher-Order Differentials and the Derivative in the Leibnizian Analysis," *Archive for History of Exact Sciences* 14 (1974/75): 1–90.

47. Youschkevitch, "Concept of Function," 60, n. 41.

48. Ibid., 61.

49. Ibid., 38–39, 70.

50. Many instances are found in Newton's *Arithmetica Universalis*, which apparently derives from his teaching at Cambridge in the period 1673–1675, was assembled into a book around 1683, and was published in 1707.

51. "Reflexions sur les diverses manieres dont on peut représenter le mouvement de la lune," *Histoire de l'Académie Royale des Sciences et Belles-Lettres* (Berlin, 1763, published 1770), 183. Euler's memory may be faulty here.

52. See Waff, "Universal Gravitation and the Motion of the Moon's Apogee," 55–56, n. 8.

53. *Novae et Correctae Tabulae ad Loca Lunae Computanda* (Berlin, 1745).

54. Euler, *Opera Omnia*, Series secunda, 23:2: "Calculum enim mihi eousque producere licuit, ut titulos seu argumenta singularum inaequalitatum ostenderet: ac plerarumque quidem aequationum veram quantitatem sola theoria determinavit, nonnullas vero per observationes definire sum coactus."

55. Katz, "The Calculus of Trigonometric Functions," *Historia Mathematica* 14 (1987): 311–24.

56. In *Opera Omnia*, Series prima, 14:543, Euler explicitly lays claim to his originality here. Clairaut, in his *Théorie de la lune* of 1752, after giving the equivalences between products of sines and cosines of the angles A and B and the sines and cosines of the sum $A + B$ and difference $A - B$, remarks: "Le célèbre Mr. Euler, à qui les Mathématiques sont redevable de tant d'artifices de calcul, est le premier que je sache qui se soit passé des valeurs des sinus sous la forme imaginaire & qui ait pensé à avoir recours aux Theorems que je viens de citer."

57. See especially in the list of references in Katz, "Calculus of Trigonometric Functions," the items Euler 1739a (*Opera Omnia* II 10, 78–97) and Euler 1743 (*Opera Omnia* I 22, 108–149).

58. "Recherches sur le mouvement des corps célestes en générale," *Opera Omnia*, Series secunda, 25:1 ff, ¶¶ 1–16; "Recherches sur la question des inégalités du mouvement de Saturne et de Jupiter," *Opera Omnia* II 25, 45 ff., ¶¶ 1–8.

59. See chapter 2 of Waff, "Universal Gravitation and the Motion of the Moon's Apogee."

60. For a more detailed description of Clairaut's analysis than I give here, see ibid., chapter 5 of Waff. See also my "Perturbations and Solar Tables from Lacaille to Delambre," *Archive for History of Exact Sciences* 22 (1980): 133–45.

61. As d'Alembert pointed out, Clairaut is wrong to assume an orbital plane: the moon's path has double curvature and is not confined to a plane. It would have been better had Clairaut like both Euler and d'Alembert, derived his equations for the motion of the moon as projected onto the ecliptic.

62. D'Alembert later criticized Clairaut for using this observation in constructing his theory, on the grounds that he should have derived the apsidal motion purely from theory.

63. A. P. Juskevic & E. Winter, *Die Berliner und die Petersburger Akademie der Wissenschaften im Briefwechsel Leonhard Eulers* (Berlin: Akademie Verlag, 1951), 2:173–74.

64. G. Bigourdan, "Lettres inédites d'Euler à Clairaut," *Comptes rendus du Congrès des sociétés savantes de Paris et des Départements tenu à Lille en 1928 . . .*, 34–35.

65. *Opera Omnia*, Series secunda, 23:160.

66. Euler, "Reflexions sur les diverses manieres," 182.

67. In his *Recherches sur differens points importans du système du monde, Premiere Partie* (Paris, 1754).

68. See ibid., pp. 119–20.

69. D'Alembert published this derivation in his *Recherches sur la Précession* in July 1749, and it was important in correcting Newton's deeply flawed derivation, thus providing the first solid evidence that the attraction between the moon and the earth was mutual. In addition, it opened the way to a correct treatment of the mechanics of rigid bodies, which Euler then pursued. See my "D'Alembert versus Euler on the Precession of the Equinoxes and the Mechanics of Rigid Bodies," *Archive for History of Exact Sciences* 37 (1987): 233–73.

70. D'Alembert, *Recherches sur differens points importans du système du monde, Premiere partie* (Paris, 1754), 113–18.

71. Ibid., 43–44.

72. Ibid., 40.

73. Ibid., 113–14.

74. Ibid., 44, 55 ff.

75. Ibid., 115.

76. Ibid., 177.

77. Ibid., 181–82.

78. Ibid., 181.

79. Ibid., 199–201.

80. Ibid., chap. 13, pp. 89 ff.

81. Ibid., 235, 234.

82. *Novae Tabulae Motuum Solis et Lunae*, published in volume 2 of the *Commentarii* of the Royal Society of Sciences of Göttingen in 1753, and the later revision of these tables, published by the British Admiralty in 1767. Mayer's *Theoria Lunae Juxta Systema Newtonianum*, which gives the theoretical basis of his tables, was sent to the British Admiralty in late 1754 but not published till 1767. Although Mayer learned much from Euler's lunar theory, his own theory was different in important respects, particularly in conflating inequalities so as to reduce the number of tables required to compute the moon's place.

83. See Nicholas Kollerstrom, "Halley and the Saros Synchrony," chap. 13 in "The Achievement of Newton's 'Theory of the Moon's Motion' of 1702" (Ph.D. diss., University of London, 1994), and Nick Kollerstrom & Bernard D. Yallop, "Flamsteed's Lunar Data, 1692–95, Sent to Newton," *Journal for the History of Astronomy* 26 (1995): 244.

7

Newton's Perturbation Methods for the Three-Body Problem and Their Application to Lunar Motion

Michael Nauenberg

In the preface to the first edition (1687) of the *Principia*, Newton wrote, "But after I began to work on the inequalities of the motion of the moon, and then also began to explore other aspects of the laws and measures of gravity and of other forces . . . I thought that publication should be put off to another time, so that I might investigate these other things and publish all my results together. I have grouped them together in the corollaries of Prop. 66 the inquires into lunar motion (which are imperfect), so that I might not have to deal with these things one by one in propositions and demonstrations, using a method more prolix than the subject warrants . . ."[1] Subsequently, in a remarkable series of twenty-two corollaries to Proposition 66 in Book I of the *Principia*, he summarized almost entirely in prose with only a single diagram the result of his researches on the effects of the sun's gravitational force on the moon's motion around the earth. About one of these results, the calculation of the bimonthly variation in lunar speed discovered by Tycho Brahe, Laplace wrote that "the method which he has followed seems to me one of the most remarkable things in the *Principia*" [la methode qu'il a suivie me parait etre une des choses les plus remarquables de l'ouvrage de Principes].[2] In the *Principia* Newton described two distinct perturbation methods, one in Book I, Propositions 43–45, and the other in Book III, Propositions 25–35, which included his application to the sun's perturbations of the lunar motion. However, certain mathematical manuscripts of the Portsmouth collection of Newton's papers, first examined by a Cambridge University committee in 1872,[3,4] revealed that by 1686 Newton had developed a third perturbation method to deal with gravitational perturbations to Keplerian motion. I will show here that this method corresponds to the variation of orbital parameters method[5] first developed in 1753 by Euler[6] and afterwards by Lagrange and Laplace.[7–9] Newton's method, called here the Portsmouth method, was apparently intended for inclusion in the *Principia*, but it never found its way into any of the three editions Newton supervised. The only evidence in the *Principia* for the existence of this perturbation method appeared in two corollaries, Corollaries 3 and 4, at the end of Proposition 17, which deals with the initial value problem: given the position and velocity of a body acted on by a inverse-square force, construct the corresponding Keplerian orbit (conic section).[10] In particular, Corollary 4 states, "And if the body is continually perturbed by some force impressed from outside, its trajectory can be determined very nearly, by noting the changes which the force introduces at certain points and estimating from the order of the sequence the continual changes at intermediary places."

Evidently Newton developed this method to deal with the observed motion of the lunar apogee, because his earlier approach, described in Corollary 2 of Proposition 45, Book I, had failed to describe this motion quantitatively by a factor of two. Indeed, Newton applied his new method to this problem, and he appeared to have succeeded in calculating the motion of the lunar apogee in agreement with observation, as he announced in the Scholium to Proposition 35, Book III, in the first edition (1687) of the *Principia*:

> Thus far regarding the motions of the moon insofar as the eccentricity of the orbit is not considered. By similar computations I have found that the apogee, when it is situated in conjunction or opposition to the sun, advances each day 23′ in respect to the fixed stars. but at the quadratures regresses each day about $16\frac{1}{3}'$; and that its mean annual motion should be virtually 40°. By the astronomical tables adapted by Mr. Flamsteed to Horrocks hypothesis the apogee at its syzygies advances with a daily motion of 20′12″, and is borne in *consequentia* with a mean annual motion of 40°41′. . . . These computations, however, excessively complicated and clogged with approximations as they are, and insufficiently accurate we have not seen fit to set out [computationes autem, ut nimis perplexas & approximationibus impeditas, neque satis accuratas, apponere non lubet].[11]

However, in the two subsequent editions of the *Principia*, this Scholium was revised, leaving out these claims entirely,[12] and at the end of Corollary 2, Proposition 45 of Book I in the third edition, he inserted, after the statement ". . . therefore the upper apse in each revolution will go forward 1°31′28″" the well-known additional comment "The apse of the moon is about twice as swift."

Evidently Newton remained unsatisfied[13] with his computation of the motion of the lunar apogee in the Portsmouth manuscript. However, in 1894 Tisserand in his *Traité de Mécanique Céleste*[14] carefully reviewed

Newton's lunar theory as it appeared in the *Principia* and also compared some of Newton's results in the Portsmouth manuscript with the results of the variation of parameters perturbation theory of Euler, Laplace, and Lagrange. For an arbitrary perturbing force, Tisserand found that Newton's equation for the rotation of the ellipse's major axis was correct to lowest order in the eccentricity of the orbit, and his application to the lunar case differed only in the numerical value of one parameter, which Newton gave as 11/2, instead of the correct value of 5. In particular, Tisserand concluded that "Newton deduced correctly that the mean annual movement of the apogee is 38°51′51″, while the astronomical tables give 40°41′5″" [Newton dèduit, tout à fait correctement ... que le mouvement moyen annuel de l'apogèe est de 38°51′51″, tandis que celui qui est donne dans les Tables astronomiques es de 40°41′5″].[15] Nevertheless, recently historians of science have concluded that these calculations of Newton were a "fudge" to get the correct answer, or were otherwise badly flawed.[16–22] This is quite surprising, because Tisserand had shown that Newton's equation for the dominant contribution to the rotation of the apsis has the correct analytic form. Moreover, Newton had integrated this equation to obtain an analytic expression for the secular motion of the apogee that can also be compared with the correct higher-order calculation carried out by Alexis Clairaut and Jean d'Alembert sixty-three years later. It will be shown here that apart from the numerical discrepancy in the parameter mentioned above, Newton's results have equivalent analytic form, a fact that seems to have escaped previous attention.[16–23]

In this chapter I will analyze Newton's Portsmouth method and relate it to his earlier perturbation approaches to the three-body problem as well as to modern perturbation techniques. Newton's physical and geometrical approach leads directly to differential equations, which Newton referred to as *hourly motion*, for the parameters of a revolving ellipse describing orbital motion due to an inverse-square force in the presence of an external perturbation force. I will show that this method corresponds precisely to the modern method of variation of orbital parameters attributed to Euler, Lagrange, and Laplace.[6–9] However, in the Portsmouth manuscripts one finds only the variational equation for the rotation of the major axis of the ellipse evaluated to lowest order in the eccentricity, whereas the analogous equation for the variation of the eccentricity is absent. This second equation is essential in order to relate his new method to his earlier approaches, as I shall demonstrate below.

An essential question that I will examine here in some detail is the relation between the revolving ellipse introduced in Section 9, Book I, as an approximate orbit for a body under the action of general central forces, and the corresponding revolving ellipse introduced in the Portsmouth method to account for the effect of perturbations on an orbit under the action of a central inverse-square force. I will show that these two ellipses can be quite different, and that some of the difficulties Newton encountered in applying his Portsmouth method may have been due in part to a lack of understanding of the relation of this approach to his earlier perturbation methods. To illustrate the problems involved I consider the special case of a circular orbit, which is a trivial solution for pure central forces, and demonstrate how such a solution emerges from the Portsmouth method, which represents an orbit by a rotating ellipse. For such an orbit with small eccentricity e, a crucial parameter is the ratio e/m^2, where m is a dimensionless parameter determined by the perturbation's magnitude. For the lunar problem, m is the reciprocal of the number of lunar cycles in a sidereal year (Newton gives the value $m^2 = 1/178.725$). The rotation of the apsis of the ellipse in the Portsmouth method turns out to be the same as that of the actual orbit only in the case that $m^2/e \ll 1$. Fortunately this condition turns out to be satisfied in the case of the moon's motion.

In the chapter's first section, Newton's earliest approach to orbital dynamics, the curvature method,[24,25] is reviewed and extended to include the contributions of transverse as well as radial perturbing forces which depend on both radial and angular coordinates. These contributions are essential to evaluate higher-order solar perturbations of the lunar motion. Although Book I of the first edition of the *Principia* does not discuss this curvature method, Newton applied it in Book III, Proposition 28, to evaluate, to first order in the solar perturbation, the variational orbit of the moon.[26] This contribution was evaluated subsequently by L. Euler[27] and in full detail more than 190 years later by G. W. Hill.[28] The chapter's second section discusses Newton's Portsmouth method and compares his relation for the "motion of the apogee" to lowest order in the eccentricity of the ellipse with the exact differential equation for this motion obtained by Euler, Lagrange, and Laplace.[6–9] The third section illustrates its application by considering the simple case of a purely radial perturbation that depends linearly with distance. Newton had previously considered this perturbation using his method of mobile orbits in Book I, Proposition 45, Corollary 2, but there is no evidence that he attempted also to solve this problem with the Portsmouth method. I will show that this example illustrates clearly the relation between Newton's two distinct approaches based on rotating elliptical orbits. The fourth section then evaluates the hourly motion of the lunar apogee, as Newton called his differential equation of motion for the lunar apsis, for the complete solar perturbation including the transverse component and angular dependence of this perturbation, and compares the results with Newton's own calculation in the Portsmouth papers. (For comparison, the appendix presents Clairaut's and d'Alembert's method of solution, based on the perturbative integration of the differential equations of motion corresponding to Newton's curvature method.) The last section gives a brief summary and conclusion. The discussion in this chapter is restricted to perturbation effects in the plane of the orbit, and Newton's treatment of the precession of lunar nodes and the oscillation of the lunar plane are not considered here.

To follow Newton's methods, which he presented in a geometrical form rather unfamiliar to most contemporary readers, I have given in each section a discussion of these methods in modern mathematical language.[29] It is essential to first understand the basic physics and the mathematics of low-order perturbation methods in the restricted three-body problem before one can analyze Newton's often rather succinct accounts and understand what he was trying to do, what he eventually achieved, and why and where he failed. To understand Newton's ideas in more detail and to follow his own geometrical language, the reader will need to consult also a copy of the *Principia* and the relevant parts of the Portsmouth manuscripts (which have been translated into English and published by D. T. Whiteside).[4]

Curvature Method

It has been argued[24] that the earliest approach Newton developed to calculate orbital motion for general central forces was based on the extension of the concept of centrifugal force or acceleration that he and Huygens[30] had developed independently for the case of uniform circular motion.[25] If a body is moving with constant velocity on a circular orbit with radius ρ, then the acceleration a along the radial direction is given by

$$a = \frac{v^2}{\rho}. \tag{7.1}$$

This acceleration is also a measure of the force required to keep a body in such a motion (in this article force is equated with acceleration).[31] For motion along a general orbit, this relation remains valid for the normal component of the acceleration, provided v is the instantaneous velocity, and ρ is the radius of curvature at the corresponding point of the orbit.

Although this generalization is not found anywhere in Book I of the first edition (1687) of the *Principia*, it was formulated in Book III, Proposition 28, which starts with the statement that "the curvature of the orbit which a body describes, if attracted in lines perpendicular to the orbit, is directly as the force of attraction and inversely as the square of the velocity."[32] The term *curvature* here was nowhere defined in the 1687 *Principia*, but Newton's calculations reveal that it is indeed proportional to $1/\rho$. Assuming that the force acts in a plane and choosing a fixed origin, the normal component a_n of the acceleration can be written in terms of the radial component a_r and transverse component a_θ of the acceleration a,

$$a_n = a_r \sin(\alpha) + a_\theta \cos(\alpha), \tag{7.2}$$

where α is the angle between the radius vector and the velocity vector. Expressing the velocity v in terms of the angular momentum h, called the *moment of areas* in the *Principia*, we have

$$v = \frac{h}{r \sin(\alpha)}, \tag{7.3}$$

and substituting (7.2) and (7.3) in (7.1) leads to a general equation for orbital motion,

$$\frac{1}{\rho \sin^3(\alpha)} = \frac{r^2}{h^2}[a_r + a_\theta \cot(\alpha)]. \tag{7.4}$$

As we shall see, Newton applied this relation already in the first edition of the *Principia* at special points of the orbit where $\alpha = \pi/2$. In the second edition Newton added new corollaries to Proposition 6 of Book I, which deals with the measure of central forces, and in particular Corollary 3 is the first explicit statement of the curvature measure, "the centripetal force will be inversely as the solid $SY^2 \times PV$," where in our notation $PV = 2\rho \sin(\alpha)$, and $SY = r \sin(\alpha)$. This measure corresponds to (7.4) for the special case that $a_\theta = 0$, apart from the overall constant of proportionality h. In this case h is a constant of the motion expressing the area law or conservation of angular momentum proved in Book I, Proposition 1. However, if the transverse component of the force a_θ does not vanish, then

$$\frac{dh}{dt} = r a_\theta. \tag{7.5}$$

Setting

$$h = r^2 \frac{d\theta}{dt}, \tag{7.6}$$

where θ is the longitude, and eliminating the time variable t, (7.5) becomes

$$\frac{dh}{d\theta} = \frac{r^3}{h} a_\theta. \tag{7.7}$$

Although Newton did not explicitly write this relation in the *Principia*, one finds that Book III, Proposition 26, entitled "To Find the Hourly Increment of the Area Which the Moon, by a Radius Drawn to the Earth, Describes in a Circular Motion," contains a calculation for the change in h at the syzygies and quadratures obtained by integrating approximately (7.7) for the case in which a_θ is the transverse component of the solar perturbation.

In fact, (7.4) and (7.7) correspond precisely to the differential equations of motion obtained sixty-three years later by Alexis Clairaut and Jean d'Alembert,[33,34] who applied Leibniz's form of the calculus to Newton's laws of motion, written here in equivalent geometrical form. This correspondence can be seen from the differential equation in polar coordinates for the projection of the curvature ρ along the radial direction. Newton obtained this relation around 1671 in the form[35]

$$\rho \sin(\alpha) = r \frac{(1 + z^2)}{(1 + z^2 - dz/d\theta)}, \tag{7.8}$$

where $z = \cot(\alpha) = (1/r)\, dr/d\theta$. It follows that

$$\frac{1}{\rho \sin^3(\alpha)} = \left(\frac{d^2}{d\theta^2} + 1\right)\frac{1}{r}, \tag{7.9}$$

and substituting this relation in the left side of (7.4) yields

$$\left(\frac{d^2}{d\theta^2} + 1\right)\frac{1}{r} = \frac{r^2}{h^2}\left[a_r + \frac{a_\theta}{r}\frac{dr}{d\theta}\right], \tag{7.10}$$

as first written by Clairaut,[33,34] where according to (7.7)

$$h^2 = h_0^2 + 2\int d\theta r^3 a_\theta. \tag{7.11}$$

In this form, Newton's geometrical relation (7.4) becomes a differential-integral equation, with θ as an independent variable. Except for special cases, such an equation can generally be solved only numerically or approximately by a perturbation expansion.

To illustrate its application we consider first the special case of a purely central force, independent of θ, for which h is constant. This is the case Newton first treats in Book I, Section 9, Propositions 43–45, entitled "The Motion of Bodies in Mobile Orbits; and the Motion of the Apsis." The first problem he considered was the evaluation of the radial acceleration a_r for a body moving along an orbit corresponding to an ellipse rotating around one of the foci, with the center of force located at this focus. Newton's interest in the dynamics of such an orbit came most likely from his earlier study of Horrocks's model for the lunar orbit.[17] Newton described such an orbit in his study of motion under the action of general central forces, which he discussed in his letter of 13 December 1679 to Robert Hooke.[24] In polar coordinates, Newton's rotating elliptical orbit takes the analytic form

$$r = \frac{r_0}{1 + e\cos(\nu\theta)}, \tag{7.12}$$

where ν is a constant. The evaluation of the curvature ρ and angle α of a curve is a geometrical problem that Newton had solved,[35] and consequently he could readily obtain a_r from his equation of motion (7.4), which implies

$$a_r = \frac{h^2}{r_0}\left[\frac{\nu^2}{r^2} + \frac{(1-\nu^2)r_0}{r^3}\right]. \tag{7.13}$$

Actually, Newton derives this expression in Book I, Proposition 44, by a somewhat different but equivalent procedure. For an orbit with small eccentricity e, he then developed a perturbation method based on this exact solution by approximating a general central force to this linear combination of an inverse square and an inverse cube force in the neighborhood of a circle of radius r_0. However, for the solar perturbation to the lunar motion, the radial component V of this force depends also on the relative angle ψ between the earth-sun and earth-moon axes, and the transverse component W does not vanish. To first order in an expansion in r/R, where R is the mean solar distance, Newton derived in Book III, Propositions 25 and 26, the relations

$$V = -a_r + \mu/r^2 = \frac{g}{2} r\,(1 + 3\cos(2\psi)) \tag{7.14}$$

and

$$W = a_\theta = -\frac{3g}{2} r\sin(2\psi), \tag{7.15}$$

where μ and g are constants ($\mu = GM$, $g = GM_0/R^3$, M is the sum of the mass of the earth and the moon, M_0 is the mass of the sun, and G is

Newton's gravitational constant). To apply his method Newton approximated the solar perturbing force by averaging (7.14) and (7.15) over the angle ψ. This implies that $a_\theta = 0$ and $a_r = \mu/r^2 - (g/2)r$. Setting $r = r_0 + \delta r$ and expanding a_r to first order in δr gives

$$a_r = \frac{\mu}{r_0^2}(1-c)\left(1 - \frac{\delta r}{r_0}\frac{(2+c)}{(1-c)}\right), \tag{7.16}$$

where $c = gr_0^3/2\mu$. This constant corresponds to $c = (1/2)m^2$, where $m = T_M/T_E$, T_E is the period of the earth around the sun, and T_M is the sidereal period of the moon around the earth. Newton's values for $T_E = 365^d6^h9^m$ and $T_M = 27^d7^h43^m$ yield $m^2 = 1/178.725$ and $c = 1/357.45$. Comparing this expansion with the expansion of (7.13),

$$a_r = \frac{h^2}{r_0^3}\left(1 - (3 - v^2)\frac{\delta r}{r_0}\right), \tag{7.17}$$

yields

$$v = \sqrt{\frac{1-4c}{1-c}}. \tag{7.18}$$

This is the result Newton gives in Proposition 45, Corollary 2, of the *Principia*.[36] Substituting his numerical value for c in (7.18), Newton found that the change $\Delta\theta$ between apogee and perigee is $\Delta\theta = 180°/v = 180°45'44''$ implying that the upper apse in each revolution will go forward $1°31'28''$. As pointed out earlier, in the third edition of the *Principia* Newton finally admitted that "the Apse of the moon is about twice as swift."

Subsequently, Newton evidently considered various possible sources for this discrepancy. The two most obvious ones are (a) averaging out the dependence of the solar perturbation over the relative angle ψ, and (b) neglecting the transverse component of this perturbation. If the transverse component is included, the angular momentum h is not a constant, and the perturbation method of Section 9, discussed above, no longer applies. Evidently in this case a circle is no longer a special solution, and the orbit's geometrical form is now more difficult to determine. Newton conjectured the existence of a special solution for the perturbed lunar orbit corresponding to an ellipse of small eccentricity with the earth located at its center. Furthermore, he found that this ellipse revolves around this center in such a way that its major axis remains always normal to the earth-sun radius. Applying his curvature method in Book III, Proposition 28, Newton treated explicitly this special solution as circular motion in the absence of perturbation, that is, the solution obtained by starting with a circular solution and turning on the solar perturbing forces adiabatically. In the limit of a large number of lunar cycles per year, or small m, this solution corresponds to the periodic solution obtained later by L. Euler[27] and in full detail by G. W. Hill,[28] who solved for the moon's motion in a frame of reference rotating uniformly around the earth with the period of one year. In this case the solution is approximately an ellipse with the earth at the center,

$$r \approx r_0(1 - x\cos(2\psi)), \tag{7.19}$$

where ψ is the angle between the sun and the moon measured from the earth, $\psi = (1 - m)\theta$, and the undetermined parameter x is assumed to be small. The requirement that the solution satisfy Newton's curvature relation (7.4), with a_r and a_θ given by (7.14) and (7.15), clearly dictates the solution's form. Notice that the radial component a_r of the perturbing force, which appears on the right side of this equation, is proportional to $\cos(2\psi)$, while the contribution of the transverse force component a_θ is multiplied by $(1/r)\,dr/d\theta$, which is of order x, and therefore can be neglected in the lowest-order approximation. However, a_θ contributes to the variation in the angular momentum h, and neglecting the variation in r on the right side of (7.11) yields

$$h^2 = h_0^2 + \frac{3m^2\mu r_0}{2(1 - m)}\cos(2\psi), \tag{7.20}$$

where h_0 is the average value of the angular momentum. To lowest order in x we have

$$z = \frac{1}{r}\frac{dr}{d\theta} = 2(1 - m)\,x\sin(2\psi), \tag{7.21}$$

and from (7.9),

$$\frac{1}{\rho\sin^3(\alpha)} = \frac{1}{r_0}[1 + (1 - 4(1 - m)^2)\,x\cos(2\psi)]. \tag{7.22}$$

Substituting (7.20) and (7.22) in (7.4), one obtains the relation

$$\frac{h_0^2}{\mu r_0}\left(1 + \left[\frac{3m^2\mu r_0}{2(1 - m)h_0^2} + (1 - 4(1 - m)^2)x\right]\cos(2\psi)\right)$$
$$= 1 - \frac{m^2}{2}(1 + 3\cos(2\psi)), \tag{7.23}$$

where we neglected terms of order m^4 and m^2x (corresponding to the contribution of the transverse force a_θ on the right side of (7.4). Hence

$$r_0 = (h^2/\mu)(1 + m^2/2), \tag{7.24}$$

$$h = h_0\left[1 + \frac{3m^2}{4(1-m)}\cos(2\psi)\right], \tag{7.25}$$

and

$$x = \frac{3}{2}m^2\frac{(2-m)}{(1-m)(4(1-m)^2-1)}. \tag{7.26}$$

This analytic form for x was first presented by Laplace[37] and Tisserand[38] and was reproduced recently by Chandrasekhar.[39] Evidently x is of order m^2, which justifies the approximations made previously. However, as terms of order m^4 have been neglected, (7.26) is valid only to cubic terms in m, leading to

$$x \approx m^2 + \frac{19}{6}m^3. \tag{7.27}$$

This result corresponds to the first two terms of an expansion of Hill's solutions in powers of m.[28] In Book III, Proposition 28, Newton gave his result only in numerical form as $x = .00719$; (7.26) gives $x = .00720$.

The relation between time t and longitude θ is obtained by integrating (7.6), giving

$$t = \int d\theta\frac{r^2}{h}. \tag{7.28}$$

Substituting for r and h (7.19) and (7.25), respectively, we find

$$\frac{t}{\tau} = \theta - \xi\sin(2\psi), \tag{7.29}$$

is the sidereal period divided by 2π, and

$$\xi = \frac{1}{(1-m)}\left[x + \frac{3m^2}{8(1-m)}\right] \approx \frac{11}{8}m^2 + \frac{85}{24}m^3. \tag{7.30}$$

Numerically, $\xi = 32'31''$, corresponding to Newton's result $32'32''$ in Book III, Proposition 29, in good agreement with the observation of Tycho Brahe.

We turn now to Newton's own method of solving this problem given in Book III, Proposition 28. In this proposition, Newton evaluates x by considering the curvature equation only at two special points: the syzygies $\psi = 0, \pi$ and the quadratures $\psi = \pi/2$ and $3\pi/2$. This does not suffice to show that Newton's revolving ellipse (7.19) is actually a solution of the curvature equation of motion, but as Laplace commented,[21] "These calculational assumptions ... are permitted to inventors in such difficult researches" [Ces hypothèses de calcul ... sont permise aux inventeurs, dans des recherches aussi difficiles]. At these special points $\alpha = \pi/2$, and according to (7.4),

$$\frac{\rho_s}{\rho_q} = \frac{h_s^2 r_q^2 a_q}{h_q^2 r_s^2 a_s}, \tag{7.31}$$

where the subscripts s, q refer to the value of the corresponding quantities at the syzygies and at the quadratures respectively. Substituting for the angular momentum h, radius r and radial force a at these special points the values given by (7.25), (7.19), and (7.14), respectively, leads to the relation[40]

$$\frac{\rho_s}{\rho_q} = \frac{(1 + 3m^2/2(1-m))(1+m^2)}{(1 - 3m^2/2(1-m))(1-2m^2)}, \tag{7.32}$$

and the curvature relation (7.9) implies that[41]

$$\frac{\rho_s}{\rho_q} = \frac{1 + [4(1-m)^2 - 1]x}{1 - [4(1-m)^2 - 1]x}. \tag{7.33}$$

Equating these two relations and solving for x gives again (7.26).

We have seen that by applying his curvature method[24,25] Newton found a special solution to lunar motion for initial conditions corresponding to a circular orbit in the absence of solar perturbation. This condition was considered later by Euler[27] and by G. W. Hill,[28] who obtained a series expansion solution for Newton's special orbit of the moon. However, obtaining other solutions by this method, that is, for general initial conditions leading to elliptic orbits in the absence of solar perturbation forces, presents a more difficult problem. In this case Newton remarked that when the solar perturbation is acting, "the eccentric orbit in which the Moon revolves is not an ellipse, but an oval of another kind"[42] which he did not describe mathematically. Newton must have realized that a simple geometrical construction could not characterize such an oval. By applying analytical methods, sixty-three years later Clairaut obtained the same curvature equation (7.4) in the form of a differential equation (7.10) and discovered that this oval was a linear combination of two distinct ellipses (as is discussed in the appendix). Apparently Newton did not follow his curvature approach further, but instead developed a different perturbation method that is discussed in the next section.

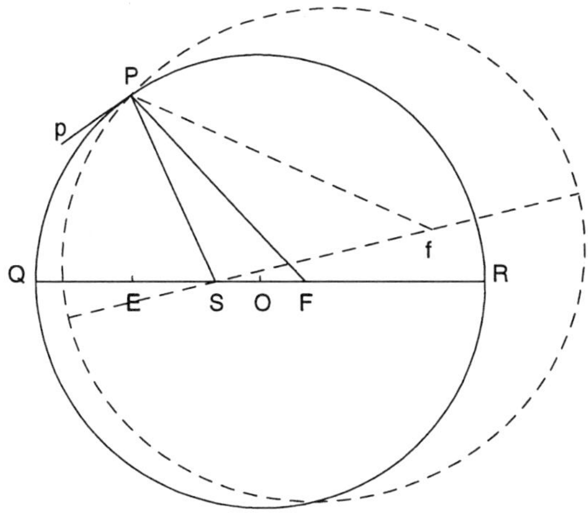

Figure 7.1
Graphical illustration of Newton's Portsmouth method for the case of a single perturbation impulse. The solid lines represent the initial elliptic orbit with foci at *S* and *F* and the broken lines the perturbed elliptical orbit after an impulse occurs when the revolving body is at *P*.

PORTSMOUTH METHOD

Certain manuscripts of the Portsmouth collection,[3,4] first examined by J. C. Adams, G. D. Liveing, H. R. Laurd and G. G. Stokes in 1872, reveal that by 1687 Newton had developed a general method for evaluating the perturbations due to the sun's gravitational force on the moon's motion around the earth. This method is described in the *Principia* only qualitativily[43] in Corollaries 3 and 4 of Proposition 17, Book I. As the starting point for the application of this geometrical and differential method, Newton assumed that the perturbing force consisted of a sequence of instantaneous impulses at short time intervals during which a segment of the moon's orbit can be characterized by the arc of an ellipse with the earth at one of the foci (see figures 7.1 and 7.2). Referring to Proposition seventeen in Book I of the *Principia*, which determines this ellipse's parameters (semi–latus rectum, eccentricity, and spatial orientation) from the moon's velocity and position at a given time,[44] these parameters could be evaluated just before and after the action of the perturbing impulse. According to Newton's second law of motion, the effect of this impulse is to change instantaneously the magnitude and direction of the moon's velocity without changing its position. Consequently, repeated applica-

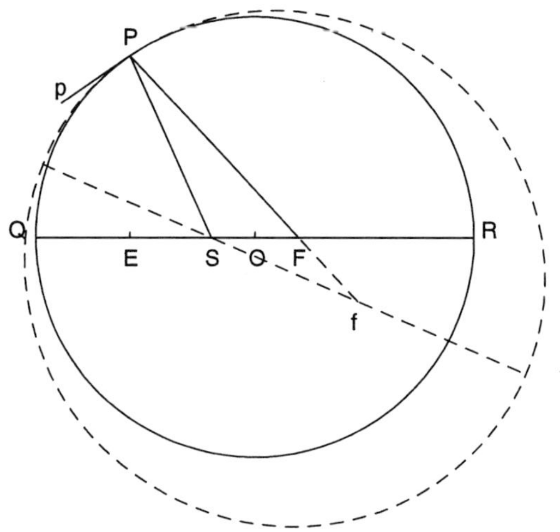

Figure 7.2
Graphical illustration of Newton's Portsmouth method for the case of a tangential perturbation impulse. Solid lines and broken lines represent the same entities as in figure 7.1. The impulse's effect is to change only the angular momentum, which changes the distance *PF* to the second focus, according to Proposition 17.

tions of Proposition 17 determine the moon's perturbed orbit as a sequence of elliptical arc segments joined together. In the limit of a vanishingly small time interval between impulses, or as Newton would put it, "conceive now that the impulses are increased in number, and their intervals diminished indefinitely so as to render the action of the forces . . . continuous," this method leads to a differential equation for the evolution of the parameters of an ellipse representing the orbit locally. In fact Newton's method corresponds[5] to what is now known as the method of variation of orbital parameters generally attributed to Euler, Lagrange, and Laplace, who rediscovered it many years later.[6–9]

In the Portsmouth manuscript Newton obtained a differential equation, which he called the *hourly motion*, for the rotation of the axis of this ellipse to lowest order in the eccentricity *e*. In a similar manner he could have also obtained the corresponding change in the eccentricity *e* and the ellipse's major axis, but there is no direct evidence in the Portsmouth manuscripts that he carried out these calculations. As we shall see, the change in eccentricity was crucial to obtain the "oval of another kind"[42] that Newton had been seeking. Indeed, it appears that some of Newton's subsequent confusing approach to solve his equation for the rotation of the ellipse's major axis can be attributed to the fact that apparently he did not carry out this analysis.

Taking the longitude θ of the moon as the independent variable, in lemma [α] Newton obtained the contribution of the radial component V of the perturbation force on the rotation of the ellipse's axis, representing the orbit (see figure 7.1) in the form

$$\frac{d\omega}{d\theta} = 2\frac{V}{P} \times \frac{SE}{SF}. \tag{7.34}$$

Here ω is the angle of the ellipse's major axis relative to some fixed direction in space, $P = \mu/r^2$ is the inverse-square force (acceleration) between the earth and the moon, SE is the component of the radius vector along the ellipse's instantaneous axis, and $SF = 2ae$ is the distance between the two foci of the ellipse with major axis a and eccentricity e. Setting $SE = -r\cos(\phi)$, where $\phi = \theta - \omega$ is the angle of the radius vector relative to the axis of the ellipse, and

$$r = \frac{(h^2/\mu)}{(1 + e\cos(\phi))} \tag{7.35}$$

is the equation for the ellipse, Newton's result (7.34) can be written in the form

$$\frac{d\omega}{d\theta} = -\frac{r^3}{\mu ae} V\cos(\phi). \tag{7.36}$$

In his derivation Newton neglected all but the lowest-order term in the eccentricity e, and therefore one should also set $r = a$ in this equation. However, it is straightforward to solve Newton's geometrical construction for the effect of the perturbation force V without approximations, which leads to the exact result[6,7,45]

$$\frac{d\omega}{d\theta} = -\frac{r^2}{\mu e} V\cos(\phi). \tag{7.37}$$

In lemma [β] Newton obtains the analogous contribution due to a transverse perturbing force W (see fig. 7.2), again only to lowest order in e,

$$\frac{d\omega}{d\theta} = 2\frac{W}{P} \times \frac{PE}{OS}, \tag{7.38}$$

where $PE = r\sin(\phi)$ and $OS = ae$. This relation corresponds to

$$\frac{d\omega}{d\theta} = 2\frac{r^3}{\mu ae} W\sin(\phi), \tag{7.39}$$

and the exact result obtained from Newton's construction is[6,7,45]

$$\frac{d\omega}{d\theta} = \frac{r^2}{\mu e}\frac{W\sin(\phi)}{(1 + e\cos(\phi))}\sin(\phi)[2 + e\cos(\phi)]. \tag{7.40}$$

To solve these equations it is necessary to obtain also the equation for the change in the eccentricity e. From Newton's construction in lemmas [α] and [β] one readily finds that to lowest order in e,[46]

$$\frac{de}{d\theta} = \frac{r^2}{\mu}[V\sin(\phi) + 2W\cos(\phi)]. \tag{7.41}$$

Although the Portsmouth manuscripts offer no evidence that Newton also obtained this equation, in Book I, Proposition 66, Corollary 9, he describes in words the variation in the eccentricity of the moon's orbit due to the sun's perturbation. In addition the change in angular momentum h is given by (7.5), which Newton also solved in Book III, Proposition 28.

These differential equations can be solved approximately by substituting directly for V and W the radial and tangential components of the solar perturbing force, (7.14) and (7.15). However, this method gives rise to a puzzle whose resolution apparently evaded Newton and may have been ultimately one of the causes of his well-known remark that the lunar problem was the only problem that had made his head ache: how can Newton's elliptic orbit solution for the moon with the earth at the center (7.19) emerge from a solution represented by a rotating ellipse with the earth at a focus? Newton remarked that for this special orbit "the perturbing force does not cause a rotation of the apsis,"[47] but this statement seems incompatible with his variational equation, (7.34) and (7.38). As I demonstrate below, to resolve this paradox it is necessary to obtain also the equation for the change in the eccentricity e, an equation not found in the manuscripts of the Portsmouth Collection.

Radial Perturbation

To illustrate the application of these equations for the orbital parameters, we consider first the simple case of a purely radial perturbation that is linear in r. This is the case that Newton treated in Proposition 45, Corollary 2, Book I of the *Principia*. It is of particular interest to compare the evaluation of the moon's orbit by Newton's Portsmouth or variation of parameters method with his earlier perturbation method developed in Proposition 45, which is also based on approximating the orbit by a rotating ellipse, but which could be applied only to the restricted case of a purely radial perturbing forces independent of angular coordinates, and to first order in the eccentricity e of the orbit. In this approximation Newton's new method is expected to give the same answer, but at first sight this seems rather puzzling, because (7.36) and (7.41) imply that both the angle ω of the ellipse's axis and its eccentricity e have periodic variations in the angle ϕ. However, Newton's approximate solution given in Proposition 45 corresponds to a revolving ellipse with a *constant* eccentricity without any periodic variations. As we shall see, the resolution of this puzzle lies in the recognition that the rotating ellipse introduced in the Portsmouth method is different from the rotating ellipse in Proposition 45, although both ellipses approximate the same orbit. It is not clear that Newton was aware of this important distinction,[48] and this may have been the source of some of the difficulties that he encountered with his variation of parameters method.

For a purely radial perturbing force $V = (g/2)r$, the angular momentum h is a constant, and (7.37) and (7.41) become

$$\frac{d\omega}{d\theta} = -\frac{g}{2\mu e} r^3 \cos(\phi) \tag{7.42}$$

and

$$\frac{de}{d\theta} = \frac{g}{2\mu} r^3 \sin(\phi). \tag{7.43}$$

To first order in e, $r^3 = (h^2/\mu)^3[1 - 3e\cos(\phi)]$, and substituting this approximation in (7.42) and (7.43) we find

$$\frac{d\omega}{d\theta} = -\frac{1}{2} m^2 \left[\frac{1}{e} \cos(\phi) - \frac{3}{2}(1 + \cos(2\phi)\right] \tag{7.44}$$

and

$$\frac{de}{d\theta} = \frac{1}{2} m^2 \left[\sin(\phi) - \frac{3}{2} e \sin(2\phi)\right], \tag{7.45}$$

where $m^2 = gh^6/\mu^4$.

The first term in (7.44) is proportional to $1/e$, and therefore this equation becomes singular in the limit of vanishing eccentricity e. This is not surprising, because in this limit of circular symmetry the angle ω is undefined. For the present we shall assume that e remains positive and finite and can be taken to be a constant $e = e_0$ on the right side of (7.44) and (7.45). Setting

$$\omega = \frac{\Delta\omega_0}{e_0} + \omega_1, \tag{7.46}$$

we then have

$$\frac{d\Delta\omega_0}{d\theta} = -\frac{m^2}{2}\cos(\phi) \tag{7.47}$$

and

$$\frac{d\omega_1}{d\theta} = \frac{3m^2}{4}(1 + \cos(2\phi)). \tag{7.48}$$

Evidently (7.48) leads to a secular as well as to a periodic change in ω_1, and setting for ω_1,

$$\omega_1 = (3m^2/4)\theta + \Delta\omega_1, \tag{7.49}$$

we obtain

$$\frac{d\Delta\omega 1}{d\theta} = \frac{3m^2}{4}\cos(2\phi). \tag{7.50}$$

To integrate (7.47) and (7.50) to first order in e_0, we keep on the right side of these equations only the secular term in ω which appears in ϕ, setting

$$\phi = \theta - \omega \approx v\theta, \tag{7.51}$$

where $v = 1 - 3m^2/4$. Neglecting $\Delta\omega_0$ and $\Delta\omega_1$ is justified here, because we find that these terms are of order m^2 and vary periodically in θ,

$$\Delta\omega_0 = -\frac{m^2}{2}\sin(v\theta) \tag{7.52}$$

and

$$\Delta\omega_1 = \frac{3m^2}{8}\sin(2v\theta). \tag{7.53}$$

Likewise, setting

$$e = e_0 + \Delta e_0 + \Delta e_1, \tag{7.54}$$

where e_0 is a constant, and substituting (7.51) in (7.45) we obtain

$$\Delta e_0 = -\frac{m^2}{2}\cos(v\theta) \tag{7.55}$$

and

$$\Delta e_1 = \frac{3}{8} m^2 e_0 \cos(2\nu\theta). \tag{7.56}$$

We consider now separately the contributions indicated by the subscripts 0 and 1. Expanding $e\cos(\phi)$ to first order in m^2, we have

$$e_0 \cos(\nu\theta) + \Delta e_0 \cos(\nu\theta) + \Delta\omega_0 \sin(\nu\theta) = e_0 \cos(\nu\theta) - \frac{m^2}{2}, \tag{7.57}$$

and substituting this expression in (7.35) we obtain

$$r \approx \frac{h^2}{\mu}\left[1 + \frac{1}{2}m^2 - e_0 \cos(\nu\theta)\right], \tag{7.58}$$

up to terms of order $m^2 e_0$. Hence the singular term in the eccentricity in the expansion of ω has the effect of changing the magnitude of the unperturbed ellipse's major axis and eccentricity. In particular, in the limit of vanishing eccentricity e_0, this term gives the correct perturbation to order m^2 of the radius of the circular orbit. The oscillating terms $\Delta\omega_1$ and Δe_1, (7.53) and (7.56), would appear to contribute periodic variations to the ellipse's eccentricity and the angle, but one finds instead that

$$\Delta e_1 \cos(\nu\theta) + \Delta\omega_1 \sin(\nu\theta) = \frac{3}{8} m^2 e_0 \cos(\nu\theta), \tag{7.59}$$

which adds only a constant contribution to the ellipse's eccentricity. Collecting our results, we find that the radial perturbation linear in r on an elliptic orbit with small eccentricity e_0 and semi–latus rectum h^2/μ has the effect of giving rise to an orbit corresponding to a rotating ellipse with eccentricity e' and semi–latus rectum r_0',

$$r = \frac{r_0'}{1 + e' \cos(\nu\theta))}, \tag{7.60}$$

where

$$e' = \left(1 - \frac{m^2}{8}\right) e_0 \tag{7.61}$$

and

$$r_0' = \frac{h^2}{\mu}(1 + \frac{1}{2}m^2). \tag{7.62}$$

Now we can see the resolution of the paradox formulated above. For a fixed value of θ, the parameters of the ellipse in the Portsmouth method differ from those of the rotating ellipse in Proposition 45. Nevertheless, both ellipses can represent the same orbit because in the Portsmouth method these parameters are not constants but instead are functions of θ. In particular this example illustrates the important fact that the angle ω is not the angle of the orbit's apogee. In Book I, Proposition 45, this angle is $(\nu - 1)\theta = (3m^2/4)\theta$, which corresponds only to the secular variation of ω (7.49), as determined by the variation of parameters method.

In the case $e_0 = 0$ we have $e' = 0$, and (7.60) represents a circular orbit with an approximate value for the perturbed radius (7.62) for a given angular momentum h. However, the expansion for ω (7.45) is only valid provided that $m^2/e_0 \ll 1$. In fact, for a circular solution, $e\cos(\phi)$ is a constant, and (7.42) and (7.43) then imply that $e\sin(\phi) = 0$. Hence in this limit the correct solution is obtained by setting $\phi = 0$ or $\omega = \theta$, and $e = -(g/2\mu)r_0^3$ in (7.35). This yields

$$r_0 = \frac{h^2/\mu}{(1 - m^2/2)}, \tag{7.63}$$

which corresponds to (7.62) in the approximation that $m^2 \approx gr_0^3/\mu \ll 1$. This demonstrates how the circular solution arises from a rotating ellipse in the Portsmouth method.

HOURLY MOTION OF THE APOGEE

We now consider the solution of the variational differential equations for the angle of the major axis ω and the eccentricity e of the ellipse when the solar perturbation with radial component V and transverse component W is given by (7.14) and (7.15). We start with a modern treatment to help shed some light on Newton's own calculations, which will be discussed afterwards.

For the variation of the angle w we must now also keep terms independent of the eccentricity e,

$$\frac{d\omega}{d\theta} = \frac{r^2}{\mu e}[-V\cos(\phi) + W\sin(\phi)(2 - e\cos(\phi)], \tag{7.64}$$

where the last term, $e\cos(\phi)$, in this equation was neglected by Newton. For the variation of the eccentricity e, which Newton did not consider, we keep terms to first order in e:

$$\frac{de}{d\theta} = \frac{r^2}{\mu}[V\sin(\phi) + W(2\cos(\phi) + e\sin^2(\phi))]. \tag{7.65}$$

Substituting in (7.64) and (7.65) the solar perturbation force with V and W given by (7.14) and (7.15), respectively, we obtain

$$\frac{d\omega}{d\theta} = -\frac{gr^3}{2\mu e}[(1 + 3\cos(2\psi))\cos(\phi) + 3\sin(2\psi)\sin(\phi)(2 - e\cos(\phi))] \tag{7.66}$$

and

$$\frac{de}{d\theta} = \frac{gr^3}{2\mu}[(1 + 3\cos(2\psi))\sin(\phi) - 3\sin(2\psi)(2\cos(\phi) + e\sin^2(\phi))]. \tag{7.67}$$

Approximating $e = e_0 + \Delta e + \delta e_1$, where δe_1 and $r^3 = r_0^3(1 - 3e_0\cos(\phi))$ are evaluated to first order in e^0, the unperturbed value of the eccentricity, we obtain

$$\frac{d\Delta e}{d\theta} = \frac{m^2}{4}[2\sin(\phi) - 3\sin(2\psi + \phi) - 9\sin(2\psi - \phi)] \tag{7.68}$$

and

$$\frac{d\delta e_1}{d\theta} = \frac{3m^2 e_0}{4}[5\sin(2\psi - 2\phi) + 4\sin(2\psi + 2\phi) - \sin(2\phi) + 5\sin(2\psi)]. \tag{7.69}$$

Likewise, setting $\omega = \Delta\omega_0/e_0 + \omega_1$ as before, we obtain

$$\frac{d\Delta\omega_0}{d\theta} = \frac{m^2}{4}(-2\cos(\phi) + 3\cos(2\psi + \phi) - 9\cos(2\psi - \phi)) \tag{7.70}$$

and

$$\frac{d\omega_1}{d\theta} = \frac{3m^2}{4}[1 + 5\cos(2\psi - 2\phi) + 3\cos(2\psi) + \cos(2\phi) - 2\cos(2\psi + 2\phi)]. \tag{7.71}$$

Like the corresponding equation for the case of a purely radial perturbation (7.48), this equation for ω_1 has a constant term on the right side equal to $3m^2/4$ implying a secular contribution to ω_1. However, as Newton demonstrated in the Portsmouth manuscript, the appearance of a cosine term on the right side of (7.71) with argument $2\psi - 2\phi$ leads to an important correction to the magnitude of this secular contribution. Neglecting all but the first term on the right side of (7.71) we have

$$\frac{d\omega_1}{d\theta} = \alpha[1 + \beta\cos(2\psi - 2\phi)], \tag{7.72}$$

where $\alpha = 3m^2/4$ and $\beta = 5$. Setting $\omega_1 = (1 - \nu)\theta + \delta\omega_1$, where $\delta\omega_1$ is assumed to be a small contribution, and expanding to first order in this quantity we obtain

$$\delta\omega_1 = -\frac{a\beta}{2(v-1+m)}\sin 2(\psi - v\theta), \tag{7.73}$$

with

$$v \approx 1 - a\left[1 + \frac{a\beta^2}{2m}\right]. \tag{7.74}$$

For the given values of the constants a and β one finds that

$$\delta\omega_1 = \frac{15m}{8}\sin 2(v-1+m)\theta \tag{7.75}$$

and

$$v \approx 1 - \frac{3m^2}{4}\left[1 + \frac{75m}{8}\right]. \tag{7.76}$$

This expression for v corresponds to the result for the secular variation in the angle of the apsis first given numerically by Clairaut and analytically by d'Alembert, as shown in the appendix.

In the approximation that $\psi = (1-m)\theta$ and $\phi = \theta - \omega = v\theta - \Delta\omega_0/e_0$, we can integrate (7.68) and (7.70) and obtain

$$\Delta e = \frac{m^2}{4}\left[-2\cos(\phi) + \frac{3}{(3-2m)}\cos(2\psi + \phi) + \frac{9}{(1-2m)}\cos(2\psi - \phi)\right], \tag{7.77}$$

where $\Delta e = e - e_0$, and

$$\Delta\omega_0 = \frac{m^2}{4e_0}\left[-2\sin(\phi) + \frac{3}{(3-2m)}\sin(2\psi + \phi) - \frac{9}{(1-2m)}\sin(2\psi - \phi)\right]. \tag{7.78}$$

Hence

$$e\cos(\phi) \approx (e_0 + \Delta e)\cos(v\theta) + \Delta\omega_0\sin(v\theta), \tag{7.79}$$

and substituting (7.77) and (7.78) we find

$$e\cos(\phi) \approx e_0\cos(v\theta) + \frac{m^2}{2}\left[-1 + \left(5 + \frac{28}{3}m\right)\cos(2\psi)\right]. \tag{7.80}$$

To first order in e we can approximate

$$r \approx \frac{h^2}{\mu}(1 - e\cos(\phi)), \tag{7.81}$$

where h is obtained by integrating (7.11) (neglecting the variation in r due to the eccentricity),

$$h = h_0\left[1 + \frac{3m^2}{4(1-m)}\cos(2\psi)\right]. \tag{7.82}$$

Thus one obtains in an expansion up to cubic powers in m,

$$r = r_0[1 - e_0\cos(\nu\theta) - x\cos(2\psi)], \tag{7.83}$$

where $r_0 = (h_0^2/\mu)(1 + m^2/2)$, and $x = m^2(1 + (19/6)m)$.

An additional term to r that turns out to be of the same magnitude as the last term in this equation is obtained if one includes also the variations $\delta\omega_1$ (7.75), and δe_1, which are zero and first order in e_0, respectively. Keeping only the first term in (7.69) we have

$$\delta e_1 = \frac{15me_0}{8}\cos 2(\psi - \nu\theta) \tag{7.84}$$

and

$$\delta e_1\cos(\nu\theta) + e_0\delta\omega_1\sin(\nu\theta) = \gamma\cos(2\psi - \nu\theta), \tag{7.85}$$

where $\gamma = 15me_0/8$. Hence

$$r = r_0[1 - e_0\cos(\nu\theta) - x\cos(2\psi) - \gamma\cos(2\psi - \nu\theta)]. \tag{7.86}$$

This is the "oval of another kind"[42] for the lunar orbit that Newton had been seeking, including effects of the solar perturbation to first order in the eccentricity parameter e_0 and to third order in the relative strength of the perturbation m.[49] In particular, for the limit of zero initial eccentricity, $e_0 = 0$, one recovers the special solution Newton first obtained by applying his curvature method in Book III, Proposition 28 [see (7.19) and (7.27)] corresponding to Euler's[27] and Hill's[28] solution for small m. In Proposition [A] of the Portsmouth manuscripts, Newton claimed that in this case, "these [solar] forces do not, when the moon is located in the orbit in which by their means it might revolve without eccentricity, contribute at all to the motion of the apogee."[50] However, he failed to distinguish the apogee of the lunar orbit from that of the ellipse associated with his geometrical construction. As I showed in the last section, the later apogee always rotates in the presence of perturbations. Consequently, not understanding this crucial point would have prevented Newton from deriving his special lunar orbit as a solution of his Portsmouth equation.

The oscillations of the apsis and the eccentricity, (7.75) and (7.84), are discussed only qualitatively in the *Principia*, in Proposition 66, Corollarics 8 and 9, of Book I. There Newton gave a physical argument, based on the solar perturbation force, for these oscillations' dependence on the angle $\omega - \theta'$, where θ' is the solar longitude.[51] These oscillations were originally proposed in 1641 by Jeremiah Horrocks in a kinematical model of the lunar orbit.[52] In the revised Scholium to Proposition 35, Book III, in the second and third edition of the *Principia*, Newton described this model as following from his gravitational theory, but evidently he fitted the amplitude of the oscillations from astronomical observation rather than from theory.[53] Without the equation for the variation of the eccentricity calculated to first order in the unperturbed eccentricity e_0, Newton could not have accounted for the quantitative success of the Horrocksian model from his gravitational theory. Newton argued further that "the motion of the apogee arises from the differences between these forces and forces which in the moon's recession from that concentric orbit decrease, if they are centripetal or centrifugal, in the doubled ratio of the (increasing) distance between the moon and the earth's center, but, should they act laterally, in the same ratio tripled—as I found once I undertook the calculations."

The precise meaning of this crucial sentence is unclear. We have shown that the secular motion of the apogee is obtained from terms that are independent of the eccentricity e in (7.64). Thus for the radial perturbation $V(r, \psi)$ this corresponds to replacing $r^2 V(r, \psi)$ in (7.64) by the difference $r^2 V(r, \psi) - r_0^2 V(r_0, \psi)$ to first order in e, where $r \approx r_0(1 - e\cos(\theta))$. Indeed, this is mathematically equivalent to Newton's own calculation. However, for the transverse perturbation $W(r, \psi)$ he applied a cubic scaling factor without justification, in effect replacing $r^2 W(r, \psi)$ by the difference $r^3 W(r, \psi) - r_0^3 W(r_0, \psi)$, which is incorrect.[54] Thus although Newton was able to obtain the correct analytic form for the differential equation for ω_1, which we will call here ω_N, some of the coefficients were wrong. Written in modern notation, Newton's differential equation for ω_N takes the form[55]

$$\frac{d\omega_N}{d\theta} = \frac{3m^2}{4}[1 + (11/2)\cos(2\psi - 2\phi) + 3\cos(2\psi) + \cos(2\phi) - (5/2)\cos(2\psi + 2\phi)]. \qquad (7.87)$$

Comparison with the correct equation (7.71) indicates that only the coefficients multiplying the terms $\cos(2\psi - 2\phi)$ and $\cos(2\psi + 2\phi)$ are incorrect. Moreover, Newton identified correctly the first cosine term on the right side of (7.87) as the term that modifies the lowest-order secular contribution to ω.[56] The reduced equation for ω_N corresponds to (7.72), which can be rewritten in the form

$$\frac{d\chi}{d\theta} = b[\cos(\chi) - n)], \tag{7.88}$$

where $\chi = 2(\psi - \phi) \approx 2(\omega_N - m\theta)$, $b = (3/2)m^2\beta$, and $n = (1/\beta)(4/3m - 1)$ (Newton's numerical values are $\beta = 11/2$ and $n = 3.0591$). Hence

$$\frac{d\omega_N}{d\chi} = \frac{\frac{1}{\beta} + \cos(\chi)}{2[\cos\chi - n]}. \tag{7.89}$$

Newton integrated (7.88) and (7.89) to obtain the changes $\Delta\omega$ and $\Delta\theta$ for half a cycle $\Delta\chi = \pi$ and found

$$\Delta\theta = \frac{1}{b}\int_0^\pi \frac{d\chi}{n - \cos(\chi)} = \frac{2\pi}{b\sqrt{(n^2 - 1)}} \tag{7.90}$$

and

$$\Delta\omega_N = \frac{1}{2}\int_0^\pi d\chi \frac{\left(\frac{1}{\beta} + \cos(\chi)\right)}{n - \cos(\chi)}$$
$$= \frac{\pi}{\sqrt{(n^2 - 1)}}\left(\frac{1}{\beta} + n - \sqrt{n^2 - 1}\right). \tag{7.91}$$

Hence

$$\frac{\Delta\omega}{\Delta\theta} = \frac{3}{4}m^2[1 + \beta(n - \sqrt{(n^2 - 1)}]. \tag{7.92}$$

The term $(3/4)m^2$ corresponds to Newton's perturbation result in Proposition 45, Corollary 2, and the factor in brackets is equal to 1.924 for Newton's numerical values for n and β. This accounts to a very good approximation for the missing factor of two in his earlier calculation (quoted in the Introduction). Annually $\Delta\omega_N = 38.86°$, in good agreement with Newton's claim in the first edition of the *Principia* that "its mean annual motion should be virtually 40°."

Expanding the square root in powers of $1/n^2$ and substituting the expression for n, one obtains to cubic order in m^3

$$\frac{\Delta\omega}{\Delta\theta} = \frac{3}{4}m^2\left(1 + \frac{3m\beta^2}{8}\right), \tag{7.93}$$

which corresponds to the previous result (7.74) obtained by a perturbation expansion. In particular, setting $\beta = 5$ yields

$$\frac{\Delta\omega}{\Delta\theta} = \frac{3}{4}m^2\left(1 + \frac{75m}{8}\right), \quad (7.94)$$

which is the result (7.76) obtained by Clairaut and d'Alembert, whose method of solution is described in the appendix.

Conclusions

In Book III, Proposition 28, of the *Principia*, Newton applied his earliest known method to compute orbital dynamics, the curvature method,[24,25] to evaluate the effect of the sun's gravitational perturbation on a possible orbit of the moon around the earth, which is circular in the absence of this perturbation. His solution corresponds to the periodic solution found later by L. Euler,[27] and in more complete detail by G. W. Hill.[28] However, there is no evidence that Newton applied this curvature method to evaluate the actual case of an elliptical orbit of the moon perturbed by the solar gravitation force. Instead, the Portsmouth mathematical papers show that by 1687 Newton had developed an alternative approach that corresponds to the method of variation of orbital parameters developed many years later by Euler, Lagrange, and Laplace.[6–9] Newton applied this method to evaluate the motion of the apogee of the lunar orbit to lowest order in the eccentricity. However, he apparently failed to realize that his equation for the motion of the apogee of the ellipse constructed by the Portsmouth method applied to the motion of the apogee of the physical orbit only in the case where the unperturbed eccentricity e of this orbit is large compared to a dimensionless parameter m^2. Here m is the reciprocal of the number of lunar cycles during a sidereal year. Thus Newton was unable to understand how the special perturbed solution of the circular orbit, which he obtained by the curvature method, emerged from his Portsmouth method. Nevertheless, he was able to deduce, by admittedly some unclear physical arguments, the correct analytic form for the equation describing the motion of the apogee and to solve this equation for the secular variation (valid to cubic order in a perturbation theory in powers of m). Indeed, Newton's analytic solution for the motion of the apogee corresponds to Clairaut's and d'Alembert's solution obtained some sixty-three years later, after they rediscovered Newton's curvature method in differential form.

The mystery remains why in his later researches on the lunar inequalities Newton apparently abandoned the further development of his remarkable Portsmouth perturbation method, which did not appear in any of the three editions of the *Principia.* The evidence presented here indicates that he could have succeeded in his quest to evaluate correctly higher-order terms to the motion of the lunar apogee had he continued to pursue his method more carefully, particular in regard to the dependence on the orbit's eccentricity.

APPENDIX: THE THEORY OF CLAIRAUT AND D'ALEMBERT

The expression for the perturbed orbit of the moon (7.83) was obtained by Clairaut and d'Alembert[33,34] as an approximate solution for small e and m^2 of the curvature equation (7.10):

$$\left(\frac{d^2}{d\theta^2}+1\right)\frac{1}{r}=\frac{\mu}{h^2}-\frac{r^2}{h^2}\left[V-\frac{W}{r}\frac{dr}{d\theta}\right], \tag{7.95}$$

$$h^2=h_0^2+2\int d\theta r^3 W, \tag{7.96}$$

where V and W are the solar perturbations obtained by Newton, (7.14) and (7.15), respectively. Substituting for r as a function of θ the approximation of a rotating ellipse, $r = r_0(1 - e\cos(\nu\theta))$, to first order in e on the right side of (7.95) and (7.96),

$$r^2V=\frac{m^2\mu}{2}\left[1-3e\cos(\nu\theta)+3\cos(2\psi)\right.$$
$$\left.-\frac{9e}{2}[\cos(2\psi+\nu\theta)+\cos(2\psi-\nu\theta)]\right], \tag{7.97}$$

$$Wr\frac{dr}{d\theta}=\frac{3m^2\mu e}{4}[\cos(2\psi+\nu\theta)-\cos(2\psi-\nu\theta)], \tag{7.98}$$

and

$$h^2=h_0^2\left[1+\frac{3m^2}{2}\left[\cos(2\psi)-\frac{4}{3}e\cos(2\psi+\nu\theta)-4e\cos(2\psi-\nu\theta)\right]\right]. \tag{7.99}$$

This implies that, to order em^2, there are also terms proportional to $\cos(2\psi+\nu\theta)$ and to $\cos(2\psi-\nu\theta)$ on the left side of the curvature equation. Thus, to obtain corresponding terms on the left side of (7.95), an improved approximation for r must have the form

$$\frac{1}{r} = \frac{1}{r_0}[1 + e\cos(\nu\theta) + x\cos(2\psi) + \delta\cos(2\psi + \nu\theta) + \gamma\cos(2\psi - \nu\theta)], \tag{7.100}$$

where δ and γ are new coefficients determined by matching the corresponding terms proportional to $\cos(2\psi + \nu\theta)$ and $\cos(2\psi - \nu\theta)$ which appear in (7.97)–(7.99). This implies that

$$\gamma = \frac{15me}{8} \tag{7.101}$$

and

$$\delta = -\frac{5m^2e}{8}. \tag{7.102}$$

The unexpected result is that the coefficient γ is of order me instead of order m^2e as in the case of δ. Therefore the contribution to r of the corresponding cosine term $\cos(2\psi - \nu\theta)$ should not be neglected in evaluating the right side of (7.95). Thus one finds that the contribution of this term gave rise to additional terms proportional to $\cos(\nu\theta)$ on the right side of (7.95), which therefore modified his previous evaluation of ν. In particular r^2V gives the added term

$$-\frac{9m^2\mu\gamma}{4}\cos(\nu\theta), \tag{7.103}$$

$Wr\,dr/d\theta$ contributes

$$-\frac{3m^2\mu\gamma}{4}\cos(\nu\theta), \tag{7.104}$$

and h^2 contributes

$$-\frac{6m^2\gamma}{\nu}\cos(\nu\theta). \tag{7.105}$$

Collecting all the terms that appear on the right side of (7.95) that are proportional to $\cos(\nu\theta)$ leads to the new relation

$$(\nu^2 - 1) = -\frac{3}{2}m^2\left(1 + \frac{5\gamma}{e}\right), \tag{7.106}$$

and substituting (7.101) for γ one obtains

$$\nu = 1 - \frac{3}{4}m^2\left(1 + \frac{75m}{8}\right), \tag{7.107}$$

which corresponds to (7.76).

Acknowledgments

I am greatly indebted to Curtis Wilson for very many valuable comments and criticisms on an earlier draft of this article. I would also like to thank Bruce Brackenridge and Alan Shapiro for useful suggestions and corrections and my colleagues in the Physics Departments at the Universities of Amsterdam and Leiden for their warm hospitality during the completion of this work. This research was partially supported by a grant from the National Science Foundation.

Notes

1. Isaac Newton, *Mathematical Principles of Natural Philosophy*, 3d ed., trans. I. B. Cohen and Anne Whitman (Berkeley and Los Angeles: University of California Press, 2000).

2. S. P. Laplace, "Sur la théorie lunaire de Newton," in *Mécanique Céleste* (Paris: Imprimiere Royale, 1846), 5:438.

3. *A Catalogue of the Porstmouth Collection of the Books and Papers Written by or Belonging to Sir Isaac Newton*. The scientific portion of which has been presented by the Earl of Portsmouth to the University of Cambridge. Drawn up by the syndicate appointed the 6th November 1872 (Cambridge: Cambridge University Press, 1888). The members of the syndicate were J. C. Adams, G. D. Liveing, H. R. Luard, and G. G. Stoke.

4. D. T. Whiteside, ed., *The Mathematical Papers of Isaac Newton*, (8 vols.) (Cambridge: Cambridge University Press, 1967–1981) 6:508–35.

5. The equivalence of Newton's Portsmouth method to the variation of orbital parameters method of Euler, Lagrange, and Laplace has not been generally appreciated. For example, in "Newton's Lunar Theory: From High Hope to Disenchantment," *Vistas in Astronomy* 19 (1976): 321, D. T. Whiteside writes that "Newton, it is clear from his private papers (the Portsmouth manuscripts) passed in some despair to introduce the additional Horrocksian assumption—one only approximately justifiable from the 3-body dynamical problem whose more accurate solution he thereby ceased to control, then and ever after.... The truth as I have tried to sketch it here is that his loosely approximate and but shadowly justified way of deriving those inequalities which he did deduce was a retrogressive step back to an earlier kinematic tradition which he had once hoped to transcend."

On the contrary, Newton's geometrical construction is not an approximate but an exact implementation of the effects of a general perturbation force on Keplerian motion. Its analytic evaluation leads to the first-order differential equations for the variation of orbital parameters formulated later by Euler, Lagrange, and Laplace. In *Newton's Principia for the Common Reader* (Oxford: Oxford University Press, 1995): 252, S. Chandrasekhar claims that Newton must have known these equations, and in *The Motion of the Moon* (Bristol and Philadelphia: A. Hilger, 1988): 57, A. Cook writes that "there are clear indications in Newton's unpublished treatment of the effects of the Sun upon the Moon that he was aware of relations equivalent to Lagrange's planetary relations." However, as I show in the next section, in the Portsmouth manuscripts one finds the derivation of the differential equation for only one of these parameters, the angle of apsis ω, which is evaluated correctly only to lowest order in the eccentricity e. Moreover the solution of this equation for the case of the solar perturbation on the lunar orbit is not quite correct.

6. L. Euler, *Leonhardi Euleri Opera Omnia*, Series secunda, *Opera Mechanica et Astronomica*, ed. L. Courvoisier and J. O. Fleckenstein (Basil: Orell Füssli Twrici, 1969), 23:286–89. Starting with the equations of motion, written as second-order differential equations in polar coordinates, Euler assumes that the solution for the orbit is described by an ellipse with time-varying orbital parameters p, e, and ω, where p is the semi–latus rectum of the ellipse, e is the eccentricity, and ω is the angle of the major axis. Then he obtains first-order differential equations for e and ω by imposing two constraints: that $p = h^2/\mu$, where h is the angular momentum, and that $E = \mu(e^2 - 1)/2p$, where E is the time-varying Kepler energy of the orbit. In modern notation, $\mu = GM$, where M is the sum of the mass of the earth and the moon and G is Newton's gravitational constant. It can be readily shown that Euler's constraints lead to the same definition of the ellipse Newton described geometrically in the Portsmouth manuscript. I thank C. Wilson for calling my attention to Euler's work.

7. P. S. Laplace, *A Treatise of Celestial Mechanics*, trans. Henry H. Harte (Dublin, 1822): 357–90. Laplace obtained the differential equations for the time dependence of the orbital parameters by evaluating the time derivate of the vector $\mathbf{f} = \mathbf{v} \times \mathbf{h} - \mu\mathbf{r}/r$, where $\mathbf{f}$ is a vector along the major axis of the ellipse with magnitude $f = \mu e$. The construction of this vector was first given in geometrical form by Newton in Book I, Proposition 17, and in analytic form by Hermann and Johann Bernoulli in the *Mémoires de l' Academie Royale des Sciences*, 1710, 521–33. Laplace's derivation of the variation of orbital parameter is in effect the analytic equivalent of Newton's geometrical approach in the Portsmouth manuscript.

8. Curtis Wilson, "The Work of Lagrange in Celestial Mechanics," in R. Taton and C. Wilson, eds., *The General History of Astronomy: Planetary Astronomy from the Renaissance to the Rise of Astrophysics* (Cambridge: Cambridge University Press, 1995), 2B:108–30.

9. Bruno Morando, "Laplace," in Taton and Wilson, General History of Astronomy, 2B:131–50.

10. It is of considerable interest that in the 1684–1685 revised treaty of *De Motu* Corollaries 3 and 4 are not included in Proposition 17 (see Whiteside, *Mathematical Papers of Isaac Newton*, 6:161), but appeared in the first edition of the *Principia*. This implies that Newton must have developed the Portsmouth perturbation method

between 1685 and 1686. Indeed, Whiteside attributes the Portsmouth manuscripts to late 1686.

11. Whiteside, "Newton's Lunar Theory," 321.

12. Alexander Koyré and I. Bernard Cohen with Anne Whitman, eds., *Isaac's Newton's Philosophiae Naturalis and Principia Mathematica*, 3d ed. with variant readings (Cambridge: Harvard University Press, 1972), 2:658–61.

13. Newton also developed a phenomenological theory of solar perturbation of the lunar motion that I will discuss here only peripherally. See *Isaac Newton's Theory of the Moon's Motion* (1702), with biographical and historical introduction by I. Bernard Cohen (Kent, England: Dawson, 1975), and N. Kollerstrom, "A Reintroduction of Epicycles: Newton's 1702 Lunar Theory and Halley's Saros Correction," *Quarterly Journal of the Royal Astronomy Society* 36 (1995): 357–68.

14. F. Tisserand, "Théorie de la Lune de Newton," in *Traité de Mécanique Céleste*, (Paris: Gauthier-Villars et Fils, 1894), 3:27–45.

15. Ibid., 45.

16. Craig B. Waff, "Isaac Newton, the Motion of the Lunar Apogee, and the Establishment of the Inverse Square Law," *Vistas in Astronomy* 20 (1976): 99–103; "Newton and the Motion of the Moon: An Essay Review," *Centaurus* 21 (1977): 64–75; and "Clairaut and the Motion of the Lunar Apse: The Inverse-Square Law Undergoes a Test," in Taton and Wilson, *General History of Astronomy*, 2B:35–46.

17. Whiteside, "Newton's Lunar Theory," 317–28.

18. Philip P. Chandler, "The *Principia*'s Theory of the Motion of the Lunar Apse," *Historia Mathematica* 4 (1977): 405–10.

19. Curtis Wilson, "The Newtonian Achievement in Astronomy," in Taton and Wilson, *General History of Astronomy*, 2A:262–67.

20. Chandrasekhar, *Newton's Principia for the Common Reader*, 450–54. For my review of this book, see *American Journal of Physics* 14 (1996): 957–58.

21. Laplace, "Sur la théorie lunaire," 419.

22. Curtis Wilson, "Newton on the Moon's Variation and Apsidal Motion: The Need for a Newer 'New Analysis,'" chapter 6 of this volume. I am indebted to Professor Wilson for sending me a copy of the chapter during the completion of my work.

23. Chandrasekhar claims that the exact solution of Newton's equation for the mean motion of the apogee either with Newton's parameter $\beta = 11/2$ or the correct parameter $\beta = 5$ "comes nowhere near resolving the discrepancy." However, this statement is based on an erroneous result given in Chandrasekhar, *Newton's Principia for the Common Reader*, equation 36, p. 453.

24. M. Nauenberg, "Newton's Early Computational Method for Dynamics," *Archive for History of Exact Sciences* 46 (1994): 221–52.

25. J. B. Brackenridge, *The Key to Newton's Dynamics: The Kepler Problem and the Principia* (Berkeley and Los Angeles: University of California Press, 1995).

26. Newton's calculation of the variational orbit by the curvature method in the 1687 edition of the *Principia* gives further evidence that by this time he had already developed the curvature measure for force (see Nauenberg, "Newton's Early Computational Method"). Previously it had been assumed that Newton introduced this measure only after he began to revise the first edition of the *Principia* [see J. B. Brackenridge, "The Critical role of Curvature in Newton's Dynamics," in P. M. Hartman and A. E. Shapiro, eds., *The Investigation of Difficult Things: Essays on Newton and the History of Exact Sciences in Honor of D. T. Whiteside* (Cambridge: Cambridge University Press, 1992)]. In fact, an explicit discussion of this measure first appeared in an added Corollary 3 to Proposition 6 in the second edition (1713) of the *Principia.*

27. Euler, *Theoria Motus Lunae*, in *Opera Omnia*, 23:64.

28. *Collected Mathematical Papers of G. W. Hill* (Baltimore: Carnegie Institution of Washington, 1905), 1:284–335.

29. A similar approach is taken in Chandrasekhar, *Newton's Principia for the Common Reader.*

30. Christiaan Huygens, *Horologium Oscillatorium* (Paris, 1673), reprinted in *Oeuvres complètes de Christiaan Huygens*, published by Société Hollandaise des Sciences, *De Vi Centrifuga*, vol. 16, 256–328. (The Hague: Martinus Nijhoff 1888–1950); Translated into English by R. J. Blackwell as *The Pendulum Clock* (Ames, Iowa: Iowa State University Press, 1966).

31. This equivalence is in accordance with one of Newton's definitions of force (definition 7) in the *Principia.*

32. It is of great interest to compare this statement with a cryptic comment in Newton's 1664 *Waste Book*: "If the body b moves in an Ellipsis, then its force in each point (if its motion in that point be given) may be found by a tangent circle of equal crookedness with that point of the Ellipsis." The term *crookedness* was Newton's earlier expression for curvature, and *tangent circle* is the osculating circle. For further details see Nauenberg, "Newton's Early Computational Method for Dynamics."

33. Curtis Wilson, "The problem of Perturbation Analytically Treated: Euler, Clairaut, d'Alembert," in Taton and Wilson, *General History of Astronomy*, 2B:103.

34. Tisserand, "Théorie de la Lune de Newton," 46.

35. Whiteside, *Mathematical Papers of Isaac Newton*, 3:170.

36. The modern method for obtaining this result is to solve directly (7.10), which takes the form of a differential equation

$$\left(\frac{d^2}{d\theta^2}+1\right)\frac{1}{r}=\frac{\mu}{h^2}\left[1-\frac{g}{2\mu}r^3\right]. \qquad (7.108)$$

A special solution of this equation is a circle with radius $\rho = r_0$, which satisfies the equation

$$\frac{h^2}{r_0} = \mu\left(1 - \frac{m^2}{2}\right), \tag{7.109}$$

where $m^2 = gr_0^3/\mu$, (valid provided $m^2 < 2$). Assuming Newton's revolving ellipse (7.12) as an approximate solution for small eccentricity e,

$$r = r_0(1 - e\cos(\nu\theta)), \tag{7.110}$$

and expanding r^3 to first order in e,

$$r^3 = r_0^3(1 - 3e\cos(\nu\theta)), \tag{7.111}$$

one finds that

$$\frac{h^2}{r_0}[1 + (1 - \nu^2)e\cos(\nu\theta)] = \mu\left[1 - \frac{m^2}{2}(1 - 3e\cos(\nu\theta)\right]. \tag{7.112}$$

Hence, r_0 is given as before by (7.109) and

$$\nu^2 = 1 - \frac{3}{2}m^2. \tag{7.113}$$

37. Laplace, "Sur la théorie lunaire de Newton," 444.

38. Tisserand, "Théorie de la lune de Newton," 39.

39. Chandrasekhar, *Newton's Principia for the Common Reader*, 244

40. In Book III, Proposition 28, Newton evaluated the dynamical condition for the curvature ratio ρ_s/ρ_q (7.31) by substituting for the ratio of the radial forces $a_s/a_q = [u(1/AT^2) - (2000/w)AT]/[u(1/CT^2) + (1000/w)CT]$, where $u = 178725$ corresponds to $u = 1000/m^2$, $AT = 1 - x \propto r_s$, $CT = 1 + x \propto r_q$, and $w = AT \cdot CT = 1 - x^2$. Actually w should be unity, but the additional term x^2 is fourth order in m and consequently negligible. The ratio h_s/h_q, evaluated in Proposition 26, is given as $(y + 50)/(y - 50) = 11073/10973$, where $y = 11023$ corresponds to $y = 200(1 - m)/(3m^2)$; Koyré and Cohen, *Isaac Newton's Philosophiae Naturalis and Principia Mathematica*, 2:624. For $w = 1$, Newton's result agrees with (7.32).

41. The evaluation of the curvature ratio from the assumed shape of an elliptical orbit rotating about its center is given in Book III, Proposition 28, in the form $(AT^3 + zCT^2AT)/(CT^3 + zAT^2CT)$, with $z = 16824/100000$ corresponding to $z = 1/(1 - m)^2 - 1$; Koyré and Cohen, *Isaac Newton's Philosophia Naturalis and Principia Mathematica*, 2:626. This result agrees with (7.33) neglecting terms of order m^2 and higher. In Cajori's revision of Motte's translation of Newton's *Principia* (Berkeley and Los Angeles: University of California Press, 1934), this relation appears on p. 447 with a sign error.

42. Whiteside, *Mathematical Papers of Isaac Newton*, 6:519.

43. In the 1690s Newton considered a revision of his demonstration of Proposition 17 that would enable him to derive the effect of a perturbation normal and tangential to the velocity more readily and without restrictions to small eccentricities. He then included an extended version of Corollaries 3 and 4 explaining in detail with a new diagram how to evaluate the perturbation's effect. He also added Corollaries 5 and 6, giving the results of his derivation for the rotation of the apsis in the case of a tan-

gential and a normal perturbative impulse (see Whiteside, *Mathematical Papers of Isaac Newton*, 6:559–63). Newton's derivation can be reconstructed, and it can be seen that his results are not quite correct.

44. In fact, Proposition 17 corresponds, in geometrical form, to the Laplace vector $\mathbf{e} = 1/\mu\mathbf{v} \times \mathbf{h} - \mathbf{r}/r$, where $\mathbf{r}$ is the position, $\mathbf{v}$ is the velocity and $\mathbf{h}$ is the angular momentum vector (see Laplace, *Treatise of Celestial Mechanics*).

45. J. M. A. Danby, *Fundamentals of Celestial Mechanics*, 2d ed. (Richmond, VA: Willmann-Bell, 1989): 327; Cook, *Motion of the Moon*, 57.

46. The exact result is

$$\frac{de}{d\theta} = \frac{r^2}{\mu}[V\sin(\phi) + \frac{W}{(1 + e\cos(\phi))}[2\cos(\phi) + e(1 + \cos^2\phi)] \tag{7.114}$$

47. Whiteside, *Mathematical Papers of Isaac Newton*, 6:517.

48. In lemmas [α] and [β] Newton refers to "the motion of the apogee ensuing from that impulse" as if the apogee of the ellipse were the same as the apogee of the moon's orbit, which is not strictly correct. Even some modern textbooks of celestial mechanics do not make the important distinction between the apogee of the physical orbit and the apogee of the ellipse in constructing the variation of orbital parameters method.

49. The relation between time t and longitude θ obtained by integrating (7.6) takes the form

$$\frac{t}{\tau} = \theta - 2e_0\sin(\nu\theta) - \xi\sin(2\psi) - \frac{2\gamma}{(1-2m)}\sin(2\psi - \nu\theta), \tag{7.115}$$

where $\tau = r_0^2/h_0$. The last term corresponds to the lunar inequality known as the *evection*, first pointed out by Ptolemy.

50. Newton's remark can be interpreted as setting $r = r_0$ on the right side of (7.64) for ω, which leads to (7.78) for $\Delta\omega_0$, with $\omega = \Delta\omega_0/e_0$. This gives a purely oscillatory contribution to ω, and together with the corresponding contribution Δe, which Newton apparently did not evaluate, this gives rise to the special lunar orbit, as can be seen by taking the limit $e_0 = 0$ in (7.83).

51. In Proposition 66, Corollary 9, Newton concludes that "therefore in the passage of the apsis from the quadratures ($\omega - \theta' = \pm\pi/2$) to the syzygies ($\omega - \theta' = 0, \pi$) it is continually augmented, and increases the eccentricity of the ellipse; and in the passage form the syzygies to the quadrature it is continually decreasing, and diminishes the eccentricity."

52. Curtis Wilson, "Predictive Astronomy in the Century after Kepler," in Taton and Wilson, *The General History of Astronomy*, 2A:197–201. Horrocks's model for the lunar orbit can be written, for small eccentricity e_0, in the analytic form $r = r_0[1 - (e_0 + \delta e)\cos(\nu\theta - \delta\omega)]$, where $\delta e = \delta e_0\cos(\xi)$, and $\delta\omega = (\delta e_0/e_0)\sin(\xi)$. Here $\xi = 2(\omega - \phi') \approx 2(\psi - \nu\theta)$, where $\phi' \approx m\theta$ is the sun's longitude. This model is justified by Newton's gravitational theory and gives an important contribution to the lunar inequalities due to the solar perturbation. According to the perturbation theory, $\delta e_0/e_0 = 15m/8$.

53. In the revised Scholium to Proposition 35, Book III, Newton states that

> by the same theory of gravity, the moon's apogee goes forwards at the greatest rate when it is either in conjunction with or in opposition to the sun, but in its quadratures with the sun it goes backwards; and the eccentricity comes, in the former case to its greatest quantity; in the latter to its least by Cor. 7,8 and 9, Prop. 66, Book 1. And those inequalities by the Corollaries we have named, are very great, and generate the principle which I call the semiannual equation of the apogee; and this semiannual equation in its greatest quantity comes to about 12°18′, as nearly as I could determine from the *phenomena* [italics added]. Our countryman, Horrocks, was the first who advanced the theory of the moon's....

In Corollaries 7 and 8, Proposition 66, Newton gave a qualitative explanation for this motion of the moon's apogee due to the perturbation of the sun, claiming it was based on results given in Book I, Proposition 45, Corollary 1. However, these results were obtained for the case of radial forces only and are therefore strictly not applicable to the solar perturbation, which is not a purely radial force with respect to the earth as a center and which depends also on the angle ψ. According to the differential equation for the motion of the lunar apogee that appears in the Portsmouth manuscript (7.72), this rate depends on the relative angle between the moon's apogee ω and the longitude θ' of the sun, where $\omega - \theta' = \psi - \phi$. It reaches a maximum value when $\omega - \theta' = n\pi$, where n is an integer, and a minimum when n is an odd integer divided by 2, in accordance with Corollary 8. In fact, substituting Newton's numerical values in (7.72), $\beta = 11/2$, one finds that the maximum rate of advance is 21.57′, and of retardation 14.83′. This is in reasonable agreement with the values 23′ and $16\frac{1}{3}'$ given in the original (1687) Scholium to Proposition 35 (see Introduction) corresponding to $\beta \approx 6$. In Corollary 9 Newton gave a qualitative argument for the variability of the eccentricity, corresponding to (7.84) (and not (7.77) as claimed by Chandrasehkar, *Newton's Principia for the Common Reader*, 251), but there is no evidence that he obtained this quantitative result from his "theory of gravity" (according to his theory the apogee's maximum variability is $15m/8 = 8°2'$, instead of 12°18′ as quoted in the Scholium to Proposition 35). Although the Horrocksian model probably inspired his Portsmouth method, in the end Newton was able to account only partially for this model from his dynamical principles.

54. For the contribution of the perturbation force's radial component to the apogee's secular motion (Whiteside, *Mathematical Papers of Isaac Newton*, 6:521), Newton took the difference $(1 - (QO/SP)^3)(2HS - PH) \approx (3x/SP)(PH - 2HS)$, where $x = SP - QO$. In our notation $SP = r = r_0(1 - e\cos(\phi))$, $QO = r_0$, $x = -r_0 e \cos(\phi)$, and $2HS - PH = (r/2)(1 - 3\cos(2\psi))$. However, for the corresponding contribution of the force's transverse component, Newton assumed the difference $(1 - (QO/SP)^4)3IH \approx (12x/SP)IH$, where $IH = (r/2)\sin(2\psi)$. The introduction here of a quartic rather than cubic power in the scale factor QO/SP is not correct. Newton considered also other powers, which has led to the claim that he was "fudging" his equation to get agreement with observation (see Whiteside, *Mathematical Papers of Isaac Newton*, 6:518, n. 26). However, I would like to stress here that Newton's insight, that the difference of properly scaled perturbation forces acting on unperturbed circular and elliptic orbits determines the motion of the apogee, is essentially correct.

55. In Newton's notation this equation appears as the ratio of

$$6HS \times SE^2 - 3PH \times SE^2 + 24IH \times PE \times SE \tag{7.116}$$

to $178.725P^2$ for "the total motion of the apogee arising from both forces to the moon's mean motion." Here $HS = r\cos^2(\psi)$, $SE = r\cos(\phi)$, $PH = r\sin^2(\psi)$, $IH = r\cos(\psi)\sin(\psi)$, and $P^2 = 1/r^2$ (Whiteside, *Mathematical Papers of Isaac Newton*, 6:521).

56. In "The Newtonian Achievement in Astronomy," C. Wilson concludes with previous commentators that Newton "fudged" his result. However, this leaves unexplained how Newton could have obtained the *correct analytic form* for this equation.

8

Force, Continuity, and the Mathematization of Motion at the End of the Seventeenth Century

Michel Blay

This is what Christiaan Huygens states in his *Discours de la cause de la pesanteur*, which, although published in 1690 in Leiden as a sequel to the *Traité de la lumière*, had been written twenty years earlier in response to a debate at the Paris Académie Royal des Sciences: "One can finally find here the reason for the Principle that Galileo chose to prove the rate of acceleration of falling bodies, which is that their velocity increases equally in equal times. For as the bodies are pushed successively by parts of matter seeking to move up to their place and acting continuously on them, with the same force, as we have just seen, at least in cases of fall which come within our experience, it is a necessary consequence that the increase of the velocities should be proportional to that of the times."[1]

Moreover, in the opening pages of the second part of the *Horologium Oscillatorium* published in 1673 in Paris, Huygens emphasizes that in order to mathematize the motion of falling bodies, the action of gravity, independently of any account of its cause, must be continuous during each of the equal intervals of time considered. For instance, Proposition I reads: "and the latter (the motion caused by gravity) being obviously the same in the second interval of time as in the first, must give to the heavy body *in the course of the second interval of time [idea decursu temporis secundi]* a velocity equal to that which it acquired at the end of the first one. . . . The force of gravity is added again *in the course of the third interval of time*."[2] On the basis of this continuist view of the action of gravity and with the help of the idea that the ratio of the spaces described remains the same whatever one chooses as successive equal times,[3] Huygens arrives at Galileo's famous law of odd numbers, which characterizes the accelerated motion of falling bodies.

One has then, on the one hand, a succession of impulses or impacts and, on the other, an uninterrupted action, like the supposed action of gravity. How does one proceed from conceptualizing action in terms of impulses or impacts to mathematizing a continuous, uninterrupted action, like that of gravity? In other words, how can one achieve mathematical rigor in the transition from the discontinuous to the continuous, with respect to the modalities of action? The main aim of Newton's *Principia* is, in my opinion, to answer these questions. How then does Newton respond to what I shall call Huygens's challenge to construct a continuous action on the basis of conceptualization in terms of impulses and impacts?

First, I shall briefly analyze Law 2 of the *Principia*. Then I shall turn to Newton's treatment of problems involving the consideration of a continuous, uninterrupted action. Finally, to bring out the significance of these inquiries, I shall consider the question of continuity in relation to Varignon's use of Newtonian texts within the framework of the Varignon's algorithmic approach.

LAW 2 IN THE *PRINCIPIA*

"The change of motion is proportional to the motive force impressed; and is made in the direction of the right line in which that force is impressed."[4] This law should not be confused with what is now called "Newton's law," which is expressed in terms of a differential equation and is written $F = ma$ or $F = md^2x/dt^2$. Actually, when Newton speaks here of a "change of motion," he gives no indication concerning the time during which this change takes place, not even a *dato tempore*. If one were to insist on stating the law in modern terms, the closest one could come would be undoubtedly the following: $F = \Delta\ (mv)$, F being the motive force impressed, m the mass, and n the velocity, where $\Delta\ (mv)$ stands for a "change of motion." On this assumption, an impressed motive force is not a force in the modern sense of the term but an impulse.

Thus in the proof of the law of falling bodies, which Newton presents as an inference from Law 2 in the Scholium that follows, the "motive force impressed" clearly appears, in the light of the procedure of totalization carried out, as an impulse: "When a body is falling, the uniform force of its gravity acting equally, impresses, in equal intervals of time, equal forces upon that body, and therefore generates equal velocities; and in the whole time impresses a whole force, and generates a whole velocity proportional to the time. And the spaces described in proportional times are as the product of the velocities and the times; that is, as the squares of the time."[5]

Moreover, two references to Law 2 Newton gives in the context of the proofs worked out in Book II confirm this interpretation of the law. Proposition 3 reads thus: "will be as the absolute forces with which the body is acted upon in the beginning of each of the times, and therefore (by Law II) as the increments of the velocities"; and Proposition 8 reads:

"for the increment PQ of the velocity is (by Law II) proportional to the generating force KC."[6] In neither case is time mentioned. Law II appears then to be associated with an impulse model of action (action by impact), a model that implies a discontinuous conceptualization of this action. How then does one solve problems that involve a continuous action? Two cases must be distinguished depending on whether in overcoming the difficulties of continuity, one is dealing with infinitesimals of the first order or those of the second order. In the first case, Newton resorts to passages to the limit that depend on manipulating intervals or particles of successive times. In the second case, he has recourse principally to Lemmas 10 and 11 of the first section of Book I.

CONTINUITY AND INFINITESIMALS OF THE FIRST ORDER

Proposition 2 of Section 1 of Book II

In the first three sections of Book II Newton examines, for instance, the motion of bodies in media in which the resistance is proportional to the velocity, or the square of the velocity, or a combination of the two. As my aim is not to present Newton's theory in complete detail, but only to bring out what is involved in the construction of mathematical physics, I shall restrict myself to Proposition 2, concerning the uniform motion of bodies in which the resistance is proportional to the velocity:[7]

> If a body is resisted in the ratio of its velocity, and moves, by its inertia only,[8] through an homogeneous medium, and the times be taken equal, the velocities in the beginning of each of the times are in a geometrical progression, and the spaces described in each of the times are as the velocities.[9]

In the opening sentence of the proof Newton seeks mainly to conceptualize the mode of action of the forces of resistance, with reference to the model of percussion:

> Let the time be divided into equal intervals *(particulas aequales);* and if at the very beginning of each interval *(ipsis particularum initiis)* we suppose the resistance to act with one single impulse *(impulse unico)* which is as the velocity, the decrement of the velocity in each of the intervals of time will be as the same velocity.[10]

Newton assumes first of all that time is divided into equal particles[11] and "at the very beginning"[12] of each particle of time the resistance acts "with one single impulse." A few lines below, Newton substitutes a continuous action for this discontinuous mode of action of resistance, with the help of a passage to the limit, relative to the number and the magnitude of the parts of a given time interval.

Moreover, as the resistance is the only force acting and by virtue of the laws of motion it is, first, "as the velocity" according to the hypothesis and, second, as "the decrement of the velocity in each of the intervals of time."[13] Since the motion lost in each particle of time is lost in the beginning of this particle of time, the velocity during each particle of time must be taken as constant and the motion as uniform. The decrement of the velocity, in the beginning of each particle of time, therefore must be proportional to the velocity of the body during the antecedent particle of time. In other words, the difference of velocity of two particles of time (their decrement) must be proportional to the velocity of the body.[14] Having proved earlier in Lemma 1, inserted between Propositions 1 and 2 of Book II, that "quantities proportional to their differences are continually proportional,"[15] Newton is easily able to establish a relation characterizing the evolution of velocity, namely that "the velocities in the beginning of each of the times are in a geometrical progression.[16] Such being the case, he adds that the particles of time comprising a certain interval of time can be diminished at will and their number can be increased infinitely, in such a way that the discontinuous action of the resistance becomes continuous without changing the characteristic relation of the evolution of velocity in the process:

> Let those equal intervals of time be diminished, and their number increased *in infinitum,* so that the impulse of resistance may become continual *(eo ut resistentiae impulsus reddatur continuus);* and the velocities at the beginnings of equal times, always continually proportional, will be also in this case continually proportional.[17]

The use of the term *impulsus* here, in the transition from the discontinuous to the continuous, brings out the conceptual difficulties involved in Newton's construction of the concept of force in the sense of a continuous action. The continuity of the action, achieved by means of a passage to the limit, is found again in Newton's treatment of the law of areas.

The Law of Areas

Proposition 1 (Theorem 1 of Book I) states:

> The areas which revolving bodies describe by radii drawn to an immovable centre of force do lie in the same immovable planes, and are proportional to the times in which they are described.[18]

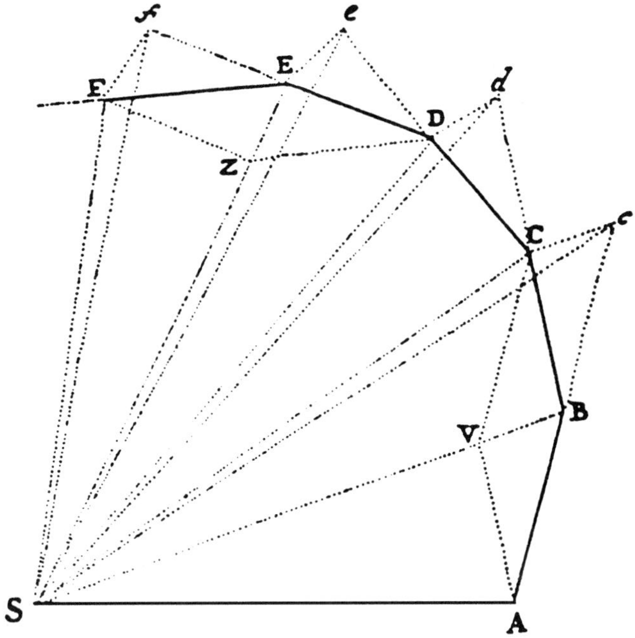

Figure 8.1

Newton gives us no indication concerning the type of curve described by the body; for instance, it is not specified whether the plane curve must be closed (in Section 3, Proposition 12, Newton examines the case of the hyperbola).

Since the body describes a curve, a force must be continually impressed on it, according to Law 1. Newton assumes that a centripetal force is acting and that its center is an immovable, mathematical point. If a material rather than a mathematical point had been taken into account, Newton would have been led to solve a different problem, one necessarily involving the third law of motion, the two-body problem.

It is essential in this Proposition 1 to prove, first of all, that the curvilinear path lies in a plane and, second, that the "radius vector" generates areas proportional to the times. The difficulty of the proof lies mainly in the mathematical treatment of the centripetal force, whose action is assumed to be continuous. How is one to account for a continuous action when the impressed force is conceptualized on the basis of a fundamentally discontinuous model of percussion or impact? (See figure 8.1.)

> For suppose the time to be divided into equal parts, and in the first part of that time let the body by its innate force describe the right line AB. In the second part of that time, the same would (by Law I), if not hindered, proceed directly to c, along the line Be equal to AB; so that by the radii AS, BS, cS, drawn to the center, the equal areas ASB, BSc, would be described.[19]

The equality of the areas of the triangles *ASB* and *BSc* follows from the fact that they have the same base ($AB = Bc$) and the same height (the perpendicular that originates at *S* and meets *ABc*). Yet, rather than continuing its rectilinear path from *B* to *c*, the body is directed toward *C*, by virtue of Law 1:

> But when the body is arrived at B. suppose that a centripetal force acts at once with a great impulse (literally, with one single but great impulse *impulsu unico sed magno*), and, turning aside the body from the right line Bc, compels it afterwards to continue its motion along the right line BC.[20]

Time being divided into equal parts, it is in the beginning of each of these parts and instantaneously that the force is acting, as in Proposition 2 of Book II. Moreover, this mode of instantaneous action implied by the underlying model of impact leads us directly back to Law 2, which, as seen before, makes no explicit allusion to time:

> Draw cC parallel to BS, meeting Be in C; and at the end of the second part of the time, the body (by Cor. I of the Laws) will be found in C, in the same plane with the triangle ASB.[21]

The coplanarity of the motion follows from the fact that *cC* is parallel to *BS* and that *c* lies in the plane defined by *ASB*. Motion under a central force lies in a plane:

> Join SC, and because SB and Cc are parallel, the triangle SBC will be equal to the triangle SBc, and therefore also to the triangle SAB. By the like argument, if the centripetal force acts successively in C, D, E, &c., and makes the body, in each single particle of time, to describe the right lines CD, DE, FE, &c., they will all lie in the same plane.[22]

Having established this, Newton goes on to express the law of areas:

> and the triangle SCD will be equal to the triangle SBC, and SDE to SCD, and SEF to SDE. And therefore, in equal times, equal areas are

> described in one immovable plane: and, by composition *(componendo)*, any sums SADS, SAPS, of those areas, are to each other as the times in which they are described.[23]

The equality of areas of the different triangles follows, as before, from the fact that they have the same base and the same height. At this stage in the rational construction the continuity of the action of the force and, correspondingly, the generation of the curve are far from established. Indeed, this succession of actions exerted in the beginning of each particle of time generates a polygon having $A, B, C, D \ldots$ as summits.

How does one proceed from this polygonal figure to a curvilinear figure generated by the action of a centripetal force assumed to be continuous?

> Now let the number of those triangles be augmented, and their breadth diminished *in infinitum;* and (by Cor. IV, Lem. III) their ultimate perimeter ADF *(ultimata perimeter)* will be a curved line: and therefore the centripetal force, by which the body is continually drawn back from the tangent of this curve, will act continually *(indefinniter);* and any described areas SADS, SAPS, which are always proportional to the times of description, will, in this case also, be proportional to those times.[24]

As the proportionality of the areas to the times is "always" actualized, this proportionality is conserved in the limit. Therefore, motion caused by a central force necessarily describes equal areas, in the same plane, in equal times. Newton introduces the continuous curve of the path by calling on Corollary 4, Lemma 3, of Section 1, concerning the consequence of series of inscribed figures,[25] thereby drawing the conclusion of an "uninterrupted" action of the centripetal force. It is nevertheless difficult to account for the exact relation between "the force acting at once with a great impulse" and the centripetal force acting "uninterrupted." The task of mathematical physics is far from being accomplished; strictly speaking, an analysis is lacking of the continuum disengaged from geometrical intuitions.

This Proposition 1 is followed up by six corollaries. I must quote Corollaries 2, 3, and 4, for they play an essential role, to which I shall return, in the proof of Proposition 6 of this section, concerning the general expression of force in the case of motion caused by a central force:

> Corollary II. If the chords AB, BC of two arcs, successively described in equal times by the same body, in spaces void of resistance, are

> completed into a parallelogram ABCV, and the diagonal BV of this parallelogram, in the position which it ultimately acquires when those arcs are diminished *in infinitum*, is produced both ways, it will pass through the center of force.[26]

This corollary merely states a geometric consequence of the mode of action of the force, emphasizing that *BV* is always "ultimately" directed toward *S*.

> Corollary III. If the chords AB, BC, DE, EF of arcs, described in equal times, in spaces void of resistance, are completed into the parallelograms ABCV, DEFZ, the forces in B and E are one to the other in the ultimate ratio of the diagonals BV, EZ, when those arcs are diminished *in infinitum*. For the motions BC and EF of the body (by Cor. I of the laws) are compounded of the motions Be, BV, and Ef, EZ; but BV and EZ, which are equal to Cc and Ff, in the demonstration of this proposition, were generated by the impulses of the centripetal force in B and E, and are therefore proportional to those impulses.[27]

Corollary 2 made it possible to study, by considering the diagonals, the direction of the force in the passage to the limit ("ultimately"). The next corollary aims to determine their magnitude, that is, the magnitude of the forces in the same passage to the limit. This result raises no difficulties, as long as one does not subject to too close scrutiny the passage to the limit. Indeed, by assuming that in equal times the motions along *BC* and *EF* are compounded of the inertial motions along *Be* and *Ef* as well as the motions along *Cc* and *Ff*, Newton is able to conclude from the fact that *BcCV* and *EfFZ* are parallelograms that *BV* and *EZ*, being equal to *cC* and *fF*, are as the impulses in *B* and *E*.

> Corollary IV. The forces by which bodies, in spaces void of resistance, are drawn back from rectilinear motions, and turned into curvilinear orbits, are to each other as the versed sines of arcs described in equal times; which versed sines tend to the center of force, and bisect the chords when those arcs are diminished to infinity. For such versed sines are the halves of the diagonals mentioned in Cor. III.[28]

The diagonals introduced in Corollary 3 are replaced here by their halves called "versed sines." Such being the case, the force is thus assumed to act in a continuous manner and without interruption. It is now time to determine its expression; here infinitesimals of the second order are involved.

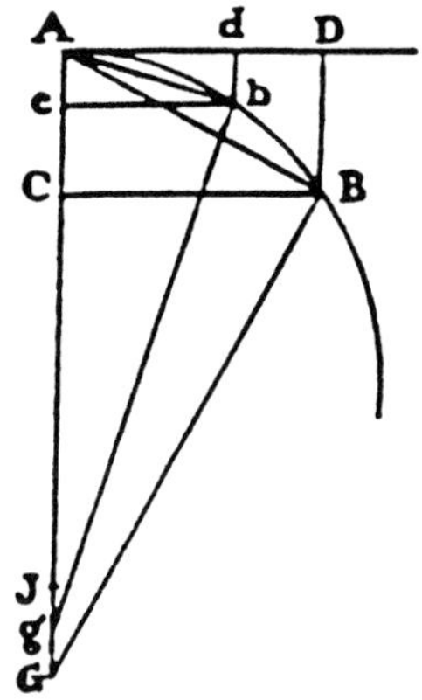

Figure 8.2

Continuity and Infinitesimals of the Second Order

In Proposition 6 of the second section of Book I, Newton gives the general expression of central force:

> Proposition 6. Theorem V. In a space void of resistance, if a body revolves in any orbit about an immovable center, and in the least time describes any arc just then nascent; and the versed sine of that arc is supposed to be drawn bisecting the chord, and produced passing through the center of force: the centripetal force in the middle of the arc will be directly as the versed sine and inversely as the square of the time.[29]

The proof that follows calls on two earlier results, on the one hand, Corollary 4 of Proposition 1 mentioned above and, on the other, Corollaries 2 and 3 of Lemma 11, Section 1 (see figure 8.2).

> Lemma XI. The evanescent subtense of the angle of contact, in all curves which at the point of contact have a finite curvature, is ultimately as the square of the subtense of the conterminous arc.[30]

The proof of this lemma, which rests mainly on the existence of an osculating circle, leads Newton to consider ultimately the last ratios of AB^2 to Ab^2 (the subtense of the arc) and of BD to bd (the subtense of the angle of contact) to be "the same." In modern terms, this result means that if the chord is taken as an infinitesimal of the first order, then BD is an infinitesimal of the second order.

From this one obtains:

> Corollary I. Therefore since the tangents AD, Ad, the arcs AB, Ab, and their sines, BC, be, become ultimately equal to the chords AB, Ab, their squares will ultimately become as the subtenses BD, bd.[31]

This corollary merely draws attention, in particular, to the fact that in the wording of Proposition 1, the chord (or subtense of the arc) can be replaced by the arc *AB* or by the tangent *AD*, or even by the sine *BC*. In Corollary 5 of Lemma 11, Newton adds what is already clear, that the curve of finite curvature is ultimately a parabola having *A* as its summit.

> Corollary II. Their squares are also ultimately as the versed sines of the arcs, bisecting the chords, and converging to a given point. For those versed sines are as the subtenses BD, bd.[32]

One is referred here not to the versed sines of the arcs *AB* and *Ab* but to those at *A* of the double arcs, as is evident from Corollary 1 of Proposition 6.[33]

Corollary 3 gives a kinematic interpretation of this result.

> Corollary III. And therefore the versed sine is as the square of the time in which a body will describe the arc with a given velocity.[34]

When a body with a given constant velocity describes an arc, the arc's length is proportional to the time of the motion. Now, the versed sine is ultimately like the square of the arc (Corollaries 1 and 2); it is then ultimately like the square of the time. This preparatory study makes it possible to understand the proof of Proposition 6:[35]

> For the versed sine in a given time is as the force (by Cor. IV, Prop. D; and augmenting the time in any ratio, because the arc will be augmented in the same ratio, the versed sine will be augmented in the square of that ratio (by Cor. II and m, Lem. XI), and therefore is as the force and the square of the time. Divide both sides by the square of the time, and the force will be directly as the versed sine, and inversely as the square of the time. Q.E.D.[36]

If one does not subject to close scrutiny the mathematical rigor of this proof, it follows from these results that if t is the time required to describe an arc, then ultimately the versed sine $a(t)$ of this arc is both like the centripetal force in the middle of the arc (see above, Corollary 4 of Proposition 1) and like the square of the time in which the arc is described (see above, Corollaries 2 and 3 of Lemma 11). Hence the centripetal force

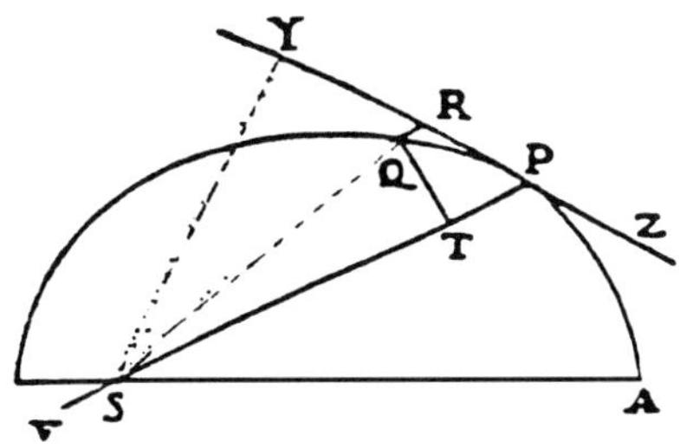

Figure 8.3

f in the middle of the arc is like the versed sine and inversely like the square of the time ($f \propto a(t)/t^2$).

Next, by Corollary 1 of this proposition, with the help of the law of areas, Newton arrives at a general expression of central force that can be manipulated geometrically (see figure 8.3):

> If a body P revolving about the center S describes a curved line APQ, which a right line ZPR touches in any point P; and from any other point Q of the curve, QR is drawn parallel to the distance SP, meeting the tangent in R; and QT is drawn perpendicular to the distance SP, meeting the tangent in R; and QT is drawn perpendicular to the distance SP; the centripetal force will be inversely as the solid (a magnitude that has the dimensions of a volume) $SP^2 \cdot QT^2/QR$, if the solid be taken of that magnitude which it ultimately acquires when the points P and Q coincide. For QR is equal to the versed sine of double the arc QP, whose middle is P: and double the triangle SQP, or $SP \cdot QT$ is proportional to the time in which that double arc is described; and therefore may be used to represent the time.[37]

In fact, ultimately, *QR* is equal to the versed sine of the double arc of *QP*, whose middle is *P*. Moreover, the double of the area of the triangle *SQP* is equal to $SP \cdot QT$. Now by Proposition 1 (the law of areas), this area is proportional to the time in which the double of the arc *QP* is described. Therefore the time can be represented geometrically by this area, and, by substitution, the centripetal force at *P* can be expressed, when *P* and *Q* coincide, by f (at P) $QR/SP^2 \cdot QT^2$; in other words this force is also "inversely as the solid $SP^2 \cdot QT^2/QR$, if the solid be taken of that magnitude which it ultimately acquires when the points *P* and *Q* coincide."

Newton now possesses an expression of the centripetal force at a point that can be manipulated geometrically. This expression makes it possible for him to solve many problems dealing with central forces, that is, in the final analysis, to express the variation of force as a function of the distance of the body in motion and the given center of force.

It is noteworthy that, in this proof pertaining to the general expression of the law of central force, Newton does not refer explicitly (via Corollary 4 of Proposition 1) to Law 2 but rather to Lemma 11, which plays a decisive role with respect to the continuous action of the force, from a local point of view.

Moreover, Newton inserts the following clause between the proof of Proposition 6 and Corollary 1: "The same thing may also be easily demonstrated by Cor. IV, Lem. X." Now, alone Lemma 10 of Section 1 deals explicitly with force: "The spaces which a body describes by any finite force urging it, whether that force is determined and immutable, or is continually augmented or continually diminished, are in the very beginning of the motion to each other as the squares of the time."[38] This relation between spaces and times, characteristic of uniformly accelerated motion, represents the force in a different way than Law 2. Already, in the first version of the *De Motu Corporum* of autumn 1684, in a text corresponding to Proposition 6 of the *Principia,* Newton refers to Hypothesis 4 according to which "the space described by a body acted on by any centripetal force is at the very beginning of its motion as the square of the times."[39] It is then essential to examine the proof that follows Lemma 10; but one must first study the antecedent lemma (see figure 8.4):

> Lemma IX. If a right line AE, and a curved line ABC, both given by position, cut each other in a given angle, A; and to that right line, in another given angle, BD, CE are ordinately applied, meeting the curve in B, C; and the points B and C together approach towards and meet in the point A: I say, that the areas of the triangles ABD, ACE, will ultimately be to each other as the squares of homologous sides.[40]

In his proof Newton considers the tangent *AFG* to the curve *ABC* as well as the abscissas *Ad* and *Ae* to the abscissas *AD* and *AE,* which are always proportional, in such a manner that, as they vanish, *AD* and *AE* remain proportional to *Ad* and *Ae.* Moreover, the ordinates *db* and *ec* are "to be drawn parallel to the ordinates DB and EC, and meeting AB and AC produced in b and c." Passing through *A*, *b* and *c* is a curve similar to *ABC*; together they make at *A* an angle of degree zero.

When the points *B* and *C* describe the curve *ABC* in such a way as to coincide, in the end, with the point *A*, the curvilinear areas *Abd* and *Ace* coincide, in the limit, with the rectilinear areas *Afd* and *Age.* By Lemma 5, these curvilinear areas have finite limits proportional to the

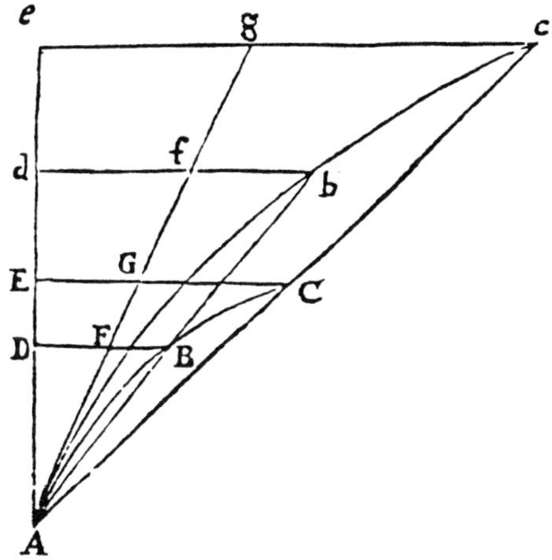

Figure 8.4

squares of sides *Ad* and *Ac*. Owing to similitude, the vanishing areas *ABD*, *ACE* are, in the limit, as the duplicate ratio of the sides *AD* and *AK*. In modern terms, the area is an infinitesimal of the second order, the abscissa being an infinitesimal of the first order.

The proof of Lemma 10 is based on the same figure:

> Let the times be represented by the lines AD, AK, and the velocities generated in those times by the ordinates DB, EC. The spaces described with these velocities will be as the areas ABD, ACE, described by those ordinates, that is, at the very beginning of the motion (by Lem. IX), in the duplicate ratio of the times AD, AE. Q.E.D.[41]

Although the concept of force does not appear in the proof, the wording of this lemma makes it clear that the force considered, whether it is constant ("determined and immutable") or variable ("continually augmented or continually diminished"), is not the force concerned in Law 2 (to which no reference is made), but rather the force taken in a sense close to our concept, or more precisely, in the present case, in the sense of a quasi acceleration (a quasi acceleration from our point of view, because the algorithm of kinematics has not been worked out; see below).

It is natural for Newton, in order to arrive at his result in Proposition 6, to call on Lemma 10 when he does not rely on Lemma 11. In both cases the force considered in Law 2 is replaced by force (acceleration) in a quasi-modern sense (Lemma 10), or principally by the equivalent technique of the osculating circle (Lemma 11). When Pierre Varignon came to give a continuist analysis of the action of force, he called on Newton's Lemma 10.

VARIGNON'S ALGORITHMIC TREATMENT

As early as 1700, in three successive memoirs, Varignon seeks to give the problem of central forces all the clarity and generality that the new methods of Leibnizian calculus along with his own research on the concept of the velocity in each instant can bring.[42] These memoirs, dated 30 January, 31 March, and 13 November 1700, are entitled, respectively: "General Manner of Determining Forces, Velocities, Spaces and Times, When Only One of These Four Factors are Given in All Sorts of Rectilinear Motions Varied at Will,"[43] "On Motion in General in All Sorts of Curves, and on Central Forces, Both Centrifugal and Centripetal, Required by the Bodies Which Describe Them,"[44] and "On Central Forces or On the Gravity Required by Planets to Describe the Orbits That They Have Been Supposed to Have until Now."[45]

As the titles of the first two memoirs imply, one encounters again here the conceptual order of the two memoirs of 1698, whereby, in particular, the concept of velocity in each instant is introduced: the first one will be devoted to "forces centrales," or the "force vers (un centre) c" or a body or again the "tendance au point c comme centre,"[46] in the case of bodies describing rectilinear trajectories. The second memoir deals with the case of "forces centrales" in which bodies describe curvilinear paths around a given fixed center. In the third memoir Varignon abandons examples of a purely mathematical nature in favor of the study of planetary orbits. I will confine my analysis here to the case of central forces causing bodies to follow rectilinear trajectories.

The Velocity in Each Instant and the Increment of the Velocity

The first memoir, dated 30 January 1700, follows as a development of that of 5 July 1698 on rectilinear motion. Varignon points this out in an opening paragraph missing from the final version published in the *Mémoires de l'Académie Royale des Sciences* but found in the *Minute Book of the Sessions of the Académie Royale des Sciences:*

> On the fifth of July 1698 at the Académie, I proved a general Rule for all sorts of motions varied at will and such that in the present figure (all the rectilinear angles being right angles) letting AH be the whole space described, the ordinates VH = VG of any curve, VB or VK the velocities at each point H of this space and the ordinates HT of any other

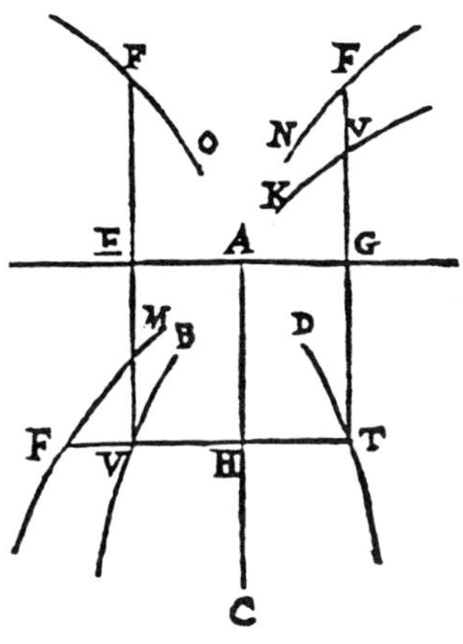

Figure 8.5

> curve TD be the times required to go from A to H. One of these three curves being given, one can always deduce the two others.[47]

Varignon then goes on to make it clear, in the same unpublished text, that this new memoir emphasizes his interpretation: he will now raise the question of the central force's intensity in association with the variation of motion:

> But thereby as I did not then include the force toward C that the body has in each point H. independently of its velocity and which I shall now call central force, because of its tendency toward C as a center.[48]

To solve this problem, Varignon assumes the different variables required and materializes their relations[49] by means of a graph whose construction, in its principle, is identical with that presented in the 5 July 1698 memoir (see figure 8.5). In the *Mémoires de l'Académie Royale des Sciences* Varignon's new study thus opens with these words:

> As the rectilinear angles in the figure produced here are right angles, let TD, VB, FM, VK, FN, FO be any six curves, of which the first three express, in virtue of their common abscissa AH, the space described by any body moved as one wishes along AC. Similarly the time required to describe it, expressed by the corresponding ordinate HT of the curve TD; the velocity of this body at each point H. by the likewise corresponding ordinates VH, VG of the curves VB, VK; the force it has toward C, at each point H. independently of its velocity (I shall now call it central force, because of its tendency toward C as a center), will

> be expressed likewise by the corresponding ordinates FH, FG, FE of the curves FM, FN, FO.[50]

Henceforth, Varignon names these six fundamental curves, *TD*, *VB*, *FM*, *VK*, *FN*, and *FO*, respectively (let us recall that the body describes any rectilinear motion *AC*):

> "the curve TD, whose ordinates HT terminate in T. will be named the *curve of the times*.
>
> The two curves VB, VK, whose equal, corresponding ordinates VH, VG terminate in V, will be named the *curves of the velocities*.
>
> Finally the three curves FM, FN, FO, whose equal, corresponding ordinates FH, FG, FE terminate in F. will be named the *curves of the forces*.[51]

The curves *VB* and *VK* correspond to the first and second diagrams of velocity, in the presentation of the 5 July 1698 memoir.

In short, one may say, in a slightly modernized language, that the curve *TD* expresses the variations of the space *AH* in function of the time *HT*; the curve *VB* that of the space *AH* in function of the velocity *VH*, the curve *VK* that of the time *AG* in function of the velocity *VG*; the curve *FM* that of the space *AH* in function of the central force *FH*; the curve *FN* that of the time *AG* in function of the central force *FG*; and, finally, the curve *FO* that of the velocity *EA* in function of the central force *FE*.

Varignon then associates with each of the four variables, space, time, velocity and force, a definite algebraic symbol:

> let AH = x be the spaces described, the times required to describe them HT = AG = t, the velocities at H (which I shall call final) HV = AE = GV = v, the corresponding central forces HF = EF = GF = y.... Thereby one will have dx for the space described, as with a uniform velocity v at each instant; dv for the increase of the velocity occurring; ddx for the space described as a result of this increase of velocity; and dt for this instant.
>
> In this case, the velocity being only a ratio of the space described in uniform motion to the time required to describe it, one will have immediately v = dx/dt for a first Rule, which will give dv = ddx/dt by making dt constant.[52]

As a result of this procedure of algebraization, Varignon is able to set up, in the first place, the kinematic concepts already considered in the 5 July

1698 memoir and necessary for his subsequent investigation of central forces in the case of rectilinear motions.

Varignon is thus now in possession of two differential expressions pertaining to the concept of velocity: "on the one hand, that of the velocity in each instant: $v = dx/dt$ on the other, that of the increment of this same velocity during the same instant dt: $dv = ddx/dt$."[53] Actually, varied motion can be interpreted as the result at each instant of a uniform motion of velocity v, equal to the total velocity acquired during the previous interval of time, and a motion equally uniform of velocity dv, the velocity acquired at the very beginning of the time interval dt.[54] This second velocity can be neglected with respect to the first within the time interval dt.

From the Increment of the Velocity to the Central Force

Setting up the differential expressions pertaining to central force hinges on the Galilean model for falling bodies:

> Furthermore, the spaces described by a body moved by a constant and continuously applied force, such as gravity as it is usually conceived, being in the compounded ratio of this force and the square of the times required to describe them, one will also have $ddx = ydt^2$, or $y = ddx/dt^2 = dv/dt$. This yields another Rule $y = dv/dt$, which, with the former $v = dx/dt$, satisfies all that I am endeavoring here to solve.[55]

Here Varignon adopts the conceptualization of the accelerating force as proposed by Newton in Lemma 10 of Section 1 of the first book of the *Principia*. It is highly revealing to compare, in this perspective, the first sentence of Varignon's text with Newton's formulation of Lemma 10:

> The spaces which a body describes by any finite force urging it, whether that force is determined and immutable, or is continually augmented or continually diminished, are in the very beginning of the motion to each other as the squares of the times.

Corollary 3 of this lemma also stipulates:

> Cor. III. The same thing is to be understood of any spaces whatsoever described by bodies urged with different forces; all which, in the very beginning of the motion, are as the product of the forces and the squares of the times.

In consequence, as the force, supposed "constante et continuellement appliquée," during the time interval dt, produces an increment of space equal to *ddx*, one may write, while strictly following Varignon,

$$ddx = ydt^2, \text{ or } y = ddx/dt^2.$$

Given $y = ddx/dt \cdot dt$, therefore $dv = ddx/dt$, hence $y = dv/dt$.

However, as I noted above that the variation of the velocity *dv*, because of the form of its expression, must appear at the very beginning of the time interval *dt*, it is essential, in this case, to bear in mind that a force acts instantaneously at the beginning of this time interval *dt*, then ceases to act until the beginning of the next time interval. In this view the expression "force constante et continuellement appliquée" betrays certain underlying problems pertaining to the production and nature of motion. The difficulties can be brought out by returning to the expression of the increment of velocity in the following way: in the case where the force actually acts in a continuous and constant manner during the time interval *dt*, so that the increment of velocity acquired at the end of this time interval is still *dv*, the space traveled will no longer be *ddx* but $\frac{1}{2}$ddx. Yet Varignon reaches the expected result; for, when passing from a proportionality to an equality in the expression of the force, he neglects to take into account the coefficient of proportionality.

By expressing things slightly differently and adopting the standpoint of the determination of the expression of the force, one may reach the conclusion that the difficulties of Varignon's conceptualization lie mainly in the fact that the expression of the increment of the velocity *dv* and that of the force *y* imply, as it were, an ambiguous modelization of the force's mode of action, In the sense that the latter is supposed to act, according to the expression that Varignon is trying to determine, either at the very first instant (such is the implicit consequence of the solely mathematical manipulations that lead to the expression for *dv*), or in a constant and continuous manner during the entire time interval *dt* (such is the explicit hypothesis involved in calculating *y*).

In the last analysis, one may take Varignon to task, from the standpoint of the coherence of his conceptualization, for not having analyzed in a rigorous manner the construction of the concept of the increment of velocity and for having undoubtedly once again gotten carried away, taking exclusively into account the mathematical operation "faire dt constante" because he indulged in setting the machinery of the new calculus going.

It then appears here that although he reaches a satisfactory expression for *y*, Varignon unfortunately leaves unanswered several questions both technical and conceptual. The general expression for central force in the case of rectilinear motions now being, as it were, acquired, Varignon is able to state what he calls the "Règles générales des mouvements en lignes droites":[56]

$$1^\circ.\ v = dx/dt$$

$$2^\circ.\ y = dv/dt \cdot (ddx/dt^2).$$

These "Règles générales" show us precisely that the concepts of velocity in each instant and of accelerating force in each instant, which Varignon has just successively constructed, can actually be deduced from one another by means of a simple calculation using Leibnizian algorithms, and in consequence, as Auguste Comte writes, "d'après ces formules, toutes les questions relatives à cette théorie préliminaire du mouvement varié se réduiront immédiatement à de simples recherches analytiques, qui consisteront ou dans des différentiations, ou, le plus souvent, dans des intégrations."[57]

After stating his general rules, Varignon determines their use in the very next lines:

> Use. I now state that from any one of the six curves above, one can always deduce the five others by means of these two Rules, assuming the required resolutions and integrations of the equalities concerned.[58]

Then, having analyzed two quite simple examples, Varignon sums up the generality of his method: "The same things will be found in the same manner under any hypothesis; the only difference will be the difficulty in the calculation, which would have merely distracted here. Thus these two examples suffice."[59]

The depth of Newton's work reveals itself. If Law 2 is not in the modern sense of the term "Newton's law"—and there is no reason it should be except at the cost of a retrospective reading—Newton nevertheless provides some remarkable solutions to questions concerning the continuity of action. In the final analysis, as my reading of Varignon's texts suggests, the real driving force of Newton's dynamics, the coherence of the *Principia,* resides in the lemmas of Section 1.[60]

Notes

Translated from the French by Anastasios Brenner.

1. L'on peut enfin trouver ici la raison du Principe que Galilée a pris pour démontrer la proportion de l'accélération des corps qui tombent: qui est que leur vitesse s'augmente également en des temps égaux. Car les corps étant poussés successivement par les parties de la matière qui tache de monter en leur place, et qui, comme on vient de

voir, agissent continuellement sur eux avec la même force, du moins dans les chutes qui tombent sous notre expérience; c'en est une suite nécessaire que l'accroissement des vitesses soit proportionnel à celui des temps. *Discours de la cause de la pesanteur* (Leiden, 1690; reprint, Paris: Dunod, 1992), 175–76. For the debates concerning the cause of gravity at the Académie Royale des Sciences, see for instance Michel Blay, *Les raisons de l'infini* (Paris: Gallimard, 1993 and Chicago University Press, 1998 with the title *Reasoning with the Infinite*), 63ff.

2. *Horologium Oscillatorium Sive de Motu Pendulorum ad Horologia Optato Demonstrationes Geometricae* (Paris, 1673), (in *Oeuvres complètes de Christiaan Huygens*, 22 vols. (La Haye: Société Hollandaise des Sciences, 1888–1950), 18:128–30.

3. In his argumentation Huygens groups by pairs intervals of time and studies the respective ratios of the spaces described. This result can be generalized.

4. A. Motte and F. Cajori, trans., *Sir Isaac Newton's Principia*, Motte's translation revised by Cajori (Berkeley and Los Angeles: University of California Press, 1934, 1962).

5. Ibid.

6. Ibid.

7. For a more detailed study of these questions, see Michel Blay, "Le traitement newtonien du mouvement des projectiles dans les milieux résistants," *Revue d'Histoire des Sciences* 40 (1987): 325–55, and "Varignon ou la théorie du mouvement des projectiles 'comprise en une Proposition générale,'" *Annals of Science* 45 (1988): 591–618.

8. Motte-Cajori, *Sir Isaac Newton's Principia*. The translation is not exact, for the Latin has "et idem sola vi *insita* per medium moveatur" (italics added). Therefore the body is moving here by virtue of its inertial force in the Newtonian sense.

9. Ibid.

10. Ibid.

11. The expression "particles of time" designates, as the sequel shows, not time of short duration, but parts of time meant to tend toward zero.

12. The action of the force thus takes place instantaneously at the very beginning of the particles of time.

13. I refer the reader to my analyses concerning the ambiguity of Newton's concept of force.

14. If $v_1, v_2, v_3, v_4, \ldots$ designate the velocities of the body "in each particle of time" taken to be equal, and $kv_1, kv_2, kv_3, kv_4, \ldots$ the resistance the medium exerts at the beginning of each particle of time, it follows that $v_2 = v_1 - kv_1, v_3 = v_2 - kv_2$, $v_4 = v_3 - kv_3, \ldots$.

15. The object of this Lemma is to establish that if

$$\frac{A}{A-B} = \frac{B}{B-C} = \frac{C}{C-D} = \ldots, \text{ then } \frac{A}{B} = \frac{B}{C} = \frac{C}{D} = \ldots.$$

16. Motte-Cajori, *Sir Isaac Newton's Principia*. According to the result established in the previous note, we have:

$$\frac{v_1}{v_1 - v_2} = \frac{v_2}{v_2 - v_3} = \frac{v_3}{v_3 - v_4} = \ldots \text{ by Lemma 1: } \frac{v_1}{v_2} = \frac{v_2}{v_3} = \frac{v_3}{v_4} = \ldots$$

and

$$v_2 = (1 - k)v_1, v_3 = (1 - k)^2 v_1, \ldots, v_{n+1} = (1 - k)^n v_1, \ldots.$$

17. Ibid.

18. Ibid.

19. Ibid.

20. Ibid.

21. Ibid.

22. Ibid.

23. Ibid.

24. Ibid.

25. Ibid. "And therefore these ultimate figures (as to their perimeters acE) are not rectilinear, but curvilinear limits of rectilinear figures."

26. Ibid.

27. Ibid.

28. Ibid.

29. Ibid.

30. Ibid.

31. Ibid.

32. Ibid.

33. See below.

34. Ibid. Corollaries 2 and 3 of Lemma 11 do not appear in the 1687 edition; this explains certain modifications in the wording of Proposition 6 in 1713 and 1726 with respect to the version of 1687. These modifications, which enhance the importance of the lemmas, are in harmony with my analyses concerning the status of "force" in Newton.

35. For a detailed study of this proposition, see Georges Barthélemy, *Concepts et méthodes de la mécanique rationnelle dans les Principia de Newton* (Ph.D. diss., Sorbonne, 1985), part 2.

36. Motte-Cajori, *Sir Isaac Newton's Principia*.

37. Ibid.

38. Ibid.

39. D. T. Whiteside, ed., *The Mathematical Papers of Isaac Newton,* 8 vols. (Cambridge: Cambridge University Press, 1967–1981), 6:33.

40. Motte-Cajori, *Sir Isaac Newton's Principia.*

41. Ibid.

42. For a detailed study of these questions and Varignon's manuscripts, see Michel Blay, *La naissance de la mecanique analytique. La science du mouvement au tournant des XVIIe et XVIIIe siècles* (Paris: Presses Universitaires de France, 1992).

43. Manière générale de déterminer les forces, les vitesses, les espaces, et les temps, une seule de ces quatre choses etant donnée dans toutes sortes de mouvemens rectilignes variés à discrétion, *Histoire de l'Académie Royale des Sciences avec les Mémoires de Mathématique et de Physique pour la même année* (hereafter, *AM*). *Tirés des registres de cette Académie,* 1700 (1703), 22–27; Archives de l'Académie des Sciences. Registres manuscrits des Procès Verbaux des séances de l'Académie Royale des Sciences (hereafter, *Ac. Sc. Registres*), t. 19, ff. 31r–37r.

44. Du mouvement en général par toutes sortes de courbes; et des forces centrales, tent centrifuges que centripètes, nécessaires aux corps qui les décrivent. *AM*, 1700 (1703), 83–101; *Ac. Sc. Registres*, t. 19, ff. 133v–141v.

45. Des forces centrales, ou des pesanteurs nécessaires aux planètes pour leur faire décrire les orbes qu'on leur a supposés jusqu'icy. *AM*, 1700 (1703), 218–37; *Ac. Sc. Registres*, t. 19, ff. 360v–364v.

46. *AM*, 1700 (1703), 22. The expression "force centrale" seems somewhat ambiguous since Varignon at no point explicitly introduces any consideration of the mass.

47. Le cinq juillet de 1698 je demontray a l'Académie une Règle générale pour toutes sortes de mouvemens variés à discretion, et telle qu'en prenant sur la figure presente (dont tous les angles rectilignes sont droits) AH pour tout l'espace parcouru, les ordonnées VH = VG d'une courbe quelconque, VB ou VK pour les vitesses à chaque point H de cet espace, et les ordonnées HT d'une autre courbe aussi quelconque TD pour les temps employés a venir de A en H; une de ces trois courbes etant donnée, l'on en deduira toujours les deux autres, *Ac. Sc. Registres*, t. 19, f. 31r.

48. Mais comme en cela je ne comprenois point alors la force vers C qu'a le corps en chaque point H. indépendamment de sa vitesse et que j'appelleray dorenavant *force centrale* a cause de sa tendance au point C comme centre. Ibid.

49. Ibid.

50. Tous les angles rectilignes étant droits dans la figure que voicy, soient six courbes quelconque TD, VB, FM, VK, FN, FO, dont les trois premières expriment par leur abscisse commune AH, l'espace parcouru par un corps quelconque mû comme l'on

voudra le long de AC. Soit de même le temps employé à le parcourir, exprimé par l'ordonné correspondante HT de la courbe TD; la vîtesse de ce corps en chaque point H. par les ordonnées aussi correspondantes VH, VG, des courbes VB, VK; ce qu'il a de force vers C, à chaque point H. indépendamment de sa vîtesse (je l'appelleray dorénavant *Force centrale* à cause de sa tendance au point C comme centre), s'exprimera de même par les ordonnées correspondantes encore FH, FG, FE, des courbes FM, FN, FO. *AM*, 1700 (1703), 22.

51. La courbe TD, à laquelle les ordonnées HT se terminent en T. s'appellera la *courbe des temps*.

> Les deux courbes VB, VK, auxquelles les ordonnées correspondantes et égales VH, VG, se terminent en V, s'appelleront les *courbes des vitesses*.
>
> Enfin les trois courbes FM, FN, FO, auxquelles les ordonnées correspondantes encore et égales FH, FG, FE, se terminent en F. s'appelleront les *courbes des forces*. Ibid.

52. ... soient les espaces parcourus AH = x, les temps employés à les parcourir HT = AG = t, les vîtesses en H (que j'appelleray *finales*) HV = AE = GV = v, les forces centrales correspondantes HF = EF = GF = y.... "De là on aura dx pour l'espace parcouru comme d'une vitesse uniforme v, à chaque instant; dv pour l'accroissement de vitesse qui s'y fait; ddx pour ce qui se parcourt d'espace en vertu de cet accroissement de vîtesse; et dt pour cet instant.

A ce compte, la vîtesse ne consistant que dans un rapport d'espace parcouru d'un mouvement uniforme, au temps employé à le parcourir l'on aura déjà v = dx/dt pour une première Règle, laquelle donnera dv = ddx/dt en faisant dt constante. Ibid., 23.

53. Varignon by "faisant dt constante" merely makes use of procedures current at the time of Leibnizian calculus. On this issue, see H. J. M. Bos, "Differentials, Higher-Order Differentials and the Derivative in the Leibnizian Calculus," *Archive for History of Exact Sciences* 14 (1974), 3–90.

54. See also E. J. Aiton, "The Inverse Problem of Central Forces," *Annals of Science* 20 (1964), 86, n. 17.

55. De plus les espaces parcourus par un corps mû d'une force constante et continuellement appliquée, telle qu'on conçoit d'ordinaire la pesanteur, étant en raison composée de cette force et des quarrés des temps employés à les parcourir; l'on aura aussi ddx = ydt^2, ou y = ddx/dt^2 = dv/dt. Ce qui fait encore une Règle y = dv/dt, qui avec la précédente v = dx/dt, satisfait à tout ce qu'on propose icy de résoudre. *AM*, 1700 (1703), 23.

56. Ibid.

57. Auguste Comte, *Philosophie premiere, cours de philosophic positive, leçons I à 45* (Paris: Hermann, 1975), 268.

58. Usage. Je dis présentement qu'une des six courbes cy-dessus, étant donnée à discrétion, on pourra toujours en déduire les cinq autres par le moyen de ces deux

Règles, supposé les résolutions et les intégrations nécessaires des égalités en question. *AM*, 1700 (1703), 23.

59. Les mêmes choses se trouveront de la même manière dans toute autre hypothèse; il n'y aura de différence que la difficulté du calcul, laquelle n'auroit fait qu'embarasser icy. Ainsi ces deux exemples suffisent. Ibid., 26.

60. I have given a more detailed analysis of some questions treated here in my book *Les 'Principia' de Newton* (Paris: Presses Universitaires de France, 1995).

9

THE NEWTONIAN STYLE IN BOOK II OF THE *PRINCIPIA*

George E. Smith

At first glance Book II looks very different from the rest of the *Principia.* In a work famous for the thesis that gravity is an inverse-square force, Book II in all but a couple of places treats gravity as constant above the surface of the earth, in the manner of Galileo. In a work noted for eschewing hypotheses of the speculative sort Descartes had put forward, Book II frames just such hypotheses about the microstructure of fluids. In a work dominated by a sustained line of argument that starts with Proposition 1 of Book I on central forces and the area rule and ends with Proposition 41 of Book III on comet trajectories, Book II appears to be merely a digression, save for a not entirely successful effort in its final section to argue that Cartesian vortices are incompatible with Keplerian motion. Indeed, in a work legendary for its successes, Book II is best known for its failures. This is what lies behind the situation that Clifford Truesdell ridiculed when he said of Book II, "This is the part of the *Principia* that historians and philosophers, apparently, tear out of their personal copies."[1]

Notwithstanding all of this, I am going to argue that the impression that Book II is different from the rest of the *Principia* is wrong. Newton was engaged in exactly the same sort of enterprise in Book II as he was in Books I and III. The differences in Book II do not stem from his having adopted a different method of empirical inquiry. That is, they do not stem from his having compromised or abandoned what I. Bernard Cohen has called "the Newtonian style." Book II is different because the empirical world deflected his inquiry down a different sort of path. Moreover, precisely because the empirical world did not cooperate in the same way, Book II sheds a good deal of added light on the Newtonian style.

Cohen's account of Newton's style in Books I and III emphasizes the way in which the mathematical theories put forward form a sequence of successive approximations. Newton starts with a one-body idealization—central forces directed toward a fixed point in space—in Sections 2 through 10 of Book I, and then in Section 11 proceeds first to a two-body idealization, in which an interacting body replaces the central point, next to an attempt at a three-body idealization, and finally in Sections 12 and 13 to mutual forces among all particles. Each successive idealization extends the one preceding it by dropping an assumption that simplifies the mathematics.[2]

The claim that Newton saw empirical science as progressing by successive approximations seems to me exactly right. I want to embellish Cohen's account in only two minor respects. First, even the simplest of the idealizations was more than just a mathematical stepping stone to the next. Newton used the idealizations to draw conclusions from phenomena in Book III. Often these conclusions took the form of inferred measures of quantities. For example, he inferred the masses of the sun, Jupiter, and Saturn and the value of the exponent of *r* in the law of celestial centripetal force by means of the one-body idealization.[3] In these cases, Newton was using his mathematical theories in much the way Huygens had previously used theory to infer the strength of gravity from pendulum measurements. The stability, mutual agreement, and precision of the inferred measures provided empirical support for the idealization employed in the inference. Sometimes the idealizations opened the way to conclusions in which phenomena provided an *experimentum crucis*, or crossroads experiment. For example, the contrasting versions of Kepler's 3/2 power rule in the one-body and the two-body idealizations allowed phenomena to select between singly directed and mutually interactive central forces, and the pattern of deviation from Kepler's other rules implied by the three-body idealization allowed phenomena to select between pairwise interaction between central and orbiting bodies, on the one hand, and mutual interaction among all bodies in the planetary system, on the other. Regardless of the kind of inference he was drawing, Newton was using the idealizations as instruments in marshaling evidence.

My second embellishment of Cohen's account concerns an empirical requirement that the idealizations were expected to satisfy. Not just any approximation will allow empirical investigation to proceed by successive approximations. Some of the residual discrepancies between each idealization and observation must themselves be convertible into compelling evidence for the next, more refined idealization. In Newtonian terms, some of the residual discrepancies must themselves amount to phenomena[4]—higher-order phenomena, as it were—from which conclusions can be drawn, especially conclusions establishing the next idealization in the sequence of successive approximations. Lunar inequalities provide the most obvious example in which residual discrepancies become phenomena from which Newton drew evidential conclusions; the variations of surface gravity with latitude and the discrepancies between these variations and the idealization of a uniformly dense earth provide two more.[5] So long as the sequence of idealizations forming the successive approximations meet this requirement, the sequence can be empirically driven. The idealizations are thus required to be a special sort of approximation. Not just any approximation will yield discrepancies when compared with observation that can then be marshaled into evidence for the next approximation. Most approximations can only lead down what will ultimately prove to be a garden path on which further progress becomes stymied when the residual discrepancies no longer define phenomena that can be converted into evidence.

When I speak of the "Newtonian style," then, I am referring to an approach involving a sequence of idealizations, each of which is used to draw conclusions from phenomena, and which together comprise successive approximations in which residual discrepancies between theory and observation at each stage provide an evidential basis for the next stage. I claim that this style is no less central to Book II than it is to the rest of the *Principia*. The fact that Book II was less successful than the rest brings out ways in which this style requires the empirical world to cooperate.

The Central Concern of Book II: From Phenomena of Motion to Investigate the Forces of Resistance

According to the preface to the first edition, Newton's central concern in Books I and III was to find the centripetal forces acting on celestial bodies from phenomena of celestial motion. The analogous central concern of Book II was to find the resistance forces acting on bodies from phenomena of motion in fluid media. The investigations of Book II started from the same point of departure as those in Books I and III, namely, uniformly accelerated motion, whether in the form of uniform circular motion or Galilean "naturally accelerated" local motion. Deviations from uniform circular motion provided the primary basis for conclusions about celestial forces, and deviations from Galilean local motion were expected to provide the primary basis for conclusions about resistance forces.

If the goal of the mathematical theory of motion in fluids developed in Book II was to enable conclusions about resistance forces to be drawn from phenomena, then the fact that Book II does not begin with, or virtually ever consider, inverse-square gravitational forces is in no way anomalous.[6] The difference between inverse-square gravity at the surface of the earth and 300 feet above it is only 1 part in 100,000. Surface gravity was being measured at the time to, at most, 1 part in 10,000.[7] Taking inverse-square variations in gravity into account was accordingly a meaningless complication when trying to draw conclusions from differences between local motions in resisting fluids and their idealized, resistance-free counterparts. Even if reason were to arise for taking this variation into consideration, it is such a small fraction of the resistance forces that it could just as well be compensated for by introducing a small correction to the constant value used for the strength of gravity.[8] We must not lose sight of the fact that the fifty-three demonstrated propositions comprising the mathematical theory of Book II, just like the ninety-eight comprising the mathematical theory of Book I, are all "if-then" propositions. They are only inference tickets, the first use of which, Newton tells us in the preface, is to reach conclusions about forces from phenomena of motion.

In the case of celestial forces, the general mathematical framework that Newton adopted as a working hypothesis was that the forces are centripetal and their strength is some sort of function of distance from the center, $f(r)$. Although one section examines centripetal forces that vary as $(a/r^2 + b/r^3)$, and another examines forces that vary arbitrarily with r, Book I focuses on forces that vary as $1/r^2$, as r, and (on occasion) as $1/r^3$. In the case of resistance forces, the general mathematical framework Newton adopted as a working hypothesis was that the forces are directed oppositely to the motion, and their strength is some sort of function of velocity, $f(v)$—more precisely, of the relative speed between the body and the fluid.[9] The mathematical efforts of Book II focus on motion under resistance forces that vary as v and as v^2, although Newton allows for resistance forces that vary as v^0 as well, pointing out that they merely produce a special case of uniformly accelerated motion.

Even though he developed mathematical solutions for each separate type of resistance force, Newton was far more concerned with the superposition of the different types of force in the resistance case than he was in the celestial case. His approach in Book II is accordingly best described as proceeding under the working hypothesis that resistance forces can be represented within an idealized mathematical framework that superposes three distinct terms:

$$F_{\text{RESIST}} = \mathbf{a}_0 + \mathbf{a}_1 v + \mathbf{a}_2 v^2 \tag{9.1}$$

As we shall see, at one point he allows this framework to be expanded to include other terms involving v to a power, integral or otherwise. The primary framework, however, is given by (9.1).

The empirical problem in Book II is to draw conclusions about $\mathbf{a}_0$, $\mathbf{a}_1$, and $\mathbf{a}_2$ from phenomena of motion. For any single body in any specific fluid medium, each of these three should have a unique value.[10] Thus one conclusion to draw is the magnitude of each of the three for specific combinations of body and medium. The value of each of the three, however, is sure to vary from one combination of body and medium to another. In modern terminology, $\mathbf{a}_0$, $\mathbf{a}_1$, and $\mathbf{a}_2$ are undoubtedly functions of properties of the body and properties of the fluid medium. The problem is to identify what these properties are in the case of each $\mathbf{a}_i$ and how it varies with these properties. Newton's solution in the analogous instance of celestial forces was that they vary as m_1, m_2, and r^2, so that $F_{\text{CENTRAL}} = Gm_1m_2/r^2$, where G is strictly a constant. A similar solution is wanted for resistance forces. The empirical problem of Book II is to establish this solution from phenomena of motion.

In the last two propositions of the tract *De Motu*, in which he took an initial stab at this problem, Newton considered only resistance forces that vary as v, with acceleration serving as the measure of force.[11] He concluded that the deceleration of spheres under such resistance varies with the density of the medium and the sphere's surface area, and inversely with its weight.[12] In other words,

$$(dv/dt)_{\text{RESIST}} = c_1\rho_f A_{\text{surf}}v/W_s, \tag{9.2}$$

where $(dv/dt)_{\text{RESIST}}$ is the deceleration from fluid resistance, ρ_f is the density of the fluid medium, A_{surf} is the surface area of the sphere, W_s is its weight, and c_1 is a constant. Newton offered no argument for (9.2). He did, however, provide a method of inferring c_1 from the observed incidence angle of a spherical projectile, given its angle and velocity of projection. So long as the same value of c_1 is obtained for different spheres with different initial conditions in different fluids, the empirical world can provide the missing argument.

Newton reconsidered resistance forces that vary as v in the first section of Book II. He there dropped all mention of how the leading coefficient for the v term, our $\mathbf{a}_1$, varies from body to body and fluid to fluid. Indeed, nowhere in any edition of the *Principia* does he ever say anything about how this particular leading coefficient might vary.[13] The only step in this direction is a slightly more tractable method for inferring the magnitude of $\mathbf{a}_1$ for a given body and medium, using the difference in range for two different angles of projection instead of the angle of incidence. No sooner does he give this method, however, than he denies its applicability, for the Scholium following it announces "that the resistance encountered by bodies is in the ratio of the velocity is a hypothesis belonging more to mathematics than to nature. In mediums wholly lacking in rigidity, the resistances encountered by bodies are as the squares of the velocities."[14] I know of no document that indicates whether Newton reached this conclusion through experiment, as Huygens had some twenty years earlier (without at the time publishing), or through qualitative physical reasoning.[15]

The physical reasoning Newton offered in the *Principia* concludes that the fluid's inertia produces a resistance force that varies as the square of the velocity. The easiest-to-follow version of this reasoning construes the fluid as composed of uniform particles. Inertial resistance arises from the impacts between these particles and the body. The number of such impacts in a unit of time varies as the number of particles per unit volume in the fluid, the frontal area of the body, and the velocity. The force transmitted per impact varies as the product of the mass of the particles and the velocity. Putting these together, the resistance associated with the fluid's inertia varies as the fluid's density, the square of the velocity, and the body's frontal area, which in the case of a sphere varies as the square of the diameter.

Based on this reasoning, Newton's working hypothesis in the *Principia* was that resistance forces can be represented, at least to a first approximation, within a mathematical framework involving the following three terms:

$$F_{\mathrm{RESIST}} = \mathbf{a}_0 + \mathbf{a}_1 v + \mathbf{b}_2 \rho_f A_{front} v^2, \tag{9.3}$$

where ρ_f is again the density of the fluid medium, A_{front} is the frontal area, and the coefficient $\mathbf{b}_2$ may perhaps vary with shape. In the case of spheres, (9.3) can be rewritten as

$$F_{\mathrm{RESIST}} = \mathbf{a}_0 + \mathbf{a}_1 v + \mathbf{c}_2 \rho_f d^2 v^2, \tag{9.4}$$

where c_2 is presumed to be strictly a constant. The empirical problem Newton set himself was first to confirm the $\rho_f d^2$ constituents of the v^2 term and then to characterize $\mathbf{a}_0$, $\mathbf{a}_1$, and $\mathbf{c}_2$—if not $\mathbf{b}_2$—from phenomena of motion.

Engineers still frequently employ a mathematical framework very similar to that given in (9.3) and (9.4) when modeling resistance. They normally set $\mathbf{a}_0$ to 0 on the basis of the "no slip" hypothesis, that is, the hypothesis that there is no relative motion between the fluid and the body at the body's surface and hence no friction in the strict Coulomb sense of the term. The v term is then sometimes spoken of as "friction" or "internal friction," though more often and more correctly it is called the "viscous" term.[16] Stokes's solution for the purely viscous flow about a sphere is generally used to characterize $\mathbf{a}_1$. Finally, by the product of a constant and an appropriate value of an empirically determined coefficient, called the "drag coefficient", for bodies of the shape in question replaces $\mathbf{c}_2$ or $\mathbf{b}_2$. For a sphere, then, engineers often model resistance as

$$F_{\text{RESIST}} = 3\pi\mu\, dv + (\pi/8)c_D\rho_f\, d^2v^2, \tag{9.5}$$

where d is the sphere's diameter, μ is the fluid viscosity, and c_D is an empirically determined value of the drag coefficient appropriate to the range of conditions in question. The v^2, or inertial, term in such a model corresponds entirely to Newton's working hypothesis. Newton, however, appears to have had no inkling of the specifics of the viscous term. For that matter, we should scarcely have expected him to, for these specifics are not at all apparent from qualitative physical reasoning. Stokes obtained them through an explicit solution of an idealized version of the Navier-Stokes partial differential equations for fluids, a version in which the fluid lacks all inertia.[17]

Even though engineers still use this two-term model for resistance, it has been known for most of this century to be not only inexact, but also misleading. The approach that now prevails models resistance simply as

$$F_{\text{RESIST}} = (1/2)C_D\rho_f A_{\mathit{front}}v^2, \tag{9.6}$$

where C_D, the drag coefficient, is a nondimensional parameter that, for a body of a given shape, is a function only of the Reynolds number, R_E. The Reynolds number represents the ratio of inertial to viscous effects in a flow. As came to be realized 200 years after Newton, the Reynolds number is the fundamental nondimensional parameter in fluid mechanics. Subject to qualifications that we can ignore here, any two flows involving similar geometries that have the same Reynolds number are entirely similar, regardless of scale.[18] When the geometry consists of flow around a sphere (or, what amounts to the same thing, a sphere moving in an unbounded fluid) the Reynolds number is given by $\rho_f\, dv/\mu$.

We still have no way of deriving C_D as a function of R_E from first principles for any shapes at all, even for the simple case of a sphere. We must rely on empirical results. Figure 9.1 displays the empirically determined drag coefficient for spheres as a function of Reynolds number. Think of the curve as dividing into three domains: (1) a low–Reynolds number domain, where to a first approximation the drag coefficient varies inversely with velocity, so that the resistance force varies as v; (2) an intermediate, or transition, domain in which the Reynolds number is between, say, 400 and 200,000 and the drag coefficient remains within a narrow band, so that the resistance force to a first approximation varies as v^2; and (3) a high–Reynolds number domain in which the flow becomes fully turbulent and the drag coefficient first drops abruptly, then rises.

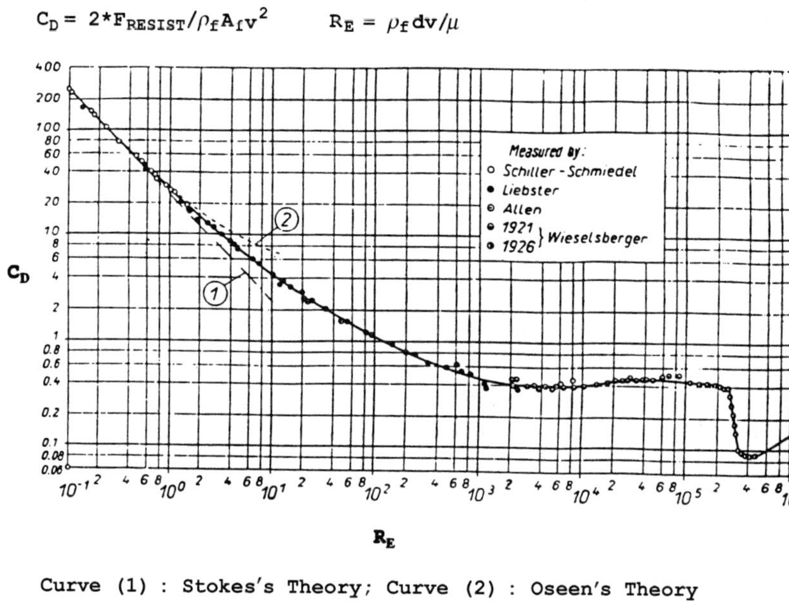

Figure 9.1
Drag coefficient versus Reynolds number for spheres. From Hermann Schlichting, *Boundary-Layer Theory*, trans. J. Kestin, 7th ed. (New York: McGraw-Hill, 1979), 17.

No two-term engineering model of resistance can begin to represent the abrupt drop in resistance at the critical Reynolds number that marks the beginning of this third domain. The situation over the other two domains, however, is less straightforward. Figure 9.2 compares the drag coefficient given by the two-term engineering model for spheres (9.5) with the empirically determined values. At first glance, the approximation is not that bad.[19] Notice, however, how the empirically determined curve dips to a local minimum at a Reynolds number around 4,000 and then rises to a local maximum a little below 100,000. No two-term model in v and v^2 can reproduce this variation. Although the variation looks small on a log-log plot, it amounts to almost 50% in magnitude, from a low around 0.38 to a high near 0.55. This is the regime of everyday macroscopic objects moving at everyday velocities. As such, it was the regime in which Newton was looking for phenomena that would allow him to draw conclusions about resistance forces with the help of his three-term mathematical framework. Because of this, although he had no way of knowing it, he was in trouble from the outset.

To understand why the C_D curve in figure 9.2 varies in the peculiar way shown above $R_E = 1000$, one needs to appreciate that the other

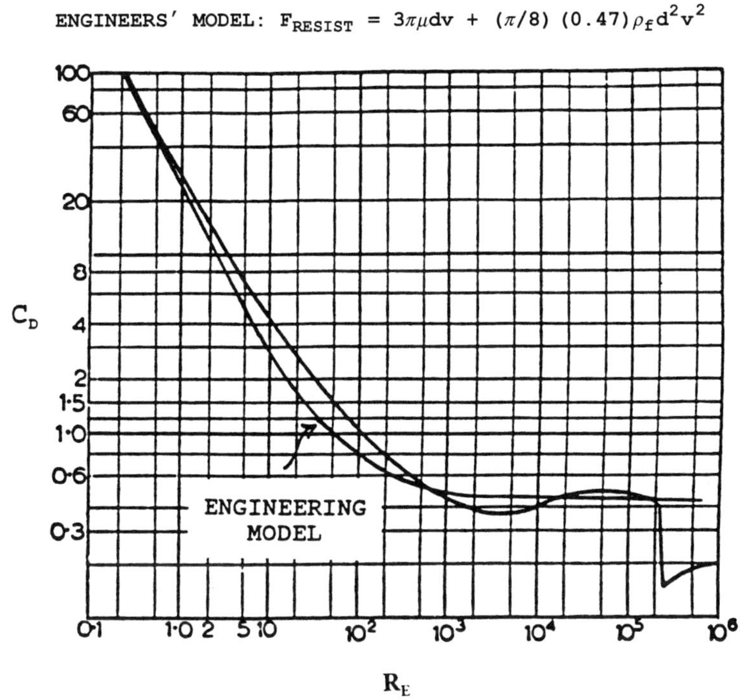

Figure 9.2
Comparison between engineers' model for C_D on spheres and real data.

factor controlling resistance, besides $\rho_f d^2 v^2$, is the character of the flow immediately downstream of the sphere. This in turn is governed by the development of the boundary layer over the front portion of the sphere and the location where this boundary layer separates from the surface, creating the wake behind it. Figure 9.3, taken from Feynman's lectures,[20] illustrates this in the case of cylinders. (The case of spheres is not so easy to depict in two dimensions.) Boundary layer development and separation are viscous flow phenomena. Thus the variation of C_D above $R_E = 1000$ in figure 9.2 indicates that there is no such thing as a purely inertial mechanism of fluid resistance. The resistance arising from a fluid's inertia fluid depends fundamentally on its viscosity too![21]

THE FIRST EDITION: NEWTON'S PENDULUM DECAY EXPERIMENTS

Newton needed two things to extract answers about resistance forces from phenomena of motion. First, he had to have a mathematical solution

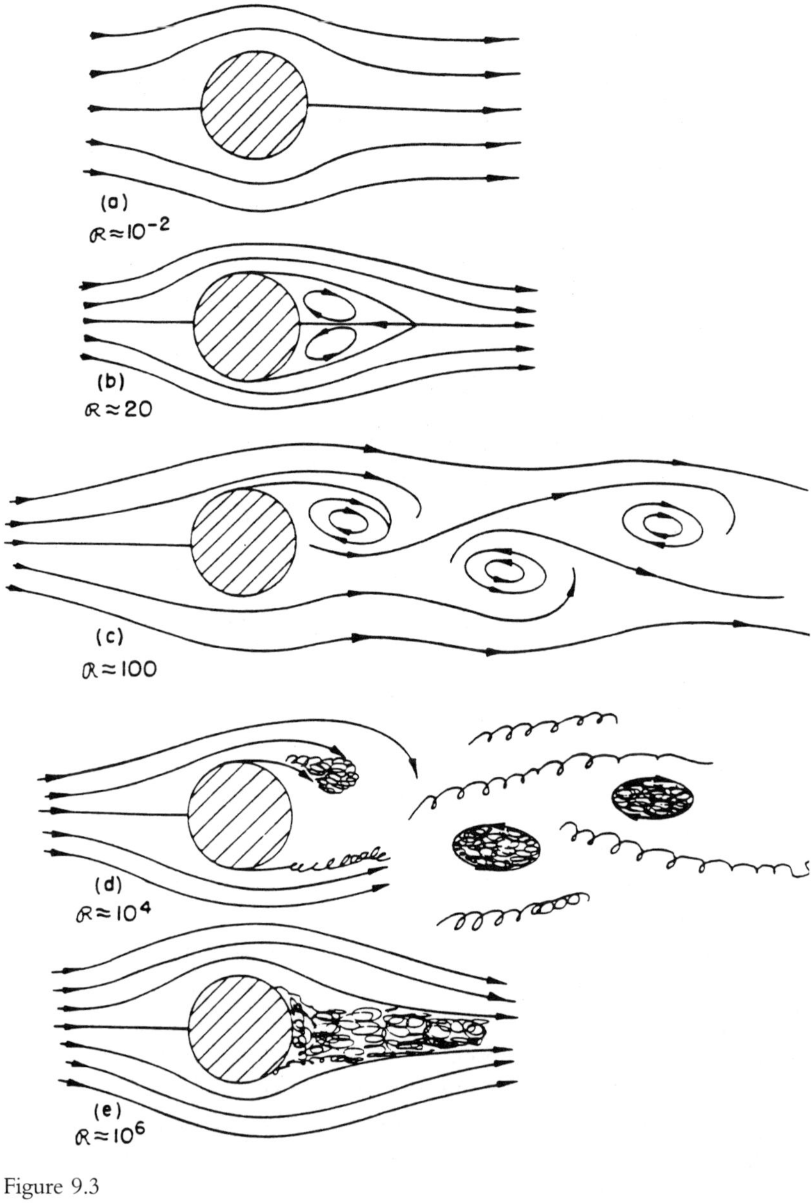

Figure 9.3
Flow past a cylinder for various Reynolds numbers. From Richard P. Feynman, Robert B. Leighton, and Matthew Sands, *The Feynman Lectures on Physics* (Reading, MA: Addison-Wesley, 1964), 41–48.

for the motion in question that would allow him to correlate features of it with the parameters in his resistance framework. Second, if two or three distinct physical mechanisms contribute to resistance, as Newton's framework presupposed, then the phenomena had to allow the separate contributions to be disaggregated at least to the point where conclusions could be reached about each separate term in the framework. Direct vertical fall takes place too quickly to have much promise of meeting this second requirement; imprecision in measuring the time of fall was likely to introduce too much variance when trying to split the total effect into separate parts. Furthermore, Newton's inability to solve the projectile problem for resistance varying as v^2 eliminated any possibility of using projectile motion, as he had proposed in *De Motu*.

Newton offered an ingenious solution to this problem in the first edition of the *Principia*. Air resistance's primary effect on a pendulum is a gradual decay of the length of the arc through which it is moving.[22] In Propositions 30 and 31 Newton managed to derive a systematic relationship between the amount of arc lost per swing, nondimensionalized with respect to the pendulum's length, and the ratio of the resistance force on the bob to its weight. Thus, if part of the lost arc is proportional to a resistance-as-v effect, part proportional to a resistance-as-v^2 effect, and so on, the total lost arc in a single swing can be represented as a sum:

$$\delta_{\text{arc}} = A_0 + A_1 V_{\max} + A_2 V_{\max}^2 + \cdots + A_n V_{\max}^n, \tag{9.7}$$

where the A_i are constants for a given pendulum. Under the assumption that the resistance is weak enough for the ratio between the velocity along the arc and the maximum velocity to remain virtually the same as in the unresisted case, the solution in Proposition 30 allows the resistance force at the point of maximum velocity to be represented by a corresponding sum:

$$F_{\text{RESIST}}/\text{WEIGHT} = [(1/2)A_0 + (2/\pi)A_1 V_{\max} + (3/4)A_2 V_{\max}^2 + \cdots](1/l). \tag{9.8}$$

And in general, if part of the lost arc is proportional to v^n, then, whether n is an integer or not, the corresponding contribution to the total resistance is given by[23]

$$\left[0.5 \Big/ \int_0^1 (1 - x^2)^{n/2}\, dx\right] A_n V_{\max}^n. \tag{9.9}$$

So, once the A_i are determined, the separate components of the resistance force associated with different exponents of velocity can be determined.

Now, V_{max} in a cycloidal pendulum, ignoring resistance, is proportional to the arc length. Hence by starting the pendulum at different points and measuring the arc lost in each case, the values of the A_i can be inferred via simultaneous algebraic equations: for example,

$$\begin{aligned} \delta_{arc} &= A_0 + A_1 V_{max} + A_2 V_{max}^2, && \text{for a 4-inch arc;} \\ \delta_{arc} &= A_0 + 4A_1 V_{max} + 16A_2 V_{max}^2, && \text{for a 16-inch arc;} \qquad (9.10) \\ \delta_{arc} &= A_0 + 16A_1 V_{max} + 256A_2 V_{max}^2, && \text{for a 64-inch arc;} \end{aligned}$$

where V_{max} is now the maximum velocity for a 4-inch arc in descent. Because $A_i V_{max}^i$ always has units of length, setting V_{max} to 1.0 amounts to nothing more than a temporary choice of units for the A_i. Thus so long as the resistance does not significantly alter the velocity ratios along the arcs, Newton had a mathematical solution for the resisted motion of pendulums that allowed him to infer values of the A_i and hence the magnitude of each component of the resistance force from the amounts of arc lost by a given pendulum. The phenomenon of pendulum decay appeared to satisfy the two requirements given earlier for inferring resistance forces from phenomena of motion.

Newton used a twelve-foot-long circular pendulum in some preliminary trials.[24] The primary pendulum, again circular, for which data are given in the *Principia* was 126 inches long, with a $6\frac{7}{8}$ inch diameter spherical wooden bob weighing $57\frac{7}{22}$ ounces.[25] He set up a curved ruler 5 inches above the center of oscillation to measure the arc lengths. Because the amount of arc lost in any one swing was too small to measure, he counted the number of arcs until one-eighth or one-quarter of the initial arc was lost and divided this total by the number of swings to obtain the average arc lost per swing. Table 9.1 gives Newton's published data for this pendulum, first with one-eighth of the total arc lost, and then with one-quarter. As Newton remarked, the ratios of the lost arcs in the consecutive cases, given in the far right column of the table, show that at high velocities (i.e., large arcs) the total resistance is tending to vary nearly as the square of the velocity; at lower velocities, by contrast, there is a noticeable additional contribution, with a velocity exponent below two. In other words, the ratios of the arcs lost in the different trials are qualitatively in accord with Newton's mathematical framework; moreover, no resistance component involving an exponent of velocity above two is needed.

So far so good. The problem comes when one tries to infer values of the A_i from these data.[26] Because the data are redundant, values of the A_i

Table 9.1
Newton's Data for the Basic Pendulum Experiment in Air

Arc (in.)	Lost arc (in.)	Number of oscillations	Average δ_{arc} (in.)		Relative V	Consecutive ratios
2	1/4	164	1/656	(1.524E − 3)	1/2	
						2.71
4	1/2	121	1/242	(4.132E − 3)	1	
						3.51
8	1	69	1/69	(1.449E − 2)	2	
						3.89
16	2	35.5	4/71	(5.634E − 2)	4	
						3.83
32	4	18.5	8/37	(2.162E − 1)	8	
						3.82
64	8	9.67	24/29	(8.276E − 1)	16	
2	1/2	374	1/748	(1.337E − 3)	1/2	
						2.75
4	1	272	1/272	(3.677E − 3)	1	
						3.34
8	2	162.5	4/325	(1.231E − 2)	2	
						3.90
16	4	83.3	12/250	(4.800E − 2)	4	
						4.00
32	8	41.67	24/125	(1.920E − 1)	8	
						3.68
64	16	22.67	48/68	(7.059E − 1)	16	

can be obtained from different combinations of them and then compared to ascertain whether the experiment is yielding strictly constant values for the given pendulum. Newton ended up trying this with four mathematical frameworks:

$$A_0 + A_1 V + A_2 V^2 \tag{9.11a}$$

$$A_1 V + A_2 V^2 \tag{9.11b}$$

$$A_{1/2} V^{1/2} + A_1 V + A_2 V^2 \tag{9.11c}$$

$$A_1 V + A_{3/2} V^{3/2} + A_2 V^2 \tag{9.11d}$$

None worked. In the case of the first framework, the values of A_2 vary by less than $\pm 4\%$, but A_0 turns out to be negative with many data combinations, and the values of A_1 vary by more than a factor of 5.[27] The second, or two-term framework, which Newton preferred initially, does not yield any negative values of A_i. The values of A_1, however, vary by more than a factor of 2.2 over combinations from the data in which one-eighth of the arc was lost, and by more than a factor of 6 over combinations in which one-quarter of the arc was lost. Furthermore, although the values of A_2 vary by less than $\pm 3\%$ in the case of the one-eighth-arc-lost data, and by less than $\pm 6\%$ in the case of the one-quarter-arc-lost data, there is a 14% difference in the values between the two sets of data. The values Huygens and others had inferred for the strength of gravity from different pendulum measurements varied in the fourth significant figure. The values Newton had inferred for the masses of Jupiter and the sun varied in the third significant figure, depending on which orbiting body was used in the inference. By contrast, the inferred values of A_2 were varying in the second significant figure, and the values of A_1 in the first. Something was wrong.

Newton never announced this in the *Principia*. Instead, he adopted the fourth framework, with the $V^{3/2}$ power term, and gave values of the A_i based on the second, fourth, and sixth data points in the one-eighth-lost-arc data. I do not know why he chose to do this. He never gave any reason, nor did he offer any explanation at all for the $V^{3/2}$ term. Most of the data combinations imply a negative value for $A_{3/2}$, and the values of both A_1 and A_2 vary more than they do in the case of the two-term framework. The only redeeming feature of his choice that I have been able to ascertain is that the quoted values of A_1 and A_2 lie very near the middle of the different values inferred from the different data combinations and different frameworks. No less oddly, Newton subsequently remarked that the one-quarter-lost-arc data are more accurate and invited the reader to carry out the calculation of the supposed constants for them. Any reader who did so discovered that the value of A_2 obtained from the corresponding second, fourth, and sixth data points from these data is almost 13% lower and the value of A_1 is 44% lower than the values Newton obtained from the one-eighth-lost-arc data.

Regardless of why Newton presented the findings in the way he did, the basic facts are clear. The results of the pendulum experiments do not begin to allow reliable conclusions to be drawn about contributions to resistance involving an exponent of v less than 2. Not surprisingly, therefore, in the remainder of the discussion of the pendulum experiments, Newton focused on the v^2 term. Even here, however, his results were disappointing. When he replaced the $6\frac{7}{8}$ inch diameter wood bob with a 2 inch diameter lead bob, the inferred v^2 resistance force decreased by a factor of $7\frac{1}{3}$ instead of a factor of $11\frac{13}{16}$, as it should have if this force varies as d^2. Newton decided from this that resistance acting on the string was having a large effect in the case of the smaller bob. To support this, he tried an $18\frac{3}{4}$ inch diameter bob, finding the v^2 resistance 7 times larger, instead of the 7.438 that a d^2 variation implied.

Similarly, when he tried a $134\frac{3}{8}$ inch long pendulum with the bob first in air and then submerged in a trough of water, he found that the total resistance was 571 times larger in water at high velocities, versus the 850 times larger that he expected it to be under the assumption that the v^2 resistance effect that dominates at high velocities varies with the fluid's density. He attributed the discrepancy here to the fact that the string was not immersed in water, so that the two experiments were not parallel. He also initially found that the resistance force in the water experiments was increasing in a proportion greater than v^2. He attributed this discrepancy to his having used too narrow a trough for the size of the bob.[28] A further experiment with a smaller bob supported this explanation.

In sum, in spite of the ingenuity of their conceptual design, Newton's pendulum experiments were largely a failure. Newton concluded from them that the v^2 component of resistance is dominant at higher velocities in both water and air, and that once suitable allowances are made for shortcomings in the experiments, this component varies as ρd^2, at least for spheres. Even this conclusion, however, has to be restated in a more qualified manner all too familiar to those who find themselves having to rely on simple hypothesis testing. The correct statement is that the pendulum experiments *did not clearly falsify* the claim that the dominant component of resistance on spheres at high velocities can be expressed in the form $c\rho d^2v^2$. The results of the experiments nevertheless failed to yield a stable value of the constant c. Nor did they yield any conclusions at all about the other component (or components) of resistance that become more prominent at low velocities.

What went wrong? Frankly, there are so many difficulties in the experiments that I still have not been able to pinpoint the effects of each and thereby to reproduce Newton's data.[29] As remarked above, in the first edition of the *Principia* Newton pointed to resistance on the string as the main problem. By the second edition, however, he had come to realize (from experiments discussed below) that the resistance forces on the bob inferred from the pendulum experiments were much too large—not just the total resistance, but the v^2 component as well. This led him to conclude that the pendulum's motion was generating a to-and-fro fluid flow, so that the relative velocity between the bob and the fluid tended to be significantly greater than supposed in the undisturbed-fluid assumption adopted in the experiment's conceptual design. (A 30% higher relative velocity would increase the v^2 resistance effect by a factor of 1.7.)[30]

Some factors not evident to Newton undercut the experiment's conceptual design. The string had to be affixed to the bob. Even a small protuberance on the sphere could have been enough to trip the boundary layer locally, altering the flow in the wake and thereby making the resistance artifactually larger than it would be on a perfect sphere.[31] The boundary layers on the bob in the smaller arc trials fall into a flow regime that is especially sensitive to such an effect.

This may well explain one of the most misleading aspects of Newton's data, which appear to be indicating a component of resistance that varies with v (or at least a power of v less than 2) at low velocities. One might naturally associate this component with viscous effects. In fact, however, the $6\frac{7}{8}$ inch diameter bob was much too large for the purely viscous contribution to resistance to be detectable at velocities above 2 cm/sec; the average velocity in the case of the smallest (2-inch) arc was above 5 cm/sec. As Stokes showed 150 years later, inertial effects mask the purely viscous contribution to resistance unless the fluid has an exceptionally high viscosity or the sphere's diameter is extremely small. Because this is not at all evident from qualitative physical reasoning, Newton had no way of realizing that any reasonable choice of bob size was going to make the purely viscous contribution to resistance far beyond empirical access in his pendulum experiments.

Accordingly, in spite of the ingenuity of their conceptual design, Newton's pendulum experiments never had any chance of yielding information about $\mathbf{a}_0$ or $\mathbf{a}_1$ in his mathematical framework. There were so many confounding factors that the experiments also had virtually no chance of yielding reasonably precise values for the constant c in the $c\rho\, d^2v^2$ term. In spite of the great effort Newton put into these experiments between 1685 and 1687, he himself appears to have given up on them after the first edition of the *Principia*. Unless I am mistaken, he never put any subsequent effort into trying to save them. If so, he showed good judgment.

Book II, Section 7: From the First Edition to the Second

In the first edition the General Scholium presenting the results of the pendulum experiments came at the end of Section 7. In the later editions, although the General Scholium still covered all the results (with some numbers reworked a little), Newton eliminated 20% of it, mostly on points of detail, and shifted it to the end of Section 6. Its main contribution there was to corroborate that the v^2 component of resistance is dominant even at relatively modest velocities and that this component appears to vary as the fluid's density and, in the case of spheres, as the square of their diameter. Newton rewrote the last half of Section 7 in the second edition and added a Scholium comparable in length to the revised

General Scholium at its end, presenting the results of a new set of experiments on resistance, involving the phenomenon of vertical fall.

My primary concern here is with these experiments and the new propositions lying behind them that were introduced in the rewrite of Section 7. Lest I be accused, however, of owning a copy of the *Principia* in which all the scandalous parts have been removed, I had best say something about Section 7 generally. Section 7, in all three editions, contains a large fraction of what became most notorious in the *Principia*: such efforts, falling under Truesdell's categories of "Conjecture, Error, and Failure," as the solid of least resistance, the efflux problem, and various wildly wrong models of the microstructure of fluids. Nothing in the *Principia* provides more grounds than Section 7 for arguing that Newton did not have a new or distinctive way of doing science. I had best counter this mistaken impression before turning to my main point.

No part of the *Principia* did Newton rewrite more extensively from the first to the second edition than Section 7; for all practical purposes, he replaced the second half of the section in its entirety. In all editions Section 7 raises questions that the experiments presented immediately following it were expected to answer. But key questions concerning the action of the fluid on the rear of moving bodies that the pendulum decay experiments were taken to be answering in the first edition gave way to very different questions for the vertical-fall experiments to answer in the subsequent editions, and consequently Newton simply dropped the portions of the General Scholium on pendulum decay that answered these questions when he shifted it to the end of Section 6.[32] Although the material on fluid resistance that disappeared in the second edition can be pieced together from the Latin[33] with some effort, it has never been available in one location, much less in English. Translations have therefore been provided in the appendix to this chapter of the replaced half of Section 7 and the conclusions eliminated from the General Scholium. In some respects this material from the original version of Book II is better than the later versions in bringing out the way in which Newton was using theory to develop questions for experiments on fluid resistance to answer.

Section 7 of Book II is the counterpart to Sections 12 and 13 of Book I. In both books, Newton first obtains a wide range of results for motion under different types of macroscopic force, then turns to how these macroscopic forces might arise out of microstructural effects. Sections 12 and 13 present some twenty-three if-then propositions, generally of the form, "if the forces among particles of matter vary as the inverse square of the distance between them (or any inverse power, or proportional to the distance), then the central force acting toward the body composed of these particles has such and such a property." With one notable exception, the efflux problem, Section 7 similarly presents a series of if-then propositions of the form "if the fluid is composed of particles of matter of a certain type, then the resistance force on a body arising from the inertia of the fluid has such and such a property." Section 7 deals only with this one type of resistance force. It includes no propositions concerning forces arising from the "tenacity and friction" of the fluid—for example, ones that vary as v (or v to a power less than 2).

How are such propositions about microscopic action producing macroscopic effects different from Descartes's propositions about aetherial matter? Newton gives the answer in the scholium immediately preceding Section 12 of Book I:

> By these propositions we are directed to the analogy between centripetal forces and the central bodies toward which those forces tend. For it is reasonable that forces directed toward bodies depend on the nature and the quantity of matter of such bodies, as happens in the case of magnetic bodies. And whenever cases of this sort occur, the attraction of the bodies *must be reckoned by assigning proper forces to their individual particles and then taking the sum of these forces. . . .*
>
> Mathematics requires an investigation of those quantities of forces and their proportions that follow from any conditions that may be supposed. Then coming down to physics, these proportions must be compared with the phenomena, so that it may be found out which conditions of forces apply to each kind of attracting bodies. *And then, finally it will be possible to argue more securely concerning the physical species, physical causes, and physical proportions of these forces.* Let us see, therefore, what the forces are by which spherical bodies, consisting of particles that attract in the way already set forth, must act upon one another, and what sorts of motions result from such forces.[34]

Thus the hope is that the if-then propositions will enable inferences to be drawn from macroscopic phenomena about the forces among microscopic particles.[35] Inferences reached this way will be more than mere conjectures, more than Cartesian "fictions."[36]

In Proposition 8 of Book III Newton gives a second reason for pursuing if-then propositions relating macroscopic centripetal forces to microstructural ones: Propositions 75 and 76 of Section 12 and their corollaries were the basis for his concluding that the gravitational forces between two suitably homogeneous spheres vary *exactly* as the inverse square of the distance between their centers, and not just approximately so.[37] This is not the place for me to go into the logic underlying the inference claimed here. We should nevertheless note that one of Section 7's central concerns is whether the resistance arising from the inertia of fluids varies exactly as $\rho\, d^2 v^2$ or only approximately so.

Section 7, in all three editions, distinguishes among three theoretically defined types of fluid. A *rarified* fluid consists of particles spread out in space, with inertial resistance arising from impacts between these particles and macroscopic bodies, as if the body were moving through debris in empty space. An *elastic* fluid is a rarified fluid with repulsive forces among the particles and thus between the particles and the body as well.[38] A *continuous* fluid consists of particles so packed together that each particle is in contact with its neighbors.

The first half of Section 7, which remained essentially the same in the first and later editions, gives results for elastic and rarified fluids. Newton derives the requirement that the repulsive forces in elastic fluids have to satisfy for the resistance to vary exactly as $\rho\, d^2 v^2$; he shows that the resistance automatically varies exactly as $\rho\, d^2 v^2$ in the absence of such repulsive forces, that is, in rarified fluids; and he argues that even if the repulsive forces do not satisfy the requirement in an elastic fluid, the resistance still varies nearly as $\rho\, d^2 v^2$ at reasonably high velocities. His result on the solid of revolution of least resistance is for rarified fluids only and is an afterthought to a proposition that the resistance on a cylinder moving in the direction of its axis in such a fluid is twice the resistance on a sphere of the same diameter, this owing to less force being transmitted when a particle collides with a body in a glancing blow than when it collides head on. Notice that should Newton's solid of least resistance turn out not to be truly the one of least resistance in a given fluid, then that fluid does not fall into his rarified category. Newton clearly put his proposal for the solid of least resistance forward in the hope that it would prove useful. Nevertheless, to take a contrary empirical finding as refuting his entire theory is, from a strictly logical standpoint, to ignore the if-then structure of the claim.

The second half of Section 7, following the Scholium on the solid of least resistance, acquired a somewhat different aim in the second edition from what it had in the first. The main results in both editions derive magnitudes of inertial resistance forces from idealized assumptions about the microstructural mechanisms giving rise to them in the different types of fluids. The aim in the first edition is to allow experimental measurements of macroscopic resistance forces to shed light on microstructural and fluid flow mechanisms entering into inertial resistance. In the second edition the aim becomes one of obtaining the magnitude of the inertial component of the resistance a priori, from theoretical considerations alone. The impression that Newton is engaged in a Cartesian-style exercise of hypothesizing microstructural mechanisms and confirming them by experiment is consequently much stronger in the two later editions.

In the case of rarified fluids, the microstructural mechanism underlying inertial resistance is motion transferred from the moving body to the individual fluid particles with which the body collides. The conclusion, as stated in the second edition, is that if the particles rebound perfectly elastically, then a "sphere encounters a resistance that is to the force by which its whole motion could be either destroyed or generated, in the time in which it describes 2/3 of its diameter by moving uniformly forward, as the density of the medium is to the density of the sphere." If, instead, the impinging particles are not reflected, the inertial resistance is half this amount, and if the particles are reflected between these extremes, the resistance is between these two values. In modern terms, this is equivalent to a drag coefficient C_D of 2.0 if the particles are maximally reflective, 1.0 if they do not reflect at all, and between these values otherwise. The force on a cylinder of the same diameter moving in the direction of its axis is twice this amount. These results are based purely on the impact between the particles and the body; in particular, they presuppose that no action is occurring among the particles, altering the fluid ahead of the body or, for that matter, behind or to the sides of it.

The same mechanism is assumed for rarified fluids in the first edition, and, although the result obtained is stated in a different manner, it is fully equivalent to this one. In the first edition, however, the claim is that the resistance on the body will be of this magnitude *approximately*. A corollary indicates that the actual resistance will be somewhat greater, especially at slow speeds, owing to non-inertial effects. A further corollary then suggests a program: first determine the extent of the elastic rebound, and hence the rule giving the magnitude of the inertial resistance, by experiments at high speeds in rarified fluids, then turn to the difference between the magnitudes given by this rule and observed resistance at slower speeds to obtain a rule for the non-inertial resistance.

Continuous fluids pose a more challenging problem, for resistance in their case cannot be treated simply in terms of individual particles' impact. Instead, Newton treats inertial resistance in continuous fluids as arising from the moving object's having to push a certain amount of fluid in front of it, or, with the body still and the fluid moving, as the force of a certain amount of fluid impinging, like a weight, on the object. The efflux

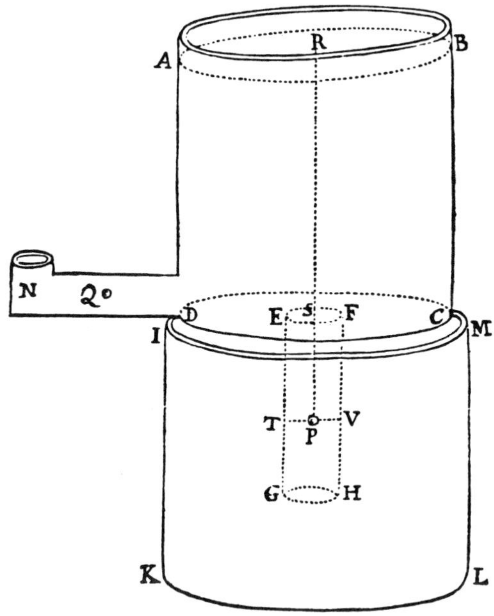

Figure 9.4
The efflux problem diagram from the first edition of the *Principia*.

problem enters in conjunction with this. Picture the fluid passing around the sphere in the cylindrical channel in figure 9.4, in which Newton displays a forerunner of a modern wind tunnel.[39] Part of the fluid forms a "stagnation" column above the sphere, so that the sphere supports the fluid's weight, which amounts to a resistance force. The goal of the solution for this force in the first edition was to determine the part of the inertial resistance force acting on the body's forward surface. As a step to this end, Newton solved the efflux problem, determining the relationship between the velocity in the cylindrical channel with no sphere present and the fluid's height. The solution he gave for the efflux problem in the first edition, however, was mistaken.[40] He corrected the solution in the second edition. (Truesdell did not exaggerate when he said that the argument Newton offered for the solution in the second edition has more the character of bluff than proof.)[41]

In the first edition Proposition 38 gives the magnitude of the inertial force on the forward surface of a moving sphere in a continuous fluid. In a corollary, Newton remarks that this magnitude is the total inertial resistance force if the sphere "while moving is urged from the rear by the same force as when it is at rest.... But if while moving it is urged from behind less, it will be more retarded; and conversely, if it is urged more, it will be less retarded." In a passage in the General Scholium giving the results of the pendulum experiments, a passage deleted in the second edition, he then indicates that the magnitude of the force obtained in Proposition 38 is around three times greater than the magnitude of the v^2 term obtained in the pendulum experiments in water.[42] Based on this, the passage concludes that the fluid acting on the moving sphere's back contributes an important inertial force counteracting that resisting the motion so that the net inertial resistance comes from two opposing effects.

Accordingly, the result given in the first edition for the inertial resistance acting on the body's forward surface represents an idealized first approximation to the overall inertial resistance. The idea was to compare this force with the inertial component of the force obtained in the pendulum experiments in water to shed light on the effects of further inertial mechanisms in continuous fluids, just as a similar comparison with experiments in a rarified fluid might shed light on the extent of the fluid particles' elastic rebound. Newton's theory of inertial resistance in the first edition was merely a stepping stone to getting important information from the pendulum experiments. Keep in mind, however, that the pendulum experiments originally had the promise of disaggregating the inertial and the other components of resistance.

The extraordinary change in the second half of Section 7 in the second edition lies in Newton's attempt to derive the precise magnitude of the inertial resistance force in continuous fluids from theoretical considerations alone. Proposition 38 gives the result: "the resistance to a sphere moving uniformly forward in an infinite and non-elastic compressed fluid is to the force by which its whole motion could either be destroyed or generated, in the time in which it describes 8/3 of its diameter, very nearly as the density of the fluid to the density of the sphere." This is equivalent to saying that the drag coefficient C_D amounts to 0.5, "very nearly." Furthermore, the force on a cylinder of the same diameter moving in the direction of its axis is the same, for the sphere and cylinder have the same frontal area, and this is the only feature of the shape that matters in the case of the impinging column of fluid ahead of the body in a continuous fluid.[43]

In spite of the great trouble to which Newton went to obtain this result for continuous fluids, his derivation is far from rigorous. A crucial step, for example, involves determining the weight of the fluid resting on the circular disk shown in figure 9.5. Newton argued first (in Corollary 7

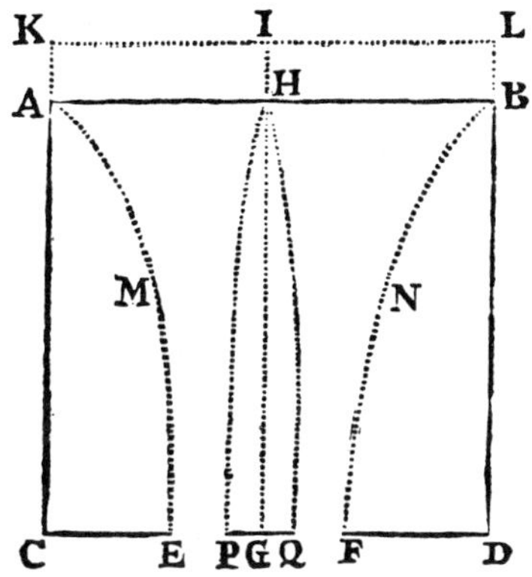

Figure 9.5
The column of "continuous" fluid borne by the moving body (Corollaries 7–9, Proposition 36, Book II, in the second and third editions).

of Proposition 36) that this weight is greater than one-third of the cylindrical column above the disk; next (in Corollary 8) that it is less than two-thirds of it; and then concluded (in Corollary 9) that it must be "very nearly" one-half of it.[44] We learn a few pages later that he already had some preliminary results for vertical fall of spheres in water, so this choice was most likely empirically informed. The same cannot be said for the arguments he offers that various other factors affect the force on the disk so little that they can be ignored in deriving the magnitude of the force —specifically, the disturbance of the flow upstream of the disk caused by its presence, the three-dimensional effects arising from the obliquity of the flow going around it, and the circulation of flow immediately downstream of it. All three of these factors are in fact important, especially the third. That Newton should dismiss it in the second edition after emphasizing it in the first is striking.

Why did Newton go to all this trouble in the second edition in order to add an a priori rule for calculating the magnitude of the inertial resistance force on spheres? Why was he willing to compromise his usual standards of rigor just to obtain specific values of this force without having to infer them from the phenomena? He gives the answer in the middle of Proposition 40 when he remarks, "This is the resistance that arises from the inertia of matter of the fluid. And that which arises from the elasticity, tenacity, and friction of its parts can be investigated as follows." Given the value of c in the $c\rho\, d^2v^2$ term, the inertial component of resistance can be calculated as a function of v beforehand. Proposition 9 of Book II can then be used to calculate the precise time a sphere takes to fall from a given height through a fluid medium, if this is the only form of resistance acting on it.[45] The resistance arising from mechanisms other than the fluid's inertia can then be investigated by contrasting the actual time of fall with this calculated time, for the difference between them stems from these other effects. Vertical fall takes place too rapidly to try to disaggregate the different mechanisms of resistance by means of a simultaneous-equation solution of the sort tried with the pendulum. If, however, the dominant v^2 term can be calculated precisely beforehand, and the sphere "encounters another resistance in addition, the descent will be slower, and the quantity of this resistance can be found from the retardation."

The problematic revision to Section 7 in the second edition was accordingly Newton's response to the failure of the pendulum experiments to provide any empirical basis for reaching conclusions about $\mathbf{a}_0$ and $\mathbf{a}_1$. The revision opened a pathway to using the phenomena of vertical fall to reach such conclusions by way of a sequence of successive approximations. I consider whether this was a reasonable and sound way for getting at the non-inertial components of resistance below. But it should already be clear that the revision of Section 7 in the second edition did not involve a total departure from the Newtonian style.

The Second and Third Editions: Newton's Vertical-Fall Experiments

The Scholium introduced at the end of Section 7 in the second edition gives the results of thirteen vertical fall experiments that Newton carried out in the years between the two editions, twelve in water plus one in air in which globes were dropped from the top of the inside dome of the virtually completed St. Paul's Cathedral. A fourteenth experiment was added in the third edition, a second in air, again in St. Paul's, but carried out by Newton's protegé, J. T. Desaguliers, who dropped globes from the lantern atop the dome, as before to the floor below. Figure 9.6 suggests that the experiments in St. Paul's involved considerable effort.

A David Gregory memorandum of 1694 provides some background to these vertical-fall experiments: "He [Newton] does not sufficiently trust the observations obtained from a pendulum to determine the ratio of the resistance, but he will subject everything afresh to the test of falling weights. He is choosing the place for contriving his experiments from the top of Trinity College Chapel into his own garden on the right as one enters the College."[46] Whether Newton carried out such an experiment

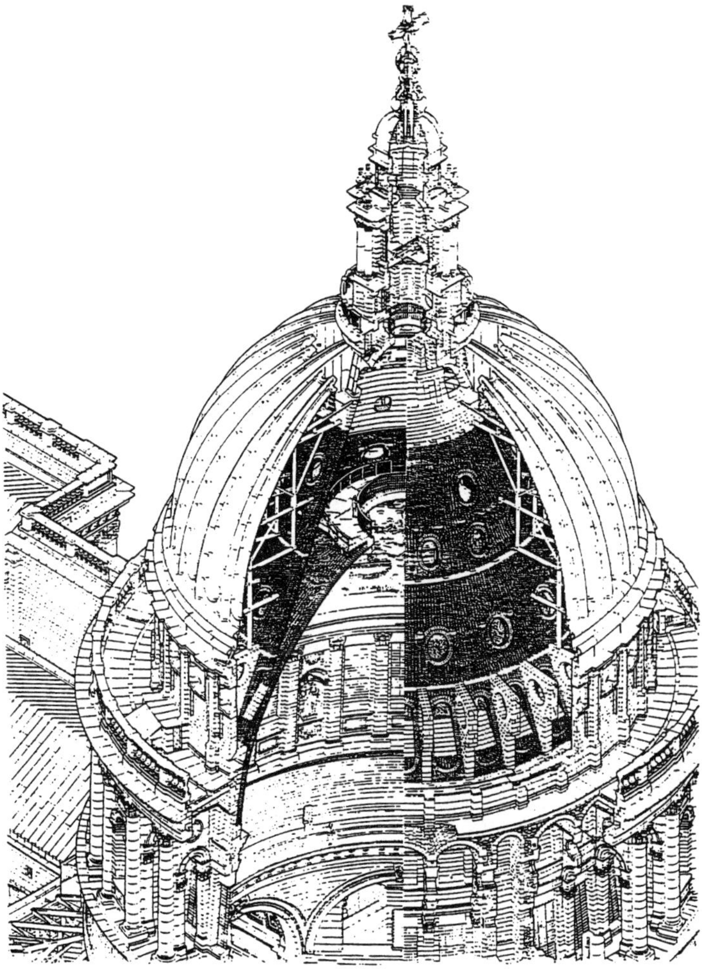

Figure 9.6
A cutaway drawing of the dome of St. Paul's. Spheres were dropped from the platform at the top of the inner dome in the Newton-Hauksbee experiment of 1710 and from the base of the lantern, where the railing is shown, in Desagulier's experiment of 1719.

in Cambridge is doubtful. The first experiment in St. Paul's was carried out in 1710, with Francis Hauksbee actually conducting it and reporting it to the Royal Society.[47] The experiments in water fall into two distinct groups, using separate equipment. I have been unable to determine when these experiments were carried out. The first group was almost certainly carried out before Newton left for London in 1696,[48] and the other, after 1707, while putting together the second edition.[49]

Like his pendulum experiments, Newton's vertical fall experiments merit a more extended discussion than is appropriate here.[50] The main concern here is with the data and the conclusions Newton drew from them.

Newton tells us that he carried out the first three experiments in water "in order to investigate the resistances of fluids before formulating the theory set forth in the immediately preceding propositions." They were done in a trough $9\frac{1}{2}$ feet deep and 9 inches square. The nine subsequent experiments were done in a $15\frac{1}{2}$ foot trough, $8\frac{2}{3}$ inches square. The descending spheres were made of wax with lead inside. They were on the order of 1 inch in diameter, varying in size and especially in weight from experiment to experiment. Descent was timed using a half-second pendulum. The greater density of water gives it two advantages over air for purposes of such experiments: buoyancy reduces relative gravity and hence the effective weight of objects falling in it, and the resistance forces acting on the objects are enhanced. As a consequence, the terminal velocity is lower and the time of fall is therefore longer. Water, however, does not have a further advantage that one might expect it to have: it does not make the viscous resistance forces more obtrusive. Although water is more viscous than air, its ratio of viscosity to density is 17 times less than that of air. Because the falling objects Newton used in his water experiments were comparatively small and the terminal velocity was lower, the Reynolds numbers in these experiments were lower than in his experiments in air, but not so low that the inertial contribution to resistance no longer masked the direct viscous contribution.

Table 9.2 summarizes the data from the experiments in water. In three of the experiments, Newton measured the time of descent and then compared the actual distance with the distance the object should have fallen in that time according to the theory. In the others he compared measured times with the times of descent according to the theory. The spreads in the time data shown in the table stem from the multiple trials of the experiments. The agreement between theory and observation in the first seven experiments in table 9.2 is quite good. In Experiments 6, 9, and 10, however, in which the velocities are comparatively high, and hence

Table 9.2
Newton's Data for the Vertical-Fall Experiments in Water

Experiment	Diameter (in.)	Weight (grns.)	Buoyant weight (grns.)	Actual time (sec.)	Theoretical fall (in.)	Actual fall (in.)
1	0.892	156.3	77.0	4.0	113.06 112.08*	112.0
2	0.813	76.4	5.1	15.0	114.07 113.17*	112.0
4	0.999	139.3	7.1	25.0 (23.5–26.5)	184.25 181.86*	182.0

Experiment	Diameter (in.)	Weight (grns.)	Buoyant weight (grns.)	Actual fall (in.)	Theoretical time (sec.)	Actual time (sec.)
5	1.00	154.5	21.5	182.0	14.5	14.8–16.5
8	0.999	139.2	6.5	182.0	26.0	25.0–26.0
11	0.693	48.1	3.9	182.5	23.3	21.8–23.0
12	1.010	141.1	4.4	182.0	32.3	30.5–32.5
6	1.00	212.6	79.5	182.0	7.5	7.5–9.0
9	0.999	273.5	140.8	182.0	5.7	6.0–6.5
10	1.258	384.4	119.5	181.5	7.8	8.9–9.5
7	1.247	293.7	35.9	181.5	14.0	14.8–16.5
3	0.967	121.0	1.0	112.0	40.0	46, 47, 50

* After correction based on Proposition 39.

the inertial resistance should be even more dominant, the theory underpredicts the time by significant amounts. In other words, the resistance appears to be greater at high velocities than the theory would have it. Newton attributes this to (1) oscillation of the spheres during descent and (2) "the swifter the balls, the less they are pressed by the fluid in back of them." Finally, the last two experiments in the table Newton discounts entirely: Experiment 3 because the weight of the spheres in water was so small that the experiment was too sensitive to confounding factors, including inaccuracies in the measured weight, to be reliable, and Experiment 7 because the large spread in time convinced him that some of the balls were oscillating.[51]

In the experiment of 1710 in St. Paul's "glass balls were dropped simultaneously in pairs, one filled with quicksilver, the other full of air," from a height of 220 feet. An elaborate device was installed allowing the fall to be triggered from the ground, in the process starting a pendulum timer. Even so, Newton found it necessary to correct the raw data to allow for a small delay in the balls becoming free of the platform on which they were resting near the ceiling. In Desaguliers's experiment of 1719, hogs' bladders formed into spheres and filled with air were dropped, simultaneously with lead spheres as controls, from a height of 272 feet. Both half-second and quarter-second pendulums, as well as a quarter-second spring clock, were used to measure the difference in the time of descent between the two types of balls; the time required for the lead balls to fall was then measured separately and added to this difference, thus eliminating the timing fault in the earlier experiment in air. Also, the longer time of fall in Desaguliers's experiment reduced the sensitivity to timing errors. As I confirm below, Desaguliers's experiment was of very high quality, by any standards.

Table 9.3 is Newton's summary of the results of the two experiments in air. The times listed for his experiment are after the correction; the observed times ranged from 8 to $8\frac{1}{2}$ seconds. The theoretical values presuppose that air is a continuous fluid. The discrepancies between the theoretical distances and the actual 220 feet of fall in this experiment, which are less than 5%, amount to less than one-quarter of a second in time. In Desaguliers's experiment the agreement between theory and fact is spectacularly good in the case of the three heavier bladders and still within 4% for the two lighter ones. Newton tells us that the fifth bladder "was somewhat retarded by" wrinkles in its surface and hence should be discounted. We should also note that each bladder was dropped twice, and the times shown in the table represent the descent in which the bladders fell most in a straight line, without oscillating "to and fro."

How good were these vertical fall data? The best way to determine this is to ignore Newton's theory and instead determine the values of the drag coefficient C_D implied by the results. Figure 9.7 plots as functions of Reynolds number the drag coefficients for all but the five experiments in water that Newton distrusted. The two results in air, with a Reynolds number of 40,000 and C_D ranging from 0.499 to 0.518 for Desaguliers's and a Reynolds number of 78,000 and C_D ranging from 0.504 to 0.538 for Newton's, show remarkably good agreement with modern values.[52] The results in water are a little on the high side, perhaps due to sidewall effects from the fairly narrow troughs or from free-surface effects at the beginning of descent. Regardless, the C_D values range between 0.462 and 0.519. Even more significant, the lowest value, 0.462, occurs at a Reynolds number where the modern measured value reaches a mini-

Table 9.3
Newton's Data for the Vertical Fall Experiments in Air

Experiment 13: Newton (1710)

Globorum pondera.	Diametri.	Tempora cadendi ab altitudine pedum 220.		Spatia describenda per theoriam.		Excessus.	
510 gran.	5,1 dig.	8″	12‴	226 ped.	11 dig.	6 ped.	11 dig.
642	5,2	7	42	230	9	10	9
599	5,1	7	42	227	10	7	10
515	5	7	57	224	5	4	5
483	5	8	12	225	5	5	5
641	5,2	7	42	230	7	10	7

Experiment 14: Desaguliers (1719)

Vesicarum pondera.	Diametri.	Tempora cadendi ab altitudine pedum 272.	Spatia iisdem temporibus describenda per theoriam.		Differentia inter theor. & exper.	
128 gran	5,28 dig.	19″	271 ped.	11 dig.	−0 ped.	1 dig.
156	5,19	17	272	$0\frac{1}{2}$	+0	$0\frac{1}{2}$
$137\frac{1}{2}$	5,3	$18\frac{1}{2}$	272	7	+0	7
$97\frac{1}{2}$	5,26	22	277	4	+5	4
$99\frac{1}{8}$	5	$21\frac{1}{8}$	282	0	+10	0

mum. The experiments themselves were therefore of high quality. One can make a case that no better data for resistance on spheres over this range of Reynolds numbers were published before the twentieth century.[53]

This assessment is from our modern vantage point. The only indication Newton had that the data were good was the close agreement between theory and observation in the times and distances of fall. Based on this agreement, he announced the following conclusion: "Therefore almost all the resistance encountered by balls moving in air as well as water is correctly shown by our theory, and is proportional to the density of the fluids—the velocities and sizes of the balls being equal."[54] In particular, as he went on to say, these data provide a much better basis for this conclusion than the data from the pendulum experiments do. They also demonstrate that the resistances measured in the pendulum experiments were excessive.

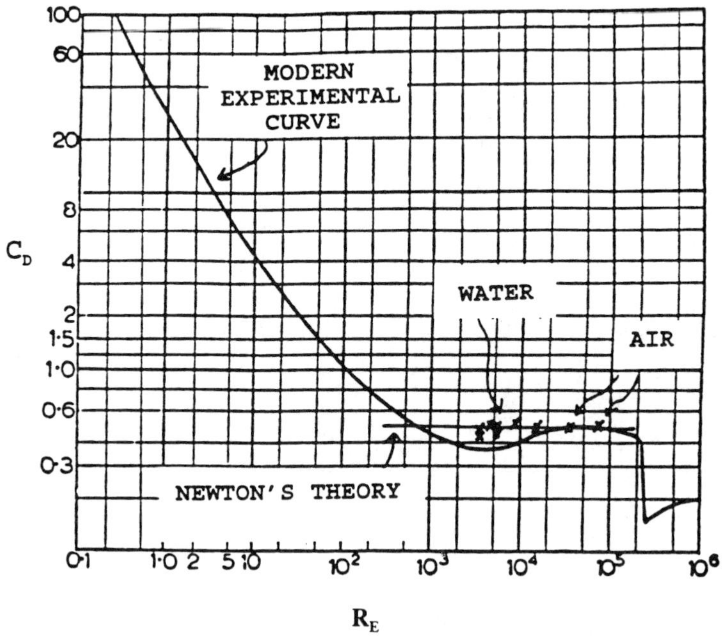

Figure 9.7
A modern assessment of the vertical-fall data from the second and third editions of the *Principia*.

Although this is not what he said in the passage just quoted, one might naturally take Newton to be concluding that the vertical-fall data confirm his theoretical account of inertial resistance in continuous fluids, namely as arising from the impinging column of fluid in front of the moving sphere. This was surely the way many at the time read him.[55] We, however, should not be so hasty in adopting this reading. Newton had to have realized just how tenuous the "proofs" of Propositions 36 through 39 were, and hence also just how loose the derivations of the theoretical values for the times and distances of fall were. Furthermore, he was never inclined to put much weight on evidence consisting of a successful hypothetico-deductive test of a theory, because he worried that too many theories could fit the same body of observations.[56] Finally, and most important of all, from the optics papers of the 1670s on, Newton had insisted on a sharp distinction between the claims that he thought his experiments established and conjectural hypotheses that went beyond the evidence in proffering explanations of these claims. This distinction was at the heart of Newton's entire approach to empirical inquiry.[57] Nothing in the way that he stated the conclusion he drew from the vertical-fall experiments gives any reason for thinking that he had abandoned this distinction here.

Consider what his vertical fall results do show. First, at least over the range of conditions the experiments covered, the resistance force on a sphere varies to a predominant extent as v^2 in both air and water. Second, it likewise varies as the fluid's density. Third, at least to a first approximation, the resistance force on spheres in both air and water varies to a predominant extent as $c\rho d^2v^2$, where the constant c is at least close to Newton's value for continuous fluids of $\pi/16$. Fourth, so far as the resistance force is concerned, air is not so different a sort of fluid from water as the contrast between Newton's rarified and continuous fluids would suggest; in particular, resistance in air does not act in the way that it would if air were a rarified medium.[58] Fifth, because the resistance force varies to a predominant extent in the same way in both air and water, two fluids that appear to have very different tenacity and internal friction, and in both cases it varies as the density, the resistance force on a sphere over the range of conditions tested arises predominately from the fluid's inertia and not from any other aspect of it. Sixth, one candidate for the microstructural mechanism giving rise to most of the resistance force over the range of conditions tested is an impinging column of fluid ahead of the moving sphere.

Missing from this list is Newton's earlier remark that the differences between the theoretical and the observed results of the vertical-fall experiments were to provide a means to investigate the contributions to resistance made by such factors as fluid friction and tenacity. Newton never returns to this proposal. The data provide ample reasons for not returning to it. The close agreement between the theory and the data indicates that these other factors are at most secondary over the range of conditions tested. The data, however, have too much variance to begin serving as a basis for investigating secondary factors. In some cases, moreover, especially Experiments 8 and 11, the observed results are indicating a resistance *less* than that implied by the theory. Fluid friction and tenacity can scarcely be producing negative resistance forces. These anomalous results must therefore be viewed as indicating timing inaccuracies of magnitude comparable to the effects of these factors. Finally, the excessive resistances in Experiments 6, 9, and 10, in which the velocities in water were highest, raise the possibility that the effects of secondary *inertial* factors ignored in the theoretical model—in particular, the effect of the fluid acting on the sphere's rear surface—may be of magnitude comparable to the effects of the non-inertial factors. Newton's theory of inertial resistance in continuous fluids is expressly only an idealized first approximation, ignoring factors like the effects of the flow at the back of the moving body. The theoretical model may have to be refined to include secondary inertial factors, or be supplemented by corrections for them, before it can serve as a basis for reaching conclusions from vertical-fall data about the contributions of fluid friction and tenacity.

At the very least, then, more refined experiments, preferably over lower velocity ranges, were going to be necessary to turn vertical-fall discrepancies into evidence concerning the effects of fluid friction and tenacity.[59] Nevertheless, if part of the resistance arises from a single, dominant inertial mechanism, and at least on spheres this part varies strictly as $c\rho\, d^2v^2$, where c is $\pi/16$, then Newton had good reason to believe that more refined, better controlled vertical-fall experiments would allow an investigation of these other effects. One can easily imagine data that would have been far less promising in this regard.

This conclusion can be put in another way. Consider the following thesis: the resistance force on spheres consists of multiple components, one of which, given by $(\pi/16)\rho\, d^2v^2$ and dominant at reasonably high velocities, arises from the fluid's inertia. Let me call this a "working hypothesis," by which I mean that further research was to be predicated on it, not just in a heuristic sense, but in the sense that evidential reasoning in this further research would indispensably presuppose it. Then the results of the vertical-fall experiments provided strong evidence that this working hypothesis was promising. Adopting it offered a way of gaining empirical access through vertical-fall experiments to the contributions to resistance made by fluid friction and tenacity. True, this working hypothesis provided only an idealized approximation to one component of resistance. The results of the vertical-fall experiments nevertheless showed that deviations from it had the promise of providing evidence for the next approximation, in which factors beyond inertia could be incorporated. The evidence from the vertical-fall experiments thus had the logical force of confirming the viability of Newton's theory as a working hypothesis.

Notice, moreover, that predicating further research on this working hypothesis was safe. There was little risk of its leading down an extended garden path. Suppose that deviations from it were not going to yield clear, stable, convergent measures for terms representing other factors contributing to resistance. Then nothing but time and effort was going to be lost in predicating research on it. If, on the other hand, clear, stable, convergent measures were to emerge, then significant further evidence would end up accruing to the working hypothesis, tending to transform its status from provisional to entrenched.

Newton's microstructural model of a continuous fluid contributed to this working hypothesis in two ways. First, this model provided reason to think that the $(\pi/16)\rho d^2v^2$ term is something more than a mere approximation akin to a curve fit. An indefinite number of alternative approximations of this sort can hold to a given level of accuracy. The deviations from most of them, however, cannot be relied on to be providing evidence about other mechanisms contributing to resistance. The microstructural model gave a possible physical mechanism, involving nothing but fluid inertia, that would yield this term exactly. Hence deviations from it had a legitimate claim to being physically instructive.

Second, Newton's model was a candidate for being a point of departure for theorizing about microstructural mechanisms involved in fluid friction and tenacity. In the case of the inertial contribution to resistance, the mechanism must involve some sort of momentum transfer between the moving body and the fluid. But what of the mechanism or mechanisms involved in fluid friction and tenacity? Newton put forward no suggestion in the *Principia* about what sorts of actions among the particles comprising a fluid might give rise to them. And well he should not have, for he had no empirical basis for doing so.[60] The continuous-fluid impingement model can constrain theorizing about such further actions once some phenomena are isolated that enable conclusions to be drawn about the macroscopic consequences of these actions. If the continuous-fluid impingement model cannot be augmented or refined to incorporate the further actions among fluid particles needed to account for the macroscopic effects of fluid friction and tenacity, then the model will have proven to be a dead end. If, on the other hand, it can be so augmented and refined—even more, if it can be so augmented and refined in an empirically driven manner—then this fact itself will be further evidence accruing to it as a microstructural theory of the dominant mechanism governing inertial resistance.[61]

In other words, Newton's continuous-fluid model is itself best regarded as a working hypothesis. The vertical-fall results showed it too to be promising as an idealized first approximation to the physics underlying fluid resistance. It was also safe. If vertical-fall deviations turned out not to yield an empirical basis for extending it to include actions among particles that give rise to resistance forces from fluid friction and tenacity, then nothing but time and effort was going to be lost in predicating further research on it. If, however, these deviations turned out to yield an empirical basis for extending it, then as a working hypothesis it would open an evidential pathway for reaching empirical conclusions about the microstructure of fluids from phenomena of motion under resistance. As Newton remarked in the Scholium leading into Sections 12 and 13 of Book I, quoted earlier, "then, finally, it will be possible to argue more securely concerning the physical species, physical causes, and physical proportions of these forces." What remained was to carry out the next phase of the empirical investigation and see whether the results fulfilled the promise.

The *Principia* offers no clear textual support for this interpretation of the logical force of the evidence obtained in the vertical-fall experiments. In the preface to the second edition, Newton says, "In Section 7 of Book 2, the theory of the resistance of fluids is investigated more accurately and confirmed by new experiments."[62] As remarked earlier, Newton's contemporaries tended to interpret him to be saying that the vertical-fall experiments established his impingement theory of resistance. That is, they took him to be engaged in a Cartesian-like exercise of devising a hypothesis about the microstructural basis of resistance and then looking to its deductive consequences in vertical fall to establish its truth, pure and simple. On any careful examination of Section 7 and the vertical-fall data, this sort of interpretation cannot escape the conclusion that Newton was surely overreaching his evidence, if not engaging in something of a bluff. The interpretation I have offered has an important virtue beyond its avoiding this conclusion. In its generic form, it offers a coherent logic for marshaling empirical evidence if scientific inquiry is taken to be proceeding in a sequence of successive approximations. Thus regardless of whether Newton ever verbalized this logic, there are grounds for arguing that it is part of the Newtonian style.

REFUTATION, REJECTION, AND PARADOX: THE AFTERMATH OF BOOK II

The final two sections of Book II proceed independently of the results on resistance in Sections 1 through 7. Section 8 treats the disturbances propagating through fluids caused by moving bodies, especially the waves produced by oscillating bodies, yet the account given there in no way depends on the forces at the interface between the moving body and the fluid. Section 9 similarly considers the vortex motion rotating cylinders and spheres engender in a fluid. It argues that the vortex that the resistance forces at the surface of a spinning sphere produce—the vortex mechanism proposed in Descartes's *Principia*—is incompatible with Kepler's 3/2 power rule. These resistance forces, however, are ones associated with fluid friction and tenacity, ones that Newton has been unable to reach any conclusions about in the preceding sections. So he is forced to open Section 9 with a hypothesis: "The resistance which arises from the lack of lubricity or slipperiness of the parts of a fluid is, other things being equal, proportional to the velocity with which the parts of the fluid are separated from one another." In other words, the viscous force is proportional to the velocity gradient normal to the direction of the fluid's motion: the definition of what is now known as a "Newtonian fluid."[63] Having to rely on a hypothesis in delivering what was intended to be a coup de grace to Descartes must have been a great disappointment to Newton.[64]

Anyone who tried to pursue the next stage of the investigation into resistance forces outlined above would not have succeeded. By extending the conditions of the experiments on spheres, someone might conceivably have developed evidence that $\mathbf{a}_0$ is 0 and that $\mathbf{a}_1$ varies more or less as the diameter d in Newton's $\mathbf{a}_0 + \mathbf{a}_1 v + \mathbf{a}_2 v^2$ framework. In the process, he might have obtained the modern engineering model for resistance on spheres. But the discrepancies between measured resistance forces and this two-term model in the range of Reynolds numbers between 500 and 50,000, not to mention the sudden drop in resistance above the critical Reynolds number, cannot be handled within any such two-term model. Of course, one can always approximate the drag coefficient versus Reynolds number curve to any degree of accuracy desired if constraints on curve fitting are removed. Still, returning to an earlier point, not any old approximation can meet the requirements for proceeding in successive approximations in the Newtonian manner. An arbitrary curve fit is scarcely going to provide empirical conclusions concerning aspects of the microstructure of fluids and the forces among the particles composing them. The mathematical framework Newton proposed at the outset for empirically investigating resistance forces was doomed to come to a dead end at the next level of approximation beyond the one Newton achieved in Section 7.

This, however, was not why Newton's account of inertial resistance was rejected historically. De Borda performed a series of experiments in the 1760s on resistance in air and water, mostly using a rotating rig with the resisted body moving around the circumference (see figure 9.8). He first found that Newton's account of the shape of least resistance in rarified fluids does not conform in any way with water or air.[65] He later

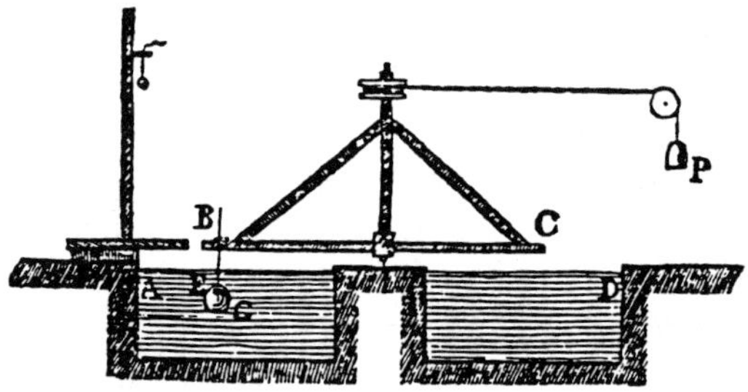

Figure 9.8
De Borda's primary rig for testing resistance forces.

found that the resistance on a hemisphere with its flat face moving directly into the fluid is more than twice that on a sphere of the same diameter.[66] According to Newton's continuous-fluid impingement model, the dominant inertial resistance should depend only on frontal area and not on shape. De Borda applied his coup de grace at the end of his 1767 paper: "The ordinary theory of the impact of fluids only gives relationships which are absolutely false and, consequently, it would be useless and even dangerous to wish to apply this theory to the craft of the construction of ships."[67] So far as I know, all interest in the magnitude of the resistance forces Newton obtained in his vertical fall experiments waned at this juncture, as, of course, would have been appropriate if those experiments' sole purpose had been to develop hypothetico-deductive evidence for his theory of the microstructural mechanism underlying resistance.

Furthermore, a far deeper problem with the general approach that Newton had pursued emerged in 1752, fifteen years before de Borda's results on the effect of shape. In his *Hydrodynamica* of 1738, Daniel Bernoulli had presented a proper solution to Newton's efflux problem, employing the principle that came to be known as Bernoulli's formula.[68] According to this formula, which he derived from the conservation of *vis viva,* the pressure along any stream tube decreases as the velocity of the fluid in it increases. Picture flow around a sphere then, with stream tubes forming symmetrically about it. The velocity along the sphere's surface should accelerate from the forwardmost point to the point of maximum breadth and then decelerate until it reaches the free-stream velocity at the rearmost point. Since the sphere is symmetric, the velocity variation around its surface must also be symmetric. So too the pressure at the sphere's surface must vary symmetrically around it. But the resistance force is simply the integral of this pressure around the surface. So in the absence of fluid friction and tenacity, the resistance force on a sphere is exactly zero.

This is a special case of a result d'Alembert published in 1752: if a body of any shape is moving through a purely inviscid, nonrotating flow—what came to be known as an *ideal* fluid—then the resistance force on it is exactly zero.[69] In other words, the correct solution to the problem Newton was trying to solve in Propositions 36 through 38 of the revised version of Section 7—namely, what is the resistance force arising from inertia alone, in the complete absence of fluid friction and tenacity?—is 0, regardless of the shape of the moving object.[70] This result is known, fittingly, as d'Alembert's paradox. It entails that there is no purely inertial mechanism of fluid resistance. As our C_D versus R_E curve for a sphere has indicated all along, resistance is never purely a function of fluid inertia, but always a combination of fluid inertia and viscosity. The forces arising directly from the fluid's viscosity are often negligible, but even then, indirect effects of viscosity govern the forces arising from the fluid inertia. Newton's two-term model of fluid resistance, in which a purely inertial idealization was to supply the first approximation, was thus doomed from the outset for deep reasons. If Newton had included such secondary inertial mechanisms as the effects of the flow at the back of the moving body, the resistance force he would have obtained from his model would have been zero.

In the beginning, while working on the first edition, Newton expected the pendulum decay experiments to disaggregate the different components of the resistance force. This freed him to devise a comparatively crude (though nonetheless ingenious) theoretical model as his first approximation to the mechanism producing inertial resistance in continuous fluids. This model considered only the force on the front of the moving body resulting from the inertia of the fluid column ahead of it. The results from the pendulum decay experiments would provide information not only about resistance's non-inertial components, but also about other mechanisms entering into the inertial component. In particular, the inertial component inferred from these experiments would tell him about the force on the rear of the body induced by the change in the fluid pressure acting on it when the body is moving. Although the pendulum decay experiments failed to yield stable values for the non-inertial resistance components, they did yield reasonably stable values for the inertial component. These results, joined with Newton's theoretical model, implied that the force induced on the rear of the moving body is significant, counteracting roughly two-thirds of the force from the column of fluid in front of it.

After the first edition was published, Newton concluded that he had to modify his solution for the force from the column ahead of the body, because his original solution for the efflux problem was wrong. His initial vertical-fall experiments in water also confirmed his suspicion that not even the inertial results from the pendulum decay experiments were reliable.

Unfortunately, vertical-fall experiments offered no prospect of disaggregating the different components of the resistance in the direct way that the pendulum decay experiments had promised to disaggregate them. Newton's best hope for gaining information about the non-inertial components was to devise a relatively accurate theoretical solution for the inertial component. The differences between observed results in vertical fall and calculated results with resistance defined by his theoretical value for the inertial component could then provide information about the other components. Based on the initial vertical-fall experiments in water and the corrected solution to the efflux problem, Newton concluded that a revised theoretical model of the inertial effect of the column of fluid ahead of the moving body could provide the close approximation he needed for the inertial resistance force. In this model the force induced on the rear of the moving body is viewed as a second-order effect.

The subsequent vertical-fall experiments did seem to show promise of ultimately allowing the non-inertial forces to be investigated in the intended manner. The only complication came from some of the experiments raising the possibility that the induced force on the rear might be comparable in magnitude to the non-inertial forces. The non-inertial components might still be separated from second-order inertial effects by varying the velocity of fall widely, provided that the second-order effects in question vary as v^2. A still better way of dealing with this complication, however, would be to devise a theoretical solution for inviscid fluid resistance that includes the induced effects at the rear of the moving body and other factors not accounted for in Newton's theoretical model. D'Alembert's solution did this, in the process revealing a deep shortcoming of the purely inviscid flow idealization. How Newton would have responded to this discovery is a fascinating question.

D'Alembert's paradox notwithstanding, the theory of inviscid fluid flow, referred to in the literature as "hydrodynamics," continued to be developed over the next century and a half. This prompted an unflattering description of the research field during this period: "fluid dynamicists were divided into hydraulic engineers who observed what could not be explained, and mathematicians who explained things that could not be observed."[71] A qualitative understanding of the mechanisms involved in fluid resistance, employing Prandtl's concept of a boundary layer, did not emerge until the beginning of the twentieth century, put forward then not by physicists engaged in research on fluid flow or the molecular structure of fluids, but by engineers concerned with heavier-than-air flight. The ideal fluid idealization, captured by Euler's equations of fluid motion, has been far from useless. But it has never been the basis for any empirical results on fluid resistance, nor for that matter on the microstructure of fluids. Here too we see that not just any idealization can serve as the starting point for an empirically driven sequence of successive approximations.

The unparalleled success of the empirical inquiry into celestial forces and motions in Books I and III of the *Principia* has been an impediment to drawing general methodological lessons from it. A close reading is needed even to see that this inquiry involved a sequence of successive approximations. The sequence from one-body to two-body then three-body and finally universal microstructural results is so tightly compacted in the *Principia* that one is easily led into viewing Newton as having devised a theory consisting of three laws of motion and the law of universal gravity, put forward as hypotheses and confirmed by their deductive consequences.[72] The subsequent history of celestial mechanics during the eighteenth and nineteenth centuries adds to this inclination, for the refinements that were introduced addressed such matters as idealizations in geometry and not Newton's four laws. The net effect of the train of successes extending from the *Principia* has been to obscure the logic of marshaling evidence when proceeding in a sequence of successive approximations. The failure of the empirical inquiry into fluid resistance in Book II and in the subsequent history of inviscid flow theory has the virtue of bringing this logic out more clearly.

CONCLUDING REMARKS

I would like to end with a conclusion and an observation. First the conclusion. My primary goal in this chapter has been to align Book II properly with the rest of the *Principia*. On my view, sometime around December 1684 Newton had an extraordinary vision of the way in which a series of mathematically derived if-then propositions promised to allow the empirical world—specifically, phenomena of planetary motion—to provide answers, at least to a high approximation, first to questions about the forces governing planetary motion and then to unresolved questions about these motions. The world cooperated, leading to the identification and first-order characterization of one of the universe's fundamental forces. Sometime a few months later, if not before, Newton had a similar vision of the way in which a series of if-then propositions promised to allow the empirical world to provide answers to questions about fluid resistance forces and motion under them, and even perhaps to questions about the microstructure of fluids. This time, however, the empirical world did not cooperate. Whereas planetary motion turned out to involve profoundly informative phenomena, fluid resistance turned out not to. There was no way for Newton to have foreseen this. All anyone could do was to pursue the indicated line of empirical inquiry, seeing where it would lead.

Finally, the observation. One often hears Newton's genius described in terms of his being peculiarly in tune with the universe and thus able to arrive intuitively at physically correct theories. One often hears this even from people who would be the first to ridicule extrasensory perception. Whatever else Book II may show, it surely shows that Newton was not able to come upon physically correct theories of the microstructure of fluids through intuition alone, nor even so much as a proposal for the purely viscous contribution to resistance. I do not think we belittle Newton's genius if we deny him any sort of extrasensory perception. Newton's genius as a physicist, I submit, was one of recognizing evidential pathways, paved by mathematically derived if-then principles, that offered a potential for vast, sustained step-by-step empirical inquiry. As such, his genius was more akin to that of a chess grand master who can see pathways of moves unfolding so far ahead of others. The other player in the chess game Newton was playing was the physical world. In the case of fluid resistance, unlike that of planetary motion, it came up with a gambit that no one had any way of anticipating.

Acknowledgments

I wish to thank Ole Knudsen, Babak Ashrafi, and India Smith for commenting on an earlier draft of this chapter, and also Eric Schliesser, I. B. Cohen, and my engineering colleagues, Willem Jansen and Gregory Oreper, for many helpful suggestions.

Notes

1. Clifford Truesdell, "Reactions of Late Baroque Mechanics to Success, Conjecture, Error, and Failure in Newton's *Principia*," in *Essays in the History of Mechanics* (New York: Springer-Verlag, 1968), 144.

2. I. Bernard Cohen, *The Newtonian Revolution* (Cambridge: Cambridge University Press, 1980), 61–78, 150ff.

3. See William Harper and George E. Smith, "Newton's New Way of Inquiry," in Jarrett Leplin, ed., *The Creation of Ideas in Physics: Studies for a Methodology of Theory Construction* (Dordrecht: Kluwer, 1995), 113–65.

4. For Newton, the term "phenomena" refers to observed, lawlike regularities.

5. See George E. Smith, "Huygens's Empirical Challenge to Universal Gravity," manuscript in preparation.

6. The exceptions, in which inverse-square gravity is taken into consideration, are Propositions 15 and 16 in Section 4, which concern the spiral motion of an orbiting body in a resisting fluid, and Proposition 22 of Section 5, which concerns the variation in the density of air above the earth's surface.

7. The principal measure of the strength of surface gravity at the time was the length of a seconds pendulum. The standard number for it was Huygens's 3 Paris ft. $8\frac{1}{2}$ lines —that is, 440.5 lines. As Newton could readily have calculated, the effect of a 300-foot elevation on this length was less than 0.015 lines.

8. In one of David Gregory's memoranda of 1694 concerning the new edition of the *Principia* that was being planned in the early 1690s, he indicates that Newton was at least considering introducing inverse-square gravity more widely in Book II: "To Proposition 10, Book II he will append another problem in which the path of a projectile is investigated according to the true system of things, that is, supposing gravity to be reciprocally as the square of the distance from a centre and the resistance to be directly as the square of the velocity—a problem which he believes to be now within his power." H. W. Turnbull, J. F. Scott, A. R. Hall, and Laura Tilling, eds., *The Correspondence of Isaac Newton*, 8 vols. (Cambridge: Cambridge University Press, 1959–1977), 3:384. In an earlier memorandum from May 1694, Gregory also mentions the possibility of replacing the pendulum measurements of resistance with ones employing "projectiles from a catapult." Ibid., 3:318. Newton therefore may well have planned to include inverse-square gravity in the projection problem because he thought that it was necessary to achieve reliable inferences from measurements of catapulted projectiles. Regardless, no such solution appeared in the subsequent published editions. Moreover, there is no reason to think that Newton had actually obtained a successful solution for the problem Gregory mentions.

9. Newton does not, to my knowledge, consider lift forces (i.e., forces arising from the relative motion of bodies and fluids that act perpendicularly to the direction of motion) anywhere in his writings.

10. These values are akin to the unique values of the "absolute measure" of force, a^3/P^2, associated with each individual force center or central body in the case of celestial forces.

11. "De Motu Corporum in Gyrum," in D. T. Whiteside, ed., *The Mathematical Works of Isaac Newton*, 8 vols., (Cambridge: Cambridge University Press, 1967–1981),

6:30–75. The title of the augmented version of this tract was changed to "De Motu Sphaericorum Corporum in Fluidis," and the heading, "De Motu Corporum in Medijs Resistentibus" was inserted just before the last two propositions, a phrase that would have been an appropriate title for Book II.

12. More exactly, Newton gave a method for determining the ratio of the resistance force at the onset of motion to the force of gravity, then added that the force of gravity can be determined from the sphere's weight. This is the closest Newton comes to introducing any notion of mass in either the original or the augmented *De Motu* tract.

13. Nor does Newton say anything in the *Principia* about how the leading coefficient of the v^0 term, $\mathbf{a}_0$, might vary. The only place where Newton subsequently remarks on either of these coefficients is in Query 28 of the *Opticks*, where he says in passing that the resistance force on a sphere arising from "the attrition of the parts of the medium is very nearly as the diameter, or, at the most, as the *factum* of the diameter, and the velocity of the spherical body together." *Opticks* (New York: Dover, 1952), 365. He offers no explanation for this claim. In the query it serves only as a step in the argument that the resistance on bodies "of a competent magnitude" arises almost entirely from the inertia of the fluid and not from the attrition of its parts; this argument in turn is a step in the larger argument of the query against the existence of an aetherial medium.

14. This and all other translations from the *Principia* in this chapter are from Isaac Newton, *Mathematical Principles of Natural Philosophy*, trans. I. Bernard Cohen and Anne Whitman with the assistance of Julia Budenz (Berkeley and Los Angeles: University of California Press, 1999).

15. Huygens had obtained solutions of motion under resistance proportional to v in the 1660s, but experiments convinced him that the resistance of real mediums varies more as v^2. He published this finding in his *Discourse on the Cause of Gravity* in 1690, along with some other points elicited by the first edition of the *Principia*. The experiments, which he never published, can be found in "Experiences de 1669 sur la force de l'eau ou de l'air en mouvement et sur les résistances éprouvées par des corps traversant ces milieux," *Oeuvres complètes de Christiaan Huygens* (The Hague: Martinus Nijhoff, 1937), 19:120–43. See also A. R. Hall, *Ballistics in the Seventeenth Century* (Cambridge: Cambridge University Press, 1952), 113–19. To my knowledge, the only explanation Newton offers for the shift away from resistance that varies purely as v is the physical explanation summarized in the next paragraph of the text. His dropping all mention of how the leading coefficient of the v term varies and his offering a new method for inferring the magnitude of this coefficient do, however, raise the possibility that he performed some experiments and found that the method does not yield a stable, converging value for $\mathbf{a}_1$ even in the case of a single body in a single medium.

16. In the Scholium at the end of Section 3 Newton speaks of the v^0 term as representing the "tenacity" of the fluid, the v term as representing its friction, and the v^2 term its inertial resistance to motion. In some places he also speaks of the "want of lubricity" [defectu lubricitatis] interchangeably with "tenacity." At the beginning of Section 9, however, "want of lubricity" seems to refer to the phenomenon we call

viscosity, and in still other places he appears to be using this phrase to designate the total non-inertial contribution to resistance, regardless of how v enters into it. (Cotes, in his preface to the second edition, similarly remarks, "the resistance of mediums arises either from the inertia of fluid matter or from its want of lubricity.") In using "viscous" to refer to the v term in this chapter, I am adopting the modern technical usage with no intention of implying that Newton had an exact counterpart to it. Our precise notion of viscosity emerged from a complex history, one element of which was the hypothesis Newton posits at the beginning of Section 9. For a review of the complications associated with this concept, see Emil Hatschek, *The Viscosity of Liquids* (New York: Van Nostrand, 1928).

17. George Gabriel Stokes, "On the Effect of the Internal Friction of Fluids on the Motion of Pendulums" (1851), in *Mathematical and Physical Papers* (Cambridge: Cambridge University Press, 1901), 3:55–62.

18. The statement holds strictly only for incompressible flows without free surfaces. Compressibility effects vary inversely with the square root of $(1 - M^2)$, where M is the Mach number (the ratio of speed to the speed of sound). In all of Newton's experiments in air, the Mach number was less than 0.10, so that compressibility effects were negligible. Free surface effects similarly involve a further dimensionless parameter, known as the Froude number, given by the ratio of v to the square root of gd.

19. Engineers often gain an improved fit in the low–Reynolds number range by using Oseen's alternative to Stokes's, indicated in figure 9.1. See Hermann Schlichting, *Boundary-Layer Theory*, trans. J. Kestin, 7th ed. (New York: McGraw-Hill, 1979), 115ff, and John Happel and Howard Brenner, *Low Reynolds Number Hydrodynamics* (Englewood Cliffs, NJ: Prentice-Hall, 1965), 40–49.

20. Richard P. Feynman, Robert B. Leighton, and Matthew Sands, *The Feynman Lectures on Physics*, 3 vols., (Reading, MA: Addison-Wesley, 1964), 2:41.8.

21. Ludwig Prandtl introduced the concept of the boundary layer, which revolutionized fluid mechanics, in the first decade of this century. He gives a lucid general account of boundary layers and the resistance forces on bodies moving in fluid mediums in the second volume of his lectures; see L. Prandtl and O. G. Tietjens, *Applied Hydro- and Aerodynamics*, trans. J. P. Den Hartog (New York: Dover, 1934).

22. If resistance's effect on the arc's amplitude is viewed as second order, then resistance's effect on the period is third order, consisting of two second-order effects that largely cancel one another, for resistance increases the time of descent, but then reduces the time of ascent. Thus the seventeenth-century practice of introducing no correction for resistance's effects on the period when using pendulums to measure surface gravity was entirely reasonable.

23. For the derivation of this expression and an analysis of its relationship to Proposition 30, see Whiteside, *Mathematical Works of Isaac Newton*, 6:448, n. 17. In the *Principia* Newton approximates $2/\pi$ by 7/11.

24. The manuscript covering the preliminary trials has been published in I. Bernard Cohen's *Introduction to Newton's "Principia"* (Cambridge: Harvard University Press,

1978), 101–103. In the *Principia* Newton argues that circular and cycloidal pendulums of the same length and the same initial arc lose virtually the same amount of arc per oscillation, and hence his theoretical relationships for arc lost by a cycloidal pendulum can be applied directly to experiments employing circular pendulums. This argument assumes (1) that the resistance force varies nearly as v^2; and (2) that the ratio of a circular pendulum's period to that of a cycloidal pendulum of the same length and same initial arc is equal to the ratio of the maximum velocity of the cycloidal pendulum to that of the circular (i.e., to the ratio of half of the total arc and hence the arc in descent to the chord of that arc). Both of these assumptions are sources of inaccuracy in the values of A_i inferred from pendulum decay experiments. (By the way, the Mott-Cajori translation of this argument has an error: instead of the correct translation, "the chords of halves of these arcs," it has, "1/2 the chords of these arcs.")

25. These are avoirdupois ounces. By contrast, ounces in the Scholium at the end of Section 7 in the second and third editions are troy.

26. I examine Newton's pendulum experiments on resistance and lay out their shortcomings in more detail in "Newton's Experiments on Fluid Resistance," manuscript in preparation.

27. I did not examine every possible combination of data points, but instead limited attention to combinations of three in which one was from the two smallest arcs, one was from the two intermediate arcs, and one was from the two largest. In other words, I followed Newton's (correct) practice of using representative combinations of data. The same is true for my comments in the text about each of the other frameworks.

28. Newton reports better results for a comparison of water and mercury but does not give the data. The sidewall effect to which he attributes the poor results for water is something that still sometimes plagues wind tunnel testing.

29. Newton remarked that he got a headache whenever he worked on the lunar orbit. I get a headache whenever I return to the problem of trying to reproduce his pendulum data analytically.

30. Stokes was careful to take this effect into account in his classic 1851 monograph on the effects of resistance on pendulum motion, cited in note 17.

31. Such unwanted fixturing effects have been another problem often plaguing wind tunnel testing.

32. I discuss in detail why Newton changed his mind about the questions fluid resistance experiments were supposed to answer in "Fluid Resistance: Why Did Newton Change His Mind?" in Richard Dalitz and Michael Nauenberg, eds., *Foundations of Newtonian Scholarship* (Singapore: World Scientific, 2000).

33. The second half of Section 7 from the first edition can be found in Alexandre Koyré and I. Bernard Cohen, eds., *Isaac Newton's Philosophiae Naturalis Principia Mathematica* (Cambridge: Harvard University Press, 1972), 775–84; and the paragraphs drawing conclusions from the pendulum decay experiments that were dropped from the General Scholium in the second and third editions, 459ff.

34. Emphasis added. The quotation is from the concluding scholium of Book I, Section 11, 588f.

35. For example, in the way that the combination of the variation of surface gravity and the earth's eccentricity picked out inverse-square gravity among the particles of matter from alternatives to it. See my "Huygens's Empirical Challenge to Universal Gravity," manuscript in preparation.

36. Newton uses the word *Figmenta* to describe Descartes's conjectures in "De Gravitatione et Aequipondio Fluidorum"; see A. Rupert Hall and Marie Boas Hall, eds., *Unpublished Scientific Papers of Isaac Newton* (Cambridge: Cambridge University Press, 1962), 92.

37. That is, spheres whose "matter is homogeneous on all sides in regions that are equally distant from their centers." Newton was almost certainly referring to Propositions 75 and 76 of Book I when he remarked to Edmund Halley in his famous letter of 20 June 1686, "I never extended the duplicate proportion lower then [*sic*] to the superficies of the earth & before a certain demonstration I found the last year have suspected it did not reach accurately enough down so low." Turnbull et al., *Correspondence of Isaac Newton*, 2:435.

38. In Proposition 23 of Section 5, Newton shows that under the assumption that the particles are static, Boyle's law requires the repulsive forces to hold only between the particles and their immediate neighbors. At issue in Section 7, however, are the forces between the impinging particles and the moving object, so that the constraint identified in Section 5 ceases to be of concern.

39. This is the figure that appears in the first edition. The figure was changed in the second.

40. Fatio appears to have had some trouble convincing Newton of the error, ultimately succeeding by persuading him to carry out an experiment. See Cohen, *Introduction to Newton's "Principia,"* 182ff.

41. Truesdell, "Reactions of Late Baroque Mechanics," 146.

42. See Koyré and Cohen, *Isaac Newton's Philosophiae Naturalis Principia Mathematica*, 460ff (and the English translation in this chapter's appendix) for the passage comparing theory and pendulum measurement in the first edition. The magnitude of the inertial force acting on the sphere's front surface given in the first edition is greater than the magnitude given in the second edition. How much greater is a matter of some confusion. The magnitude given in the second edition is equivalent to a C_D of 0.5. The text of Proposition 38 in the first edition concludes, "And therefore the resistance on a globe progressing uniformly in any very fluid medium whatever, in the time in which the globe describes 2/3 of its own diameter, is equal to the force that, uniformly impressed upon a body of the same size as the globe and having the same density as the medium, in the time in which the globe by progressing described 2/3 of its own diameter, could generate the globe's velocity in that body." If we take the phrase "by progressing described 2/3 of its own diameter" at the end of this statement to mean "progressing uniformly at its acquired velocity," then the time is $(2/3)d \div v = (2/3)(d/v)$, and the magnitude given for the resistance corresponds to a C_D of 2.0. If,

instead, we take this phrase to mean, "progressing from rest to its acquired velocity," then the time is $(2/3)d \div (v/2) = (4/3)(d/v)$, and the magnitude corresponds to a C_D of 1.0. The first corollary to the proposition says that this resistance is the same as that obtained for a rarified fluid in which the particles are inelastic and hence do not rebound; this implies that the correct reading is the one giving a C_D of 1.0. The comparison with the pendulum experiments, however, says that the measured inertial component is around one-third of the magnitude obtained in the proposition; this implies a C_D of 2.0. Complicating the matter further are a number of errata Newton listed for these two passages in one of his copies of the first edition (ibid., 782, 462); these errata imply a C_D of 1.0 unless, as is likely, they were introduced as corrections to the erroneous solution to the efflux problem in the first edition, in which case they imply a C_D of 2.0. The phrasing Newton adopted in normalizing the inertial resistance force in the second edition removes all ambiguity.

43. Proposition 39 provides corrections for when the sphere's diameter is large compared to the breadth of the flow stream. These corrections are applied in some of the experiments in water discussed in the chapter's next section.

44. In the first edition, Newton contended that the weight in question is that of the whole cylindrical column of the fluid. His argument in the second edition for bracketing the weight between one-third and two-thirds of this cylinder is that the weight must be greater than that of a cone of fluid, for the boundary defining the column of fluid acting on the disk must be convex, and it must be less than that of a hemispheroid of fluid with a defining boundary perpendicular to the disk, for the angle between the column of fluid and the disk must be acute.

45. Or equally, the precise distance an object falls in a given time if the only resistance acting on it varies as v^2. In effect, Newton derives the solution of the following differential equation:

$$m\frac{d^2s}{dt^2} = mg_b - \lambda\left(\frac{ds}{dt}\right)^2,$$

namely,

$$s = \frac{m}{\lambda}\ln\left(\cosh\left[\left(\frac{\lambda g_b}{m}\right)^{1/2} t\right]\right),$$

where g_b is the effective acceleration of gravity (after buoyancy effects have been taken into account), m is the mass of the falling object, and λ is given by

$$\lambda = C_D\rho A_f/2 (= C_D\pi\rho d^2/8 \text{ for spheres}).$$

46. Turnbull et al., *Correspondence of Isaac Newton*, 3:384. See also 3:317ff.

47. *Journal Book (Copy) of the Royal Society* 20 (7 and 14 June 1710): 243. Since Hauksbee had become the principal demonstrator of experiments in the Royal Society following Hooke's death in 1703, it was appropriate for him to carry out these experiments for Newton. The *Journal Book* does not indicate whether Newton was on the scene during them; the details he gives in describing them suggests he was.

48. An anecdote relayed by Conduit tells of the glass window at the bottom of a nine-and-a-half-foot trough breaking during an experiment, inundating Newton.

Keynes MS, 130.15, quoted in Richard S. Westfall, *Never At Rest: A Biography of Isaac Newton* (Cambridge: Cambridge University Press, 1980), 455. Westfall concludes that the initial experiments in water were conducted while the first edition was in progress. This is difficult to reconcile with Newton's failure at the time to recognize that the resistance forces inferred from the pendulum decay experiments were excessive.

49. A memorandum of Gregory's, dated 15 April 1707, remarks, "The only two things that Sir Isaac Newton further desires, to make a new edition of Princ. Math. Philos. Nat. are these; First about the Resistance of Fluids in Lib. II about which he is affray'd he will need some new experiments. The second is concerning the Procession of the Æquinoxes Lib III of which he thinks he is fully master, but has not yet written it out. He does not think to goe about these things, at least the Edition, for two years yet." W. G. Hiscock, ed., *David Gregory, Isaac Newton, and Their Circle: Extracts from David Gregory's Memoranda* (Oxford: Oxford University Press, 1937), 40. This passage may be referring to vertical fall experiments in water or in air, but it may also be referring to efflux experiments (see ibid., 41).

50. My "Newton's Experiments in Fluid Resistance," manuscript in preparation, presents a more detailed discussion.

51. Following Experiment 7, Newton intentionally weighted the balls asymmetrically and dropped them heavy side first to reduce their tendency to oscillate.

52. Newton used a ratio of the density of water to air of 860 from the table accompanying Proposition 40 in obtaining his vertical fall experiments in air. The C_D values shown for air in figure 9.7 are based on the modern ratio of 830.

53. R. G. Lunnon, "Fluid Resistance to Moving Spheres," *Proceedings of the Royal Society of London*, Series A, 10 (1926): 302–26, noted the high quality of the vertical fall data.

54. Book II, Section 7, 759.

55. See the discussion of de Borda in the chapter's next section.

56. Newton's reluctance to put much weight on evidence consisting of empirically verified deductions from hypotheses dates from early in his career. In a then unpublished portion of his letter of 6 February 1672 to Henry Oldenburg presenting his results on light and color, he remarked: "I shall now proceed to acquaint you with another more notable difformity in its Rays, wherein the *Origin of Colours* is unfolded. A naturalist would scearce expect to see ye science of those become mathematicall, & yet I dare affirm that there is as much certainty in it as in any other part of Opticks. For what I shall tell concerning them is not an Hypothesis but most rigid consequence, not conjectured by barely inferring 'tis thus because not otherwise or because it satisfies all phænomena (the Philosophers universal Topick,) but evinced by ye mediation of experiments concluding directly and without any suspicion of doubt." Turnbull, *Correspondence of Isaac Newton*, 1:96. (This passage is discussed in chapter 4.) Similarly, in the published reply to Pardies's second letter attacking the light and colors paper, Newton said: "For the best and safest method for philosophizing seems to be, first to inquire diligently into the properties of things, and establishing those properties by experiments and then to proceed more slowly to hypotheses for the

explanation of them. For hypotheses should be subservient only in explaining the properties of things, but not assumed in determining them; unless so far as they may furnish experiments. For if the possibility of hypotheses is to be the test of the truth and reality of things, I see not how certainty can be obtained in any science; since numerous hypotheses may be devised, which shall seem to overcome new difficulties." *Philosophical Transactions* 85 (July 15, 1672): 5014; translation from I. Bernard Cohen, ed., *Isaac Newton's Papers and Letter on Natural Philosophy* (Cambridge: Harvard University Press, 1958), 106.

57. Harper and Smith, "Newton's Way of Inquiry," explores at length the fundamental role that Newton's sharp distinction between hypotheses and established results played in the development of his science. The distinction is evident in the *Principia* not only in the famous remark, "hypotheses, whether metaphysical or physical, whether of occult qualities or mechanical, have no place in experimental philosophy," in the General Scholium added in the second edition, but also in the care with which Newton called attention to the conjectural character of his microstructural explanations of Boyle's law in Section 5 of Book II and Snel's law in Section 14 of Book I. A single quotation from the controversy provoked by his light and colors paper should suffice to show that he insisted on the distinction throughout his career: "In the mean while give me leave, Sir, to insinuate, that I cannot think it effectual for determining truth, to examin the several waies by which Phænomena may be explained, unless where there can be a perfect enumeration of all those waies. You know, the proper Method for *inquiring* after the properties of things is, to deduce them from Experiments. And I told you, that the Theory, which I propounded, was evinced to me, not by inferring 'tis thus because not otherwise, that is, not by deducing it only from a confutation of contrary suppositions, but by deriving it from Experiments concluding positively and directly." *Philosophical Transactions* 85 (15 July 1672): 5004.

58. The implication that the inertial resistance in both air and water acts as if both of these fluids are continuous, rather than rarified, has the corollary that Newton's result for the solid of least resistance should not be expected to hold in either air or water. (There was, of course, no basis for concluding this in the first edition, where the result on the solid of least resistance first appeared.) Newton's notion of a rarified fluid turns out not to be entirely lacking in real-world application. It has provided a starting point for resistance in hypersonic flows; see W. D. Hayes and R. F. Probstein, *Hypersonic Flow Theory*, vol. 1, *Inviscid Flows* (New York: Academic Press, 1966), 129ff. (I thank Prof. Laszlo Tisza for calling this reference to my attention.)

59. Experiment 3, for example, would have been ideal if the buoyant weights had been determined with confidence and worries about other confounding factors had been removed. Newton's remarks about this experiment display his appreciation of the problem of experimentally separating secondary viscous effects from other confounding factors: "I am uncertain whether their falling more slowly is to be attributed to the smaller proportion of the resistance that arises from the force of inertia in slow motions to the resistance that arises from other causes, or rather to some little bubbles adhering to the ball, or to the rarefaction of the wax from the heat either of the weather or of the hand dropping the ball, or even to imperceptible errors in weighing the balls in water." Book II, Section 7, 752.

60. In Archibald Pitcairne's notes (dated 2 March 1692) on Newton's short tract "De Natura Acidorum," Newton is indicated to have made the following proposal: "Viscosity [visciditas] is either just a deficiency of fluidity (which is located in the smallness, and thus the separability of parts, understood as parts of last composition) or a deficiency of slipperiness or smoothness preventing the lowest parts from sliding over others. Acid is often the cause of viscosity. Often an otherwise slippery spirit when joined to earth, such as oil of turpentine [resin of the terebinth tree] when restored to its *caput mortuum* [residue after extraction], becomes sticky." See Turnbull et al., *Correspondence of Isaac Newton*, 3:211.

61. One should not forget that Maxwell initially became persuaded of Clausius's kinetic theory of gases when he found, to his surprise, that it yielded a promising account of viscosity as a form of lateral molecular momentum transfer. See S. G. Brush, *Kinetic Theory*, 3 vols., (Oxford: Pergamon Press, 1965) 1:26ff, and James Clerk Maxwell, "Illustrations of the Dynamical Theory of Gases," included in ibid., 148–71.

62. In a draft of the preface to the second edition, Newton says, "The theory of resistance of fluids, which in the first edition was put forward imperfectly because of a shortage of experiments, is here perfected." See A. Rupert Hall and Laura Tilling, eds., *The Correspondence of Isaac Newton* (Cambridge: Cambridge University Press, 1975), 5:112ff.

63. A Newtonian fluid is rigorously defined as one in which the stress tensor and the rate-of-strain tensor are linearly related. In simple cases of the sort with which Newton was concerned in Section 9, this becomes

$$\tau = \mu \frac{\partial v}{\partial r},$$

where τ is the force per unit area and the viscosity, μ, does not vary with velocity. Strictly speaking, Newton's hypothesis does not concern the magnitude of the resistance force at the surface of the moving object so much as it concerns how the effects of this force propagate through the fluid.

64. Newton's argument against Descartes in Section 9 happens to be flawed, but not as a consequence of the hypothesis concerning viscous forces. As Stokes pointed out in 1845, Newton gets the pressure gradient in a vortex around spinning cylinders and spheres wrong because he employs a balance of forces, not a balance of torques, on the two surfaces of the rotating shells comprising the fluid; see Stokes, "On the Theories of the Internal Friction of Fluids in Motion, and of the Equilibrium and Motion of Elastic Solids," in *Mathematical and Physical Papers*, 1:103. Historically, Newton's claim that Cartesian vortices were incompatible with Kepler's 3/2 power rule was less important in their rejection than was the inability to reconcile them with comet trajectories, especially with the trajectories obtained with the benefit of Newtonian theory.

65. Le Chevalier de Borda, "Expériences sur la résistance des fluides," *Mémoires de l'Académie Royale des Sciences* (1763), 358–76.

66. Le Chevalier de Borda, "Expériences sur la résistance des fluides," *Mémoires de l'Académie Royale des Sciences* (1767), 495–503. This result of de Borda's for disks was correct.

67. Ibid., 503. Translation by René Dugas, *A History of Mechanics* (New York: Dover, 1988), 313.

68. Bernoulli's formula, in modern form, is

$$\frac{p}{\rho} + \frac{1}{2}v^2 + \phi = \text{constant},$$

where ϕ is the potential per unit mass from external forces and the constant can change from one streamline to the next.

69. Jean d'Alembert, *Essai d'une nouvelle théorie de la résistance des fluides* (Paris: David, 1752). A proper statement of the paradox imposes the further requirement that the fluid be unbounded—that is, that it not have any (nearby) free surfaces. For a modern discussion, see L. D. Landau and E. M. Lifshitz, *Fluid Mechanics: Course of Theoretical Physics*, Vol. 6, trans. J. B. Sykes and W. H. Reid (Oxford: Pergamon Press, 1959), and Garrett Birkhoff, *Hydrodynamics: A Study in Logic, Fact, and Similitude*, rev. ed. (Princeton: Princeton University Press, 1960), 12ff. An especially clear derivation and analysis of d'Alembert's paradox for spheres can be found in Prandtl and Tietjens, *Applied Hydro- and Aerodynamics*, 104–107.

70. The Introduction to d'Alembert's *Essai* provides grounds for viewing him as trying to solve Newton's problem of inertial resistance in continuous fluids but taking into account the three effects Newton discarded as secondary: the disturbance in the flow upstream of the object, the distribution of flow around the object, and the action of the fluid on the object's rear. In his introduction, d'Alembert also remarks that Newton's theory of resistance in rarified fluids is "a research of pure curiosity, . . . not applicable to nature" (*xiv*). Of Newton's continuous-fluid impingement model, d'Alembert remarks that it is "intended to elude rather than surmount the difficulty of the problem" (*xx*). For a discussion of the *Essai*, see Truesdell, "Rational Fluid Mechanics, 1687–1765," in *Leonhardi Euleria Opera Omnia*, Series II (Lausanne: Auctoritate et Impensis Societatis Scientarum Naturalium Helveticae, 1954) 12:L–LVIII.

71. Quoted from Birkhoff, *Hydrodynamics,* 4. Birkhoff obtained the quote from a summary article covering a conference on very high speed flow, where M. J. Lighthill attributed it to Sir Cyril Hinshelwood. See *Nature* 178 (1956): 343.

72. For a classic statement of this view of the *Principia*, see Pierre Duhem, *The Aim and Structure of Physical Theory*, trans. P. P. Wiener (Princeton: Princeton University Press, 1991), esp. 190–205. This view is in sharp conflict with what Newton says in the *Principia* about his methodology, especially with the reformulated beginning of Book III and the General Scholium added to it in the second edition. This conflict has led proponents of Duhem's view of the *Principia* into such remarkable extremes as Imre Lakatos's "Newton's Effect on Scientific Standards," in *The Methodology of Scientific Research Programs* (Cambridge: Cambridge University Press, 1978), 193–222.

Appendix: Newton on Fluid Resistance in the First Edition: English Translations of the Passages Replaced or Removed in the Second and Third Editions

Newton entirely replaced the second half of Book II, Section 7, in the second edition of the *Principia*. As discussed in the body of the chapter,[1] two factors prompted this revision. First, Newton became persuaded, largely from experiments, that his original solution to the efflux problem was erroneous. Second, some initial vertical-fall experiments in water convinced him that his pendulum decay experiments were yielding excessively high values for the resistance forces. In conjunction with the revision to Section 7, Newton moved the General Scholium covering the pendulum decay experiments from the end of Section 7 to the end of Section 6, in the process shortening it in ways that befit its reduced importance in the second edition. In particular, he cut three paragraphs from the General Scholium that drew conclusions from the combination of the results of the pendulum decay experiments and the propositions in the second half of Section 7, especially conclusions about the fluid medium's action on the rear of a moving sphere.

Although the passages from the first edition are available in Latin in the variorum edition of the *Principia* edited by Koyré and Cohen, they have not been available in English. The translations here are intended to correct this situation. The translation of the original version of the last half of Section 7 is by I. Bernard Cohen, Anne Whitman, and George E. Smith, and the translation of the eliminated paragraphs of the General Scholium is by Julia Budenz and George E. Smith; the annotation of both is mine.

Book II, Section 7, Propositions 36–40, in the First Edition

Propositions 32 and 33 of Book 2, Section 7 remained more or less the same in all editions. Proposition 34[2] in the first edition was eliminated

[1] This is also discussed, in greater detail, in Smith, "Fluid Resistance: Why Did Newton Change His Mind?" cited in note 32 above.

[2] Proposition 34 in the first edition asserts: *What is demonstrated in the two preceding propositions holds as well for a system of particles touching one another provided that those particles are of extremely great lubricity.* The proposition has two corollaries.

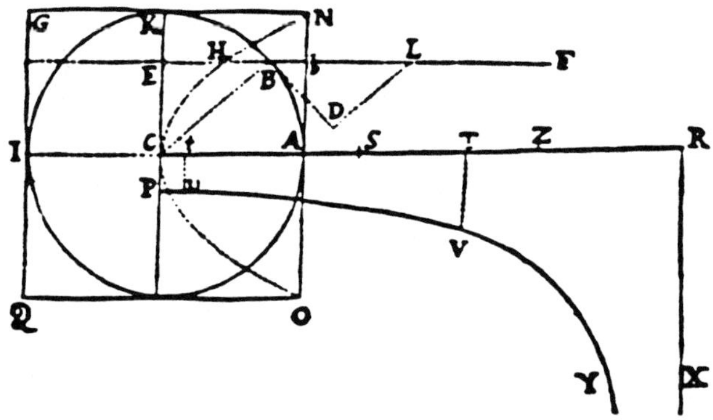

Figure 9.9

in the second and third, presumably to allow the last proposition of the section to retain the number 40. Thus, the five propositions translated below[3] were replaced by Propositions 35–40 in the second and third editions.

PROPOSITION 36 *Problem 8*
To find the resistance of a spherical body progressing very swiftly in a rare and elastic fluid. [See figure 9.9][4]

Let ABKI designate a spherical body described with center C and semidiameter CA. Extend CA first to S and then to R so that AS is a third part of CA and CR is to CS as the density of the spherical body to the density of the medium. Erect the perpendiculars PC and RX to CR, and with center R and asymptotes CR and RX describe any hyperbola PVY. On CR take CT of any length, and erect the perpendicular TV cutting off the hyperbolic area PCTV, and let CZ be a side of this area in conjunction with the straight line PC.[5] I say that the motion that the globe, in describing the space CZ, will lose from the resistance of the medium will

[3] From the Latin text in Koyré and Cohen, *Isaac Newton's Philosophiae Naturalis Principia Mathematica* 775–84 cited in note 33 above.

[4] The figure, reproduced here as figure 9.9, is the version of the figure accompanying the immediately preceding proposition (i.e., Proposition 34 of the second and third editions) in the first edition.

[5] That is, take *CZ* so that the product of it and *PC* is equal to the area *PCTV*.

be to its entire motion at the beginning as the length CT to the length CR, approximately.[6]

For (by the third law of motion) the motion that a cylinder GNOQ described about the globe would lose by impinging upon the particles of the medium is equal to the motion it would impress upon these same particles. Let us suppose that the individual particles are reflected from the cylinder and recoil from it with the velocity with which the cylinder approached them. For such will be the reflection, by corollary 3 of the laws, provided that the particles are as small as possible and are reflected by the greatest possible elastic force. Therefore the velocity with which they recoil from the cylinder, added to the velocity of the cylinder, will compose a total velocity two times greater than the velocity of the cylinder, so that the motion that the cylinder loses as a result of the reflection of each particle will be to the cylinder's whole motion as twice the particle to the cylinder. Accordingly, since the density of the medium is to the density of the cylinder as CS to CR, if Ct is the length described by the cylinder in the least possible time, the motion lost in that time will be to the cylinder's whole motion as 2Ct × CS to AI × CR. For this is the ratio of the medium's matter that is pushed forward and reflected by the cylinder to the mass of the cylinder. Hence, since the globe is 2/3 of the cylinder, and the resistance of the globe (by the above proposition) is two times smaller than the resistance of the cylinder, the motion that the globe loses in describing the length L will be to the whole motion of the globe as Ct × CS to 2/3 AI × CR, or as Ct to CR. Erect the perpendicular tv meeting the hyperbola at v, and (by book 2 proposition 5 corollary 1) if a body, in describing a length proportional to the area CtvP, loses any part Ct of its whole motion CR, it will, on describing a length proportional to

[6] Letting $dv/dt = -\lambda v^2$, then $v = v_0/(1 + v_0\lambda t)$ and the distance traversed $s = \ln(1 + v_0\lambda t)/2$. Then the fraction of the motion lost in traversing the distance s, $v/v_0 = (v_0 - v)/v_0$, is $1 - 1/e^{\lambda s}$. Let the hyperbola be $xy = a$. Taking $\lambda = PC/a$ then gives Newton's claim, $\Delta v/v_0 = CT/CR$. Using the modern nondimensionalization of resistance forces in terms of the drag coefficient, C_D, we have

$$\lambda = \frac{C_D \rho_f \pi r^2}{2m_s},$$

where ρ_f is the medium's density and m_s is the sphere's mass. Substituting Newton's geometric magnitudes then gives

$$\lambda = \frac{3}{4}\frac{CS}{CR}\frac{C_D}{2CA},$$

or $C_D = 2\lambda CR$. But $PC \times CR = a$, so that Newton's claim amounts to $C_D = 2$.

the area CTVP, lose the part CT of its motion. But the length Ct is equal to CPvt/CP, and the length CZ (by the hypothesis) is equal to CPVT/CP, and so the length Ct is to the length CZ as the area CPvt to the area CPVT. And, on that account, since the globe, in describing the least possible length Ct, loses a part of its motion that is to the whole as Ct to CR, it will, in describing any other length CZ, lose a part of its motion that is to the whole as CT to CR. Q.E.D.

Corollary 1. If the velocity of the body at the beginning is given, the time in which the body, in describing the space Ct, will lose the part Ct of its motion will be given; and from there, by saying that the resistance is to the force of gravity as that lost part of the motion is to the motion that the gravity of the globe would generate in the same time, the proportion of the resistance to the gravity of the globe will be given.

Corollary 2. Since in determining these things I have supposed that the particles of the fluid are reflected from the globe as greatly as possible by their own elastic force and that the impetus of the particles thus reflected upon the globe is two times greater than if they were not reflected, it is manifest that in a fluid whose particles are deprived of all elastic force and all other reflective force, the spherical body will suffer a resistance two times smaller, and so, by losing the same part of its velocity, will progress two times farther than in accordance with the construction of this problem offered above.

Corollary 3. And if the particles' reflective force is neither very great nor none at all, but holds some mean ratio, the resistance likewise, within the limits set in the construction of the above problem and corollary, will hold a mean ratio.

Corollary 4. Since slow bodies are resisted a little more than in accordance with the square of the velocity, they will, in describing any length CZ, lose a greater part of their motion than that which is to their whole motion as CT to CR.

Corollary 5. Moreover, when the resistance of very swift bodies is known, the resistance of slow bodies will also become known, provided that the law of the decrease of the resistance in proportion to the ratio of the velocity can be found.

Proposition 37 *Problem 9*

To define the motion of water flowing out of a given vessel through a hole.

If a vessel is filled with water and is perforated in the bottom so that water flows through the hole, it is manifest that the vessel will sustain the weight of the whole water, minus the weight of that part that projects perpendicularly above the hole. For if the hole were closed up by an obstacle, the obstacle would sustain the weight of the water perpendicularly incumbent on it, and the bottom of the vessel would sustain the weight of the remaining water. But if the obstacle is removed, the bottom of the vessel will be pressed by the same pressure of the water as before; and the weight that the obstacle was sustaining, since it no longer sustains it, will make the water descend and flow down through the hole.

From which it follows that the motion of all the effluent water will be that which the weight of the water perpendicularly incumbent on the hole would be able to generate. For each particle of water, insofar as it is not impeded, descends by its own weight, and does so with a uniformly accelerated motion; and insofar as it is impeded, it will press on the obstacle. That obstacle is either the bottom of the vessel or the water flowing down below it; and therefore that part of the weight which the bottom of the vessel does not sustain will press the water flowing down and will generate a motion proportional to itself.

In these circumstances let F designate the area of the hole, A the height of the water perpendicularly incumbent on the hole, P its weight, AF its quantity, S the space it would describe in any given time T by falling freely in a vacuum, and V the velocity that by falling it will have acquired at the end of that time; and its acquired motion AF × V will be equal to the motion of all the water flowing out in the same time. Let the velocity with which it exits the hole by flowing out be to the velocity V as d to e; and since the water with velocity V could describe the space 2S, the water flowing out in the same time, with its velocity (d/e)V, could describe the space 2(d/e)S. And therefore a column of water whose length is 2(d/e)S and whose width is the same as that of the hole could by flowing down in that time come out of the vessel, that is, the column 2(d/e)SF. Wherefore the motion $2(d^2/e^2)SFV$, which will be reckoned by multiplying the quantity of water by its own velocity, that is, all the motion generated in the time of that efflux, will be equal to the motion AF × V. And if those equal motions are divided by FV, $2(d^2/e^2)S$ will become equal to A. Whence d^2 is to e^2 as A to 2S and d to e will be as the square root of the ratio of 1/2 A to S. Therefore the velocity with which the water exits the hole is to the velocity that the water, falling and describing the space S would acquire, as the height of the water incumbent on the hole is to the mean proportional between that height doubled and that space S which a body would describe in falling in the time T.[7]

[7] That is, the efflux velocity is to V as A is to $\sqrt{(2AS)}$.

In these circumstances if those motions are turned upwards, since the water with velocity V would ascend to that height S from which it had fallen down, and (as is known) the heights are as the square of the velocities, the water flowing out would ascend to the height 1/2 A. And therefore the quantity of water flowing out will, in the time in which a body by falling could describe the height 1/2 A, be equal to the column of all the water AF projecting perpendicularly above the hole.

Since the effluent water, if its motion is turned upwards, would rise perpendicularly to half the height of the water incumbent on the hole, it follows that, if the water comes out obliquely through a channel on the side of the vessel, it will describe in non-resisting spaces a parabola whose latus rectum is the height of the water in the vessel above the mouth of the channel, whose diameter extends, perpendicular to the horizon, from that mouth, and whose ordinates are parallel to the axis of the channel.

All these things are to be understood of a fluid of a most fine texture.[8] For if the water consists of thicker parts, it will flow out more slowly than in accord with the ratio given above, especially if the hole through which it flows out is narrow.

Finally, if the water goes out through a channel parallel to the horizon, since the bottom of the vessel is intact and is pressed everywhere by the same pressure of incumbent water as if the water were not flowing out, the vessel will sustain the weight of all the water, the efflux notwithstanding, but the side of the vessel out of which it flows will not sustain all the pressure it would sustain if the water were not flowing out. For the pressure of that part where it is perforated will be removed, and this pressure is equal to the weight of a column of water whose base is equal to the hole and whose height is the same as that of the total water above the hole. And therefore if the vessel like a pendulous body is suspended from a nail by a very long string, it will, if the water flows out in any direction along a horizontal line, always recede from the perpendicular in the contrary direction. And the same is true of the motion of balls that are filled with moistened gunpowder and that, as the matter bursts out through the hole into flame, recede from the region of the flame and are carried with the impetus in the contrary direction.

Proposition 38 *Theorem 29*
To define the resistance on the anterior surface of spherical bodies in any very fluid mediums. [See figure 9.10]

[8] Newton added "and lacking all tenacity" in his copy.

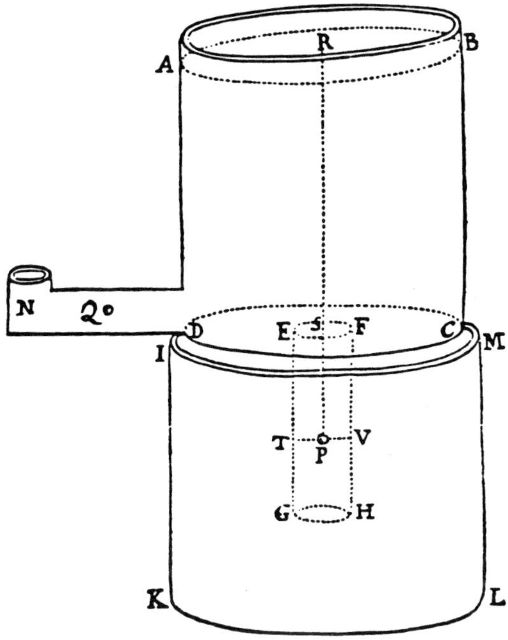

Figure 9.10

Let water flow down from a cylindrical vessel ABCD through a cylindrical channel EFGH into a lower vessel IKLM, and from there let it flow out over the vessel's edge IM. Let that edge be of the same height as the bottom CD of the upper vessel, so that the water descends through the whole channel with uniform motion; and in the middle of the channel let a globe P be placed, and let PR be the height of the water above the globe and SR its height above the bottom of the vessel. It is manifest by the previous proposition[9] that the quantity of water flowing down in a given time will be as the size of the hole through which it flows down; that is, if the globe is taken away, as the mouth of the channel; but, if the globe is present, as the space on all sides between the globe and the channel. For the velocity of the water flowing down will (by the previous proposition) be that which a body could acquire by falling and describing in its fall half the water's height SR; and so it is the same whether the globe is removed or is present. And therefore the water flowing down

[9] What actually appeared in the first edition was *proportionem* (proportion) though the word in the manuscript for the first edition was *Propositionem*.

will be as the size of the space through which it passes. If the water is not a liquid of very fine texture[10] and very fluid, its passage through a narrower space will, because of the thickness of the particles, be somewhat slower; but we here suppose the liquid to be most fluid. In these circumstances the quantity of water whose descent the globe impedes in a given time is to the quantity of water that would descend in the same time if the globe were removed, as the base of the cylinder described about the globe to the mouth of the channel, or as the square of the globe's diameter to the square of the diameter of the channel's cavity. And therefore the quantity of water whose descent the globe impedes is equal to the quantity of water that could descend in the same time through a circular hole in the bottom of the vessel equal to the base of that cylinder, and whose descent through any circular part of the bottom equal to that base is impeded.

Now, the weight of the water that the vessel and the globe jointly sustain is the weight of all the water in the vessel except that part which accelerates the water flowing out and is sufficient to generate its motion, and which, by the previous proposition, is equal to the weight of the column of water whose base is equal to the space between the globe and the channel through which the water flows down and whose height, designated by the line SR, is the same as the height of the water above the bottom of the vessel. The bottom of the vessel and the globe then jointly sustain the weight of all the water in the vessel projecting perpendicularly above them. Whence, since the bottom of the vessel sustains the weight of the water projecting perpendicularly above it, it remains that the globe also sustains the weight of the water projecting perpendicularly above it. The globe indeed does not sustain the weight of that water that is stagnant and incumbent on it without any motion, but by resisting the water flowing out impedes the effect of that much weight, and so sustains a force equal to that weight of the water flowing out. For it impedes the descent and efflux of the quantity of water that that weight would exactly effect if the globe were removed. The water by its weight, insofar as its descent is impeded, urges every obstacle; and so the obstacle, insofar as it impedes the descent of the water, sustains a force equal to the weight by which that descent would be effected. The globe moreover impedes the descent of the quantity of water that the weight of the column of water perpendicularly incumbent on it could effect, and therefore sustains the force, equal to that weight, of the water running out. The action and

[10] *Subtilissimo* is here and elsewhere being translated by "of extremely fine texture" rather than by its English cognate, "extremely subtle."

reaction of the water, by the third law of motion, are equal among themselves, and are directed in contrary directions. The action of the globe upon the descending water, impeding its descent, is directed upwards, and so impedes this descending motion; and therefore the contrary action of the water upon the globe is equal to the force that can either destroy or generate the same motion, that is, to the weight of the column of water that projects perpendicularly above the globe and whose height is RS.[11]

If now the upper mouth of the channel is blocked in such a way that the water cannot descend, the globe will indeed be pressed on all sides by the weight of the water stagnating in the channel and the lower vessel IKLM; but, that pressure notwithstanding, the globe will be at rest if it is of the same specific gravity as the water. That pressure will not impel the globe in any direction. And therefore when the channel is opened and the water descends from the upper vessel, all the force by which the globe is impelled downwards will arise from that water's descent and so will be equal to the weight of a column of water whose height is RS and whose diameter is the same as the globe's. That weight, in the time in which any given quantity of water could, with the globe removed, flow out through a hole equal to the base of a cylinder described about the globe, is sufficient to generate all its motion; and so, in the time in which the water by running down uniformly in the cylinder described 2/3 of the globe's diameter, [that weight], if uniformly continued, would generate all the motion of the part of the fluid that is equal to the globe.[12]

The things that have been demonstrated about the water in the channel are to be understood also of any running water whatsoever by which any globe at rest in it is urged. And the things that have been demonstrated about water hold also in all fluids of extremely fine texture whatsoever.

[11] The weight on the sphere's front, in modern terms, then, is $g\rho_f \pi r^2 h$, where r is the sphere's radius, h is the height of the column of water RS, ρ_f is the water's density, and g is the acceleration of gravity. The velocity of the water flowing past the sphere is $\sqrt{(gh)}$. Combining these with the modern nondimensionalization of resistance forces,

$$F = \frac{C_D}{2} \rho_f A_{\text{frontal}} v^2,$$

implies a drag coefficient C_D of 2.0.

[12] The motion mv generated by a force F continued uniformly over a time t is Ft. Since the velocity of the effluent water is $\sqrt{(gh)}$, the time required for the water to go 2/3 of the sphere's diameter is $(4/3)r/\sqrt{(gh)}$. The force, as above, is $g\rho_f \pi r^2 h$. Combining these two entails that a mass of a sphere of water of the same size as the sphere, namely $(4/3)\rho_f \pi r^3$, will indeed acquire a velocity of $\sqrt{(gh)}$ in the time in question.

Now, by corollary 5 of the laws, the force of the fluid upon the globe is the same whether the globe is at rest and the fluid moves with a uniform velocity or the fluid is at rest and the globe goes with the same velocity in the opposite direction. And therefore the resistance of a globe progressing uniformly in any very fluid medium whatever, in the time in which the globe describes 2/3 of its own diameter, is equal to the force that, uniformly impressed upon a body of the same size as the globe and having the same density as the medium, in the time in which the globe by progressing described 2/3 of its own diameter, could generate the globe's velocity in that body. The resistance of the globe on the preceding part of its surface is that great. Q.E.I.

Corollary 1. If a spherical solid moves freely in a fluid of very fine texture with the same density as its own and while moving is urged from behind by the same force as when it is at rest, its resistance will be that which we have described in Proposition 36 corollary 2.[13] Whence, if we go into the reckoning, it will be evident that the solid will lose half of its motion before it has, by progressing, described the length of its own diameter.[14] But if while moving it is urged from behind less, it will be more retarded; and conversely, if it is urged more, it will be less retarded.

Corollary 2. Those people are deluded, then, who believe that the resistance of projectiles is infinitely diminished by an infinite division of the parts of the fluid. If a fluid is extremely thick, the resistance will be diminished somewhat by the division of its parts. But after it has acquired the appropriate degree of fluidity (such as perhaps the fluidity of air or water or quicksilver), the resistance on the anterior surface of the solid will not be much diminished by a further division of the parts. For it will never be less than in accord with the limit that we have given in the above corollary.

[13] As noted in earlier notes, Newton's analysis in Proposition 38 implies an inertial resistance force equivalent to a drag coefficient C_D of 2.0. By contrast, the inertial resistance force given for inelastic particles in corollary 2 of Proposition 36 is equivalent to a drag coefficient of 1.0. The source of the discrepancy is unclear. One possibility is that the phrase "by progressing described 2/3 of its own diameter" in Newton's nondimensional specification of the force is open to two interpretations. When interpreted as *progressing uniformly at its acquired velocity*, then the time is $(2/3)d/v$ and Newton's non-dimensionalized magnitude for the resistance corresponds to a C_D of 2.0. When interpreted as *progressing from rest to its acquired velocity*, then the time is $(2/3)d/(v/2)$, and the magnitude corresponds to a C_D of 1.0. Perhaps Newton became confused between the two when he formulated this corollary.

[14] If the resistance force is equivalent to a C_D of 1.0, then a sphere will lose 53% of its motion in progressing the length of its own diameter; if instead the force is equivalent to a C_D of 2.0, then it will lose 78% of its motion.

Corollary 3. Media in which projected bodies progress very far without sensible diminution of motion are, then, not only very fluid but also far rarer than those bodies that move in them, unless perhaps someone should say that every extremely fluid medium, by a perpetual impulse made upon the back part of a projectile, promotes its motion as much as it impedes and resists on the front part. And indeed it is likely that some part of that motion that the projectile impresses on the medium, is given back to the body from behind by the medium carried circularly. For also having made certain experiments, I have found that in sufficiently compressed fluids some part is given back.[15] But that it is all given back in any case whatsoever neither seems reasonable nor squares well with the experiments I have hitherto tried. For it will be more fully evident by the two following propositions that, if fluids, however fine in texture, are dense, their force to move and resist solids is very great.

LEMMA 4
If a spherical vessel full of quiescent homogeneous fluid is moved straight forward by an impressed force, and with advancing motion always accelerated goes on in such a way that it is not moved during the time in a circle, the parts of the enclosed fluid, by participating equally in the motion of the vessel, will be at rest among themselves. The same will hold in a vessel of any shape whatsoever. This is manifest and does not need demonstration.

PROPOSITION 39 *Theorem 30*
Every fluid that progresses with accelerated motion in the manner of a wind becoming strong, and whose parts are at rest among themselves, seizes all bodies of the same density that are floating in it and carries them away with it with the same velocity.

For, by the above lemma, if a vessel that is spherical, rigid, and full of a homogeneous quiescent fluid, progresses by a motion impressed little by little, all the parts of the fluid, participating in the motion of the vessel, will always be at rest among themselves. Therefore, if some parts of the fluid are congealed, they would go on being at rest among the remaining parts. For since all the parts are at rest among themselves, it is the same whether they are fluid or whether some of them become rigid. Therefore, if the vessel is moved by some force impressed from without and impresses its motion upon the fluid, the fluid will also impress its motion upon its own congealed parts and will take them along with it. But

[15] These are presumably the pendulum experiments presented in the General Scholium at the end of Section 7 in the first edition, the relevant passages from which are translated below.

those congealed parts are solid bodies of the same density as the fluid; and the account of the fluid is the same whether it is enclosed in a moving vessel or blows, like the wind, in free space. Therefore every fluid that is borne by an advancing accelerated motion, and whose parts are at rest among themselves, takes along with it any enclosed solids of the same density that were initially at rest, and compels them to move in the same way. Q.E.D.

PROPOSITION 40 *Problem 10*
To find the resistance of spherical solids in very fluid mediums of given density.

In any given fluid whatever, find the uttermost resistance on a solid given in kind whose size is increased without limit.[16] Then say: as the motion it loses in the time in which by progressing it describes the length of its own semidiameter is to its total motion at the outset, so the motion that any given solid would lose by describing the length of its own diameter,[17] in the same fluid now made to be of extremely fine texture, is to its total motion at the outset, very nearly. For if the smallest particles of the fluid made to be of fine texture have the same proportion and the same situation with regard to the solid moving in it that the like number of smallest particles of a fluid not made to be of fine texture have to the solid increased, and the particles of both fluids are slippery to the highest degree and are entirely lacking in centrifugal or centripetal forces, and the solids begin to move similarly in these fluids in any proportional times whatever, they will go on moving similarly and so, in the time in which they describe spaces equal to their semidiameters, will lose parts of motions proportional to the wholes; and this is so even if the parts of the medium made to be of fine texture are diminished and the size of the solid moving in the medium not made to be of fine texture is increased without limit. Therefore, from the resistance of a solid increased in a medium not made to be of fine texture, the resistance of a solid not increased in a medium made to be of fine texture will be given by the above proportion. Q.E.I.

If the particles are not entirely slippery to the highest degree, it must be supposed that they are equally slippery in the fluids, so that from the defect of lubricity the resistance may be equally increased on both sides; and the proposition will still be valid.

[16] The idea here becomes clearer in the sequel: keep increasing the size of the body whose resistance is being found experimentally until the nondimensionalized resistance in the fluid in question reaches its largest value. Beyond this point, the size of the fluid particles, relative to the size of the body, ceases to matter.

[17] Changed to "semidiameter" in Newton's copy of the first edition.

Corollary 1. Therefore, if from the increased magnitude of a spherical solid its resistance is increased in the duplicate ratio, the resistance of a given spherical solid will by no means be diminished as a result of a diminished magnitude of the particles of the fluid.

Corollary 2. But if, by increasing the spherical solid, the resistance is increased in less than a duplicate ratio of the diameter, it will, by diminishing the particles of the fluid, be diminished in the ratio by which the increased resistance is wanting from the duplicate ratio of the diameter.

Corollary 3. Whence it is obvious that the resistance of a given solid cannot be much diminished by the division of the parts of the fluid. For the resistance of the increased solid will have to be approximately as the quantity of the resisting fluid matter that that solid thrusts forward by moving and propels from places it itself enters and occupies; that is, as the cylindrical space through which the solid moves, and so in the duplicate ratio of the semidiameter of the solid, approximately.

Corollary 4. Therefore, supposing two fluids, the force of resisting of one of which is far exceeded by that of the other, the fluid that resists less is rarer than the other; and the forces of resisting of all fluids are nearly as their densities, especially if the solids are large and move swiftly, and the compression of the fluids is equal.

Three Paragraphs from the General Scholium on Pendulum Decay Experiments in the First Edition That Were Eliminated in the Second Edition[18]

Paragraph cut from the first edition immediately following the table presenting the data for pendulum decay in water, p. 459 of Koyré and Cohen, Isaac Newton's Philosophiae Naturalis Principia Mathematica.

The resistance here never increases in a ratio greater than the square of the velocity. And the same result is true enough in the case of a larger pendulum provided that the vessel is enlarged in the ratio of the pendulum. But if the velocity is increased by degrees to infinity, the resistance both in air and in water will finally have to increase in a ratio a little more than the square because in the experiments described here the resistance is less than in accord with the ratio demonstrated for very swift bodies in Propositions 36 and 38 of this book. For, very swift bodies leave an empty

[18] I.e., the General Scholium that was moved from the end of Section 7 to the end of Section 6 in the second and third editions. Several other changes were made to this Scholium. What singles out the three paragraphs translated here is that they concern general conclusions drawn from the experiments and a conical pendulum experiment corroborating these conclusions.

space behind them, and for that reason the resistance they experience on their front parts will not be at all diminished by the pressure of the fluid medium on the rear parts.

Two paragraphs cut from the first edition immediately before the discussion of the attempt to measure the resistance of any aetherial medium, p. 460ff of Koyré and Cohen.

Therefore, since a globe of water moving in air experiences a resistance by which 1/3261 of its motion is taken away while it describes the length of its semi-diameter (as already has been shown previously),[19] and since the density of air to the density of water is around 800 or 850 to 1, the consequence is that this rule [that follows] holds generally. If any spherical body moves in any sufficiently fluid medium and only that part of the resistance is regarded that varies as the square of the velocity, this part will be to the force which can either take away or generate the total motion of the body while that body describes twice[20] the length of its semi-diameter in motion uniformly continued, very nearly as the density of the medium is to the density of the body. Therefore the resistance is about three times greater [*sic*][21] than in accordance with the law given in the first corollary of Proposition 38; and thus about two thirds of that total motion which the front parts of the globe impress upon the medium while moving is restored to the rear parts of the globe by the medium as it returns in a circle and rushes into the space that the globe would other-

[19] This value, which was changed to 1/3342 in the second and third editions, is put forward at the end of the discussion of the first set of data for pendulum decay in air earlier in the General Scholium.

[20] Changed to "four times" in Newton's errata to the first edition. "Twice" implies a drag coefficient of 1.33, whereas "four times" implies a drag coefficient of 0.67. Newton's fraction of the motion lost, 1/3261, implies a drag coefficient of 0.695 if the ratio of the density of water to the density of air is taken to be 850 and a drag coefficient of 0.654 if this ratio is taken to be 800.

[21] As evident from the next clause, as well as from the rest of the paragraph, Newton has misspoken here, for he must mean not "three times greater than," but "three times less than." In other words, the measured resistance force is around one-third of the resistance force on the front face given in Proposition 38. Newton changed "*major*" to "*minor*" in his annotated copies. This change was accompanied by still other changes: one annotation reads, "is less in the ratio of 2 to 3," and another, "is around one half less." Other corrections made to the passage in his annotated copies suggest that these latter fractions, two-thirds and one-half in contrast to one-third, reflect an effort to incorporate a correction for his mistaken solution of the efflux problem in Proposition 37, a correction that reduces the resistance acting on a sphere's front face given in Proposition 38 by a factor of 2.

wise leave empty behind itself. Hence, if the velocity of the globe increases to such an extent that the medium cannot so quickly rush into that space, but rather some emptiness is always left behind the globe, the resistance will finally turn out to be about three times greater than in accordance with the general rule most recently laid down.

Up to now we have used experiments with oscillating pendulums, because their motions can be observed and measured more easily and more accurately. But I have also taken into consideration the motions of pendulums propelled circularly and describing circles while returning in a curve, because they are uniform and on that account seem much more suitable for investigating the resistance corresponding to a given velocity. For, in making a circularly propelled pendulum revolve twelve times, I noted the sizes of the two circles which it described in the first and last revolution. And I gathered[22] from this the velocities of the body at the beginning and the end. Then, by positing that the body, in describing twelve mean circles with its mean velocity, would lose the difference between those velocities, I gathered the resistance by which that difference could be lost in that whole course of the body through twelve circles; and, although experiments of this kind allowed less accurate tests, the resistance that was found nevertheless agreed excellently with the preceding.[23]

[22] The Latin verb is *collegi*. Newton often used it when speaking of drawing conclusions.

[23] The idea of this experiment is straightforward. So long as the conical pendulum's deceleration from resistance is not excessive and the velocity is large, the decrement in velocity, Δv, over the course of n cycles is approximately $\lambda v_{\text{avg}}^2 n P_{\text{avg}}$, where v_{avg} is the average velocity during the n cycles, P_{avg} is the average period of the cycles, and the resistance rule is given by $dv/dt = -\lambda v^2$. The period of a conical pendulum is

$$P = 2\pi\sqrt{h/g},$$

where g is the acceleration of gravity and h is the height of the conical pendulum, namely,

$$h = \sqrt{\ell^2 - r^2},$$

where ℓ is length of its string and r is the radius of the circle that the bob describes in a given cycle. The velocity in any cycle when the deceleration from resistance is small is

$$v = r\sqrt{g/h}.$$

Combining these expressions yields

$$\Delta v = \lambda v_{\text{avg}}(2\pi r_{\text{avg}})n.$$

Thus, given the length of the pendulum and the acceleration of gravity (or equivalently Huygens's value for the distance of free fall in the first second), a determination of the radii of the first and last circles suffices to determine the velocities, λ (namely, from $\Delta r/[2\pi r_{\text{avg}}^2 n]$), and hence the magnitude of the resistance force.

Appendix

Some Recollections of Richard Samuel Westfall (1924–1996)

I. Bernard Cohen

Richard S. Westfall, known to friends and colleagues as "Sam," was a scholar of rare distinction, as declared in his publications and in many prizes and awards. He was most generous, always glad to be of help to other scholars, and he was faithful in his friendships. The chapter he wrote for the Dibner Newton symposium, printed below, seems to be the last composition that he completed for publication before his death.

I first encountered Sam in person at a meeting of the History of Science Society held in Chicago on 29 December 1959. Sam was then teaching at Grinnell College and had recently published a book on *Science and Religion in Seventeenth Century England.* I was one of two speakers at a joint session of our History of Science Society and the American Historical Association. The other speaker was Owsei Temkin, the eminent historian of medicine and biological thought.

My assignment was to present some of the recent developments in Newtonian scholarship. At the end of the session, Sam came up to the platform and introduced himself. I knew who he was, of course, and had referred to his book in my talk. We had been in correspondence but had never met. In his customary forthright manner and without a moment's delay, Sam announced to me his intention of writing a full-length biography of Isaac Newton. Since I had personally examined the vast mountains of unpublished Newton manuscripts and the extreme difficulties in such an assignment, I tried to warn him of the gigantic enterprise he was daring to mount. Some years earlier, Rupert Hall and I had been considering a two-volume *Life and Times of Isaac Newton* but had given up the project when we came to see how many years of intense concentration and labor it would require.

As a somewhat older colleague who knew the magnitude of the assignment Sam was about to undertake, I felt honor bound to warn him of his project's awesome magnitude. I told him, as bluntly as I could, that he was a young scholar (he was then 35 years of age) at the start of a promising career and that the work he was proposing would occupy a most creative and productive period of his life at a time when he needed to make his reputation and secure advancement along the academic ladder. I wanted him to be certain that he knew what he was getting into. I told him that he would be spending the next ten to fifteen years on this project.

In the event, my estimate was too short. His great biography, *Never at Rest* , did not appear until 1980, twenty-one years after our conversation. In retrospect I am of course glad that Sam was not deterred by my "realistic" appraisal of the magnitude of the assignment he had chosen. I am proud that, shortly after the Chicago meetings, I was able to help Sam gain support for his project of writing a biography of Newton. I also had the honor of being the first person to mention this proposal in print, in the published version of my talk, which appeared in *Isis* in 1960. Here I quoted from what was described as "a privately circulated preliminary statement of the project," as follows: "In a finished biography each aspect of Newton's life should, of course, receive its full due; the ultimate purpose, however, must be a fuller understanding of his scientific work. Although Newton was a complex and not uninteresting personality, few would have heard of him and fewer still would study him had he not been a great scientist. I am proposing a project in the history of science in which a fuller understanding of the man should increase our understanding of his work."

How grateful we are that Sam was willing to undertake this project and to do so in such an exemplary manner! I know of no other biography of a scientist that is in the same category of excellence for its wealth of information and insight into scientific ideas. We may be especially grateful that Sam was willing to take on this assignment, because it required someone with a rare combination of talents, interests, and understanding. Others could have coped with Newton's mathematics, physics, and astronomy, but could they also have dealt sympathetically and nonjudgmentally with Newton's explorations of theological issues, Biblical chronology, and the secrets of the Bible's prophetic books (to say nothing of problems of prophecy, of traditions of ancient wisdom, and alchemy)? And would they have been able to combine all of this with Newton's activities in the Mint in the latter part of his life?

I once told Sam that I had never opened up his book without finding something that was either new to me or that I had forgotten. In a letter to me he once wrote that "it is not clear to me that I deserve to be mentioned in the same breath with the others you list." But characteristically, he hastened to add that he "would not protest too much."

Sam was an innovative scholar, seeking an understanding of certain issues concerning the life and work of Galileo, the relations between science and the church, and aspects of scientific patronage in the seventeenth century. He and I collaborated on a number of projects, including the production of a biographical television program on Isaac Newton, the occasion of my last visit with him. We worked together as well in editing a Norton Critical Edition of Newton.

Our last full cooperation shows Sam's extraordinary generosity of spirit in a way that mere words can never do. After we (that is, Alexandre Koyré, Anne Whitman, and I) had published our Latin edition of the *Principia*, reviewers, other Newton scholars, and colleagues and teachers urged us to undertake a new English translation. Among the strongest voices urging us to do so were Michael Mahoney and Sam Westfall.

After Anne Whitman and I agreed to undertake this monumental task, Sam provided us with a set of guidelines that were of the utmost importance in our final decisions in many different ways. In particular, he stressed the need of a text that would be available to (that is, readable by) students.

The true sign of Sam's generosity and *conscience professionelle*, the full measure of his friendship, was displayed after the job had been completed, consisting of a new English version of all three "books" of the *Principia* plus a book-length guide to go along with it. Sam went through the guide and the whole of the translation page by page and sentence by sentence (sometimes even phrase by phrase). He sent me not only pages of corrections of typographical errors, but page after page of suggested improvements of my explanations in the guide and in the versions of Newton's text. He argued so valiantly and convincingly concerning certain fundamental aspects of the work that I had no choice but to accept his improvements, all of which were for the betterment of the text. In a long scholarly lifetime I have never encountered so great a display of scholarly solidarity and so magnificent a sign of true friendship.

Acknowledgment

This account of my relations with Sam Westfall was prepared for the memorial service held at Indiana University on 6 December 1996.

The Background to the Mathematization of Nature

Richard S. Westfall

Historians of science are almost unanimous in making what Alexandre Koyré called the "mathematization of nature" one of the central, perhaps the most central characteristic of the Scientific Revolution. During the sixteenth and seventeenth centuries, quantification, quantitative precision, mathematics in multifarious forms entered science in a way it never had before. I seek in this chapter to illuminate some of the background of the mathematization of nature. I will not say a word about the content of mathematics. I see nothing in my material that would explain the development of the new analysis and the calculus. Rather I seek to examine the role of mathematics in the culture of western Europe at that time in an attempt partially to illuminate why mathematical endeavor that expanded the discipline's boundaries beyond those recognized in ancient Greece should have exploded as it did during the sixteenth and seventeenth centuries.

Central to my argument is the spread through European society of that period of four technologies, all of them vital to the life of the time, and all of them heavily mathematical. The first of these, water management, which I shall call hydraulic engineering to avoid circumlocutions, although the term was not used then, was anything but new. Its long history stretched back to the beginning of civilization; it might even be said to have been a necessary precondition of civilization. The semi-arid river valleys in which the early civilizations appeared required irrigation. The same river valleys saw periodic floods as well, and if the floods could not be entirely contained, at least their worst effects could be mitigated. Without a supply of water, cities were impossible. Few remains of Rome, which flourished of course millennia after the earliest cities and in a different setting, have imprinted themselves more vividly on our imaginations than its aqueducts, and not only those serving Rome itself. Centuries later but still well before our period, in thirteenth-century Italy, men skilled in these activities were already being called "engineers." By that time, the need for them had acquired a new, industrial dimension as water mills supplied an increasing proportion of the energy that powered economic enterprise. Obviously hydraulic engineering was not born in the sixteenth and seventeenth centuries.[1]

Nevertheless, hydraulic engineering underwent a major expansion during that time, to the extent that it is not misleading to speak of its spread through society. After the long depression of the late Middle Ages, economic life began to revive at the end of the fifteenth century. Surging populations pressed upon the supply of land, and the demand for reclamation grew accordingly. The plain that stretches west-northwest from Florence to Pistoia, the breadbasket of Tuscany, largely a morass at the beginning of the sixteenth century, was drained, and somewhat later, further up the Arno, the Val di Chiana.[2] So were extensive marshes along the shores near Pisa and Venice.[3]

The Venetian lagoon, which had slowly become a problem during the late Middle Ages, now demanded constant attention. A wide swath of land north of the Po received irrigation. Problems of the Reno and Po near Ferrara called forth a succession of expedients that were never adequate in that age but did keep a cadre of hydraulic engineers employed.[4] In northern Europe, the area we now know as the Netherlands had also seen extensive hydraulic engineering during the Middle Ages; some management of water was close to the basic requirement if human beings were to live there. Lands were drained, and dikes were built along shores and river banks. When the land sank as a result of the drainage, the waters claimed anew a good proportion of what early hydraulic engineering thought it had reclaimed. With the sixteenth and seventeenth centuries, efforts at reclamation underwent a vast expansion and an equal increase in effectiveness. During those two centuries twice as much land was reclaimed, this time permanently, as during the whole Middle Ages. At this time Dutch engineers learned to pump out polders that had sunk below the level of the sea.[5] Although Italy and the Netherlands were the areas of greatest activity, they were not the sole areas. For example, England drained its great fens in the middle of the seventeenth century.[6]

Technical experts multiplied in equal proportion. In Italy, the different states created permanent boards, manned by technical experts—such as the Uffiziali dei Fiumi in Florence and the Collegio alle Acque in Venice—to manage problems with water. We can follow the careers of some of the most prominent experts. Two Dutchmen, Jan Leeghwater and Cornelius Vermuyden, for example, built international reputations and acquired international clienteles for their services.[7] Not by accident did a book of Benedetti Castelli, published in the early seventeenth century, launch the modern science of hydraulics, which built a corpus of theory that gradually replaced earlier empirical practices.[8]

Like hydraulic engineering, military engineering also had a long tradition behind it. In this case, however, one can speak without reservation of a new military engineering because of a new impulse and a new direction that it received right at the beginning of the sixteenth century. During the fifteenth century, the growing spread and effectiveness of artillery had doomed the medieval fortress. Its walls were too vast to miss but too fragile to carry cannon in defense. The late fifteenth century and early sixteenth witnessed the gradual evolution of a new style of fortification that converted the cannon, erstwhile weapon of offense, into the anchor of an impregnable defense. The Fortezza da Basso in Florence, constructed in 1534–1535, is taken as the culmination in which the new style reached full realization. Fortification in the new style was the essence of military engineering during the sixteenth and seventeenth centuries.[9]

The structures had low profiles; they were landfills with stone veneers, though the veneer was less essential than the landfill, which did not present a high target but could mount and protect canon. More important was the outline of the new style, built up from what was called the angle bastion. The heart of a fortress made up of angle bastions was the principle that every surface exposed to attack be defended by flanking fire from cannon located in and shielded by a neighboring bastion. A sketch by Simon Stevin showing lines of fire captures the idea of the angle bastion (figure A.1).[10] Anyone who visits a historic fortification can grasp the basic principle in a glance by repeating my recent experience; sighting along a face of one bastion of the Morro castle, constructed by the Spaniards in the sixteenth century to defend Havana Harbor, I found myself looking straight into the mouth of the portal in the neighboring bastion, where a cannon to defend the bastion by which I stood had once been mounted, of course. The well known Castel Sant'Angelo along the Tiber near the Vatican, where a circle of bastions constructed in the sixteenth century surrounds the medieval stronghold built in its turn around the mausoleum of Hadrian, displays the new concept in its contrast to the old in a single setting.

Designing a fortification made up of angle bastions was an exercise in applied geometry, as a plan by Stevin suggests (figure A.2).[11] The engineer had to control angles to ensure coverage and distances to correspond to the cannon's range. Anyone known to be skilled in mathematics was apt

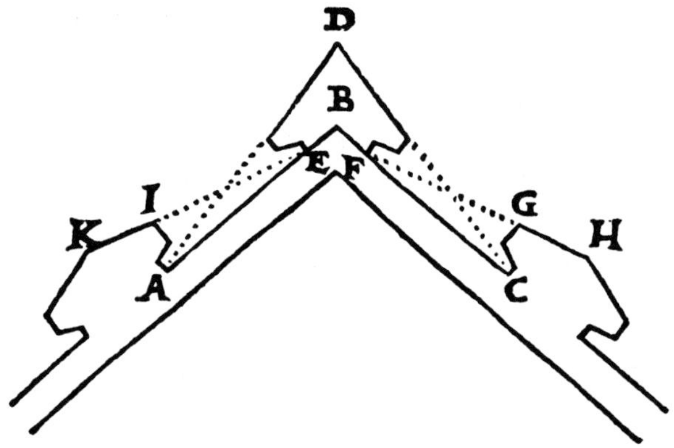

Figure A.1

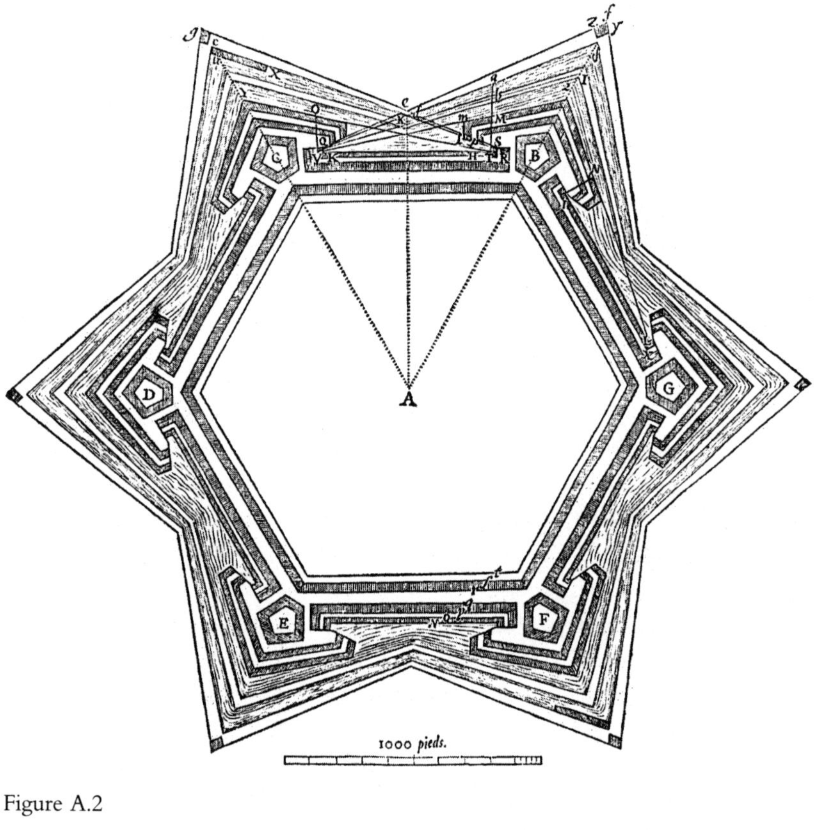

Figure A.2

to find himself dragooned into service as a military engineer. Not infrequently, they were the same men who worked as hydraulic engineers.

Like military engineering, navigation also took a new direction that emphasized its mathematical content at the beginning of the sixteenth century, as Europe ventured out beyond the closed seas of medieval commerce onto the great world oceans. Portugal had begun the process with its explorations down the coast of Africa during the fifteenth century. At the end of the century, it cleared the Cape of Good Hope and found the route to India, at almost the moment when Spain began to explore across the Atlantic, and during the sixteenth century the rest of the maritime countries of Europe entered in. The method of navigation that had worked so well in the Mediterranean and even up the Atlantic coast to England and the low countries no longer sufficed. The basic problem was the navigation charts, which distorted a spherical surface into a flat one. On the great oceans, it was impossible to lay out a course by chart and compass alone and end up anywhere near one's intended destination. A Portuguese mathematician demonstrated why this was so. There was only one solution, to learn to determine latitude and longitude and to lay out on a globe a course from one's present location to one's goal. From the beginning, the concerned authorities turned to their mathematicians and astronomers for assistance. Portuguese astronomers taught their navigators to determine latitude at any point on the globe via celestial observations. The determination of longitude proved far more difficult. Every promising suggestion came from a scientist, and although the final step in the definitive solution came from a clockmaker, his step was possible because of earlier work by scientists. None of the major advances in navigation came from practical tarpaulins who sailed before the mast; every one came from mathematically skilled scientists.[12]

All of this is well known, although I think the extent to which the scientific community was involved in these activities is not equally well known. Perhaps I am projecting my own recent ignorance about cartography onto others, but I have allowed myself to be convinced that most historians of science do not understand the extent to which cartography, as an undertaking that stepped outside the learned tradition of geography to serve practical needs, was in the full sense of the word a new enterprise in the sixteenth century. There were very few maps in Europe before 1500. The *mappaemundi*, of which everyone has heard, were more theological documents than maps, histories of the creation and the redemption of man in a form somewhat like a map. They took the shape of a circle, with Jerusalem at the center and paradise at the top. Christian symbols decorated them. One of the well-known *mappamundi* has the head of Christ at the top, his feet at the bottom, and his two hands on either side, a cartographic embodiment of the the words of John that "all things were made through him, and without him was not anything made that was made." Of the 1,100 *mappaemundi* known to exist, 900 appear near the beginning of manuscripts, where they function as illustrations rather than maps—the work of copyists rather than cartographers.[13]

Medieval portulan charts present a different state of affairs. Precisionists may dispute whether navigational charts that show only the outline of shores should be called maps. If they were not maps, they were at worst first cousins to maps, and they offer an extraordinary representation of the Mediterranean and the Continent's western coast. No one knows how they were made in the first place. Extraordinary as they are, their use was confined to mariners.[14]

Most important of all was the classical heritage of geography western Europe recently rediscovered. Ptolemy's *Geography* had been recovered initially in Byzantium at the end of the thirteenth century, and a number of copies of the Greek manuscript, some of which came to western Europe, were made during the following two centuries. Early in the fifteenth century, an Italian scholar translated it into Latin; more copies of the Latin were made. Near the end of the fifteenth century, the book was published; there were a total of seven fifteenth-century editions with more sixteenth century ones. In Ptolemy Europeans could find the concept of a graticule, a network of lines of latitude and longitude whereby the earth's spherical surface was made to imitate analogous coordinates applied by astronomers to the heavens above, which defined the location of points, such as cities, river mouths, and capes, on the earth. Europeans could find as well three different projections that Ptolemy divised to minimize the distortions that inevitably arise when the earth's spherical surface is represented on a flat sheet of paper.[15]

Historians of cartography debate whether Ptolemy himself drew maps to accompany his *Geography*, but it is well established that maps, most commonly one of the known inhabited world as Ptolemy conceived it and twenty-six others of its major regions, accompanied about a quarter of the Greek manuscripts and most of the Latin. Significantly for what it says about maps' role in fifteenth-century society, the first printed edition of the *Geography*, in 1475, did not contain maps. However, the second edition, 1477, did, and so did most subsequent editions. For most readers, they were almost certainly the first maps they had ever seen.

Ptolemy's *Geography* presented an indispensable theoretical basis for subsequent European cartography, and the maps that accompanied its editions were undoubtedly a major factor in making Europeans aware of the possibilities that maps held. Nevertheless, the manuscripts and early printed editions did not in themselves embody a tradition of cartography. The men who originally drew maps to accompany the manuscript toward the end of the thirteenth century must have understood the thrust of the text, but nearly all the subsequent maps were the work of copyists rather than cartographers. Moreover, the work concerned itself with the discipline of geography—the location of known, inhabited places on the surface of the globe—not with local, cadastral mapping, which became the primary substance of the practical cartography that developed during the sixteenth century.

Beyond these three confined categories, maps scarcely existed in medieval Europe. Historians of cartography believe that England, with its live antiquarian tradition, has explored its heritage of maps more fully than any other country. A total of three maps from the two centuries between 1150 and 1350 are known. The number then increases slightly, so that ten are known from each of the following half centuries before 1500. That is, the maps from before 1500 known in England total thirty-three. After 1500 the number begins to mount. There are two hundred from the next half century, and after 1550 it is probably impossible to count them. The numbers appear to be similar elsewhere, though undoubtedly somewhat larger for the period just before 1500 in Italy, Europe's leader in all things cultural. Historians of cartography insist that the small number of maps is not a question of survival. Other documents survived. It is instructive that no medieval language had a word exactly equivalent to our word "map." It is also instructive to look at books such as Leonard Digges's *Techtonicon*, published in 1556, one of the earliest manuals on surveying in England. Digges taught his readers how to measure pieces of property, but he did not foresee the production of maps. Before 1500, properties were described in prose, in what have been called "speaking maps."[16]

During the sixteenth century—historians of cartography are beginning to date the change more precisely to the fourth decade of the century—map consciousness, the realization of maps' capacity to summarize and convey information, literally exploded in Europe. It no longer remains feasible to count the maps produced. The Florentine archives alone contain between 4,000 and 5,000 maps from the century following 1550.[17] I myself have used some of the records of the Florentine Uffiziali dei Fiume, the Commissioners of Rivers, from the early seventeenth century. By then, when the Commissioners sent one of their technical experts out to deal with an emergency, they expected and received a map of the area in question with every report.

The birth of map consciousness coincided with the beginning of the Scientific Revolution. I have become convinced that cartography should be seen as the first modern scientific technology, one that had a precious theoretical tradition in Ptolemy but no previous empirical tradition worth mentioning. Every major figure in the early history of European cartography was a scientist of importance for other reasons as well, and all of the procedures of cartography depended directly on mathematics and science. Gemma Frisius defined the method of triangulation. Tycho Brahe carried out the first known triangulation. Early in the seventeenth century, Willibrord Snel used triangulation to measure a line from the southern end of the United Provinces to the northern and, with observations of the two latitudes, to calculate the earth's size.[18] Gian Domenico Cassini, the first astronomer royal in France, was from the beginning of his tenure in Paris deeply involved in the project to map France; Jean Picard and Philippe de La Hire were his colleagues in that project.[19]

The same type of observations that established latitude for navigators did the same, of course, for cartographers. For longitudinal differences, however, cartography was more fortunate. Jupiter's moons never worked for navigation; observations of the moons from the deck of a pitching ship proved impossible. Observation on land was another matter, and since there was no need to establish longitude immediately, the observations could be compared at leisure, not with imperfect tables, but with those of the same phenomena made on the same night at the observatory in Paris. In the latter years of the seventeenth century, for the first time ever, cartographers fixed the precise coordinates of hundreds of places both in Europe and around the globe.[20] Navigation could also benefit, because navigators now knew at least the locations of ports to which they were sailing. The same scientists who raised cartography to the level of quantitative precision also fathered a flock of improved surveying instruments. Among other things, they applied telescopic sights to the instruments, so that measurements could reach a greater degree of precision.

Cartography appears to me to be the foundation stone of the mathematical technologies. Without exhausting the content of the other three, it did support all of them. The problems of navigation, which depended on fixing latitude and longitude, had obvious similarities to those of cartography. Hydraulic engineering could not proceed without surveying, especially of level lines. The plan of a fortress was a map in miniature. The birth of map consciousness early in the sixteenth century greatly expanded the demand for mathematics.

In addition to these four, which I look upon as fundamental, mathematics entered centrally into at least two other technologies of importance, civil engineering (by which I refer to roads and bridges) and architecture. Along with hydraulic engineering, they were activities equally present in other civilizations. The same is not true of the other three. No one competed successfully with Europe in navigation of the world oceans. Europe explored the rest of the world; the rest of the world did not explore Europe. The new style of fortification has been called Europe's distinctive architecture of the early modern age, the form that it transported around the globe.[21] I have never seen a fortification built by another culture in Europe, but Europe built its angle bastions around the world. I myself have seen such fortresses, built by Spaniards in the sixteenth century in Havana and Acapulco, built by Frenchmen in the seventeenth and eighteenth centuries at Port Royale and Louisburg in Nova Scotia, built by Englishmen on the American frontier in the eighteenth century at Ticonderoga and Pittsburgh. Although I have not seen them, I am informed that they stand along the coast of Africa, in India, and in Indonesia. Most important of all, map consciousness as a widespread phenomenon appeared in Europe. Europe mapped the rest of the world; the rest of the world did not map Europe. There had never been a culture with a similar demand for mathematics.

I want to pursue here the way in which this demand impinged on the scientific community. I intend to cite numbers. I have collected data on the scientists from the sixteenth and seventeenth centuries that appear in the *Dictionary of Scientific Biography*, 631 in all. For my purposes here, 631 is misleadingly high. Three-eighths of them were physicians, and with only two exceptions, physicians did not engage in mathematical technologies. I am concerned with the 394 who were not part of the medical community. Only a small number did any civil engineering or architecture. The four major technologies were quite another matter. Fifty-six or fifty-seven in each case involved themselves in hydraulic engineering, military engineering, and navigation, overlapping but not identical groups. Nearly twice as many, 101, practiced some cartography. To avoid misunderstanding, let me explain those figures. For many of the men, the technology in question was not their primary activity. In every case, however, they did engage at least once in a real project of practical significance concerned with the technology. The numbers for each technology constitute significant percentages of 394.

The percentages become much larger if we limit ourselves to mathematicians. By a mathematician in this context I mean someone who made at least one contribution to mathematics. A number of those whom I call mathematicians according to this definition were not known primarily as mathematicians. Using this definition, I count 205 mathematicians, but again I want to reduce the number, this time by 59, leaving 146 mathematicians. For 30 of the 59, mathematics was a wholly subordinate part of their careers. Almost nothing is known about 10 more of the group, and another 10 were wealthy aristocrats who had no need to engage in practical activities. The remaining 9 of those excluded include a couple of Scholastic philosophers from the early sixteenth century, several authors of early commercial arithmetics, and so on. Of the 146, 33 to 35 in each case had something to do with hydraulic engineering, military engineering, and navigation; 56 did some cartography. Because the groups are not identical, a total of 98 engaged in one or another of the four major technologies. Eight more either invented or made relevant instruments, giving a total of 106.

It is important to my argument that the four technologies pertained primarily to the existing structure of power and wealth. Navigation, with its suggestion of capitalistic commerce, sounds different at first, but commerce did not begin with modern capitalism. Everywhere governments were heavily dependent on income from customs and made it their business to foster trade. The great Dutch drainage projects of the seventeenth century were carried out by capitalistic enterprises, but reclamation pertained mostly to established agriculture and the landed class.[22] The four major technologies were essential to the health and vitality, not of some hypothetical society yet to be born, but of existing society. They were concerned with the food supply, the defense, the commerce, and the delineation of juridictions within states and external borders where they confronted others. Can anyone doubt how vital to the existing order the technologies were?

It is no surprise then that the same men who patronized art, music, and literature also patronized those who could provide essential technical services. In an age when most careers above common labor depended on patronage, young men with aptitude in mathematics could not fail to understand that within the existing order careers could be successfully pursued in the mathematical technologies. Of the 146 mathematicians, only 7 appear to have stood outside the patronage system. Five of those seven also had no connection with the mathematical technologies. Of the 106 involved in the technologies, 102 received patronage from established authorities—76 from some court, 53 from some aristocrat, 36 from some ecclesiastical official, and the same number from some governmental official. The numbers add up to far more than 102. Patronage was hardly ever an exclusive relationship; the great majority had more than one patron. As I said, careers were available within the established order. A young man might commit himself to mathematics with confidence.

Statistics are abstract. Let me follow the same theme by observing the impact of the demand for mathematics on several concrete individual mathematicians. Oronce Finé was born in France in 1494. In 1524 he found himself in prison, apparently because of his opposition to the concordat between the French monarchy and the Papacy. Though educated as a physician, Finé had begun to make a name for himself as a mathematician, and as a mathematician he had won the protection of Admiral de Bonnivet, governor of the the Dauphiné. Bonnivet brought Finé to the attention of Francis I, who promptly ordered his release from prison in order that he might work on the fortifications of Milan, then in French hands. Finé went on to gain prominence as a cartographer, producing two world maps, a new projection of the globe onto a flat sheet of paper and a map of France. In 1531 Francis named him professor of mathematics at his new Collège Royale, a position he held until his death.[23]

Pedro Nuñez was active in Portugal at much the same time. He began his career teaching traditional subjects at the University of Lisbon. Like Finé, Nuñez had been educated in medicine but had earned a growing reputation in mathematics. He is known today as Portugal's greatest mathematician. Nuñez demonstrated that a rhumb line on a flat chart between two places does not correspond to the great circle route between them. He also invented an important instrument, called the *nonius* after the Latin version of Nuñez's name, a precursor of the vernier that made it possible to observe small fractions of angles. Nuñez was amply rewarded for his accomplishments. When the Portuguese university moved from Lisbon to Coimbra, the king ordered his appointment as professor of mathematics, effectively establishing mathematics as a university discipline in Portugal because there was no place for a professor of mathematics in any of the existing faculties. The king also appointed him royal cosmographer with a large salary. He became tutor to the royal children. Nuñez learned as well that patronage had another vital dimension. He was of Jewish origin; because of the protection of the king, the Inquisition never touched him.[24]

Egnatio Danti, a Dominican, lived in Italy a generation later. In 1562, perhaps through his brother, who was a painter at the court, he came to the attention of Cosimo I, who commissioned Danti to paint the great mural maps that still adorn the Palazzo Vecchio. Cosimo appointed Danti professor of mathematics at the University of Pisa at a salary of thirty-six scudi; since his budget was pinched, the grand duke dismissed a professor of theology to cover Danti's salary. Later, Danti mapped Perugia, catching the eye of Gregrory XIII, who had him map the papal states and execute a second set of mural maps, these for the Vatican. Danti was an innovator of surveying instruments. While in Florence, he planned a canal across Italy through the city. He left behind a manuscript on fortification. Through the patronage of Gregory he became the papal mathematician and cosmographer, and he died as bishop of Alatri.[25]

Thomas Digges, in England, was a bit younger than Danti. Digges was the son of a well-known father, whose book on surveying I have mentioned. He published on surveying in his own right. As a military engineer, he worked on the fortifications of Dover and drew a plan of the city. He also wrote on navigation. Digges differed from the previous three I have mentioned in that he came from the prosperous gentry. Even prosperous gentry aspire to more, however, and Digges dedicated his books to prominent figures in the government, such as Bacon and Burghley, and especially Leicester. Leicester's patronage secured his appointment as Muster Master General of the English forces in the Netherlands.[26]

Simon Stevin was a bit younger than Digges. From an impoverished background, he began his career as a bookkeeper in the Spanish Netherlands. When we next catch sight of him, he has raised himself, by means we will never know, to the status of a hydraulic engineer, now in the United Provinces. About 1590, Prince Maurice became aware of Stevin; the relation with the prince dominated the rest of his life. He became the prince's adviser in all matters technical, including administration. He composed several of the works for which he is known for Maurice's instruction. Stevin was deeply involved in hydraulic engineering, as I have said, and also in military engineering. He wrote on navigation. He did less with cartography, although he did improve a surveying instrument. When Stevin died, Prince Maurice expressed his valuation of Stevin's technical knowledge by continuing to support his children.[27]

Five men from five different countries spread through the length of the sixteenth century all conform to a common pattern, the tendency of those with mathematical talent to employ it in the mathematical technologies and to find patronage from the established order. All five appear in the *Dictionary of Scientific Biography*. We can find the same pattern in others who did not make the *DSB*. Richard Lee was a military engineer and early cartographer in England, stationed in the 1530s in Calais. Though of humble birth, Lee won enough prestige in his profession that he was able to marry the daughter of Sir Richard Grenville, high marshall of Calais. Henry VIII knighted him in 1544 after a campaign in Scotland and endowed him with monastic property. Through his technical expertise, Lee had ascended into the gentry, and he married his daughters to other gentry.[28]

A French-Scottish military engineer and cartographer, Jean Rotz, was active in France about the same time. In 1550, Henry II ennobled him for his services.[29] Juanelo Turriano, a transplanted Italian, served both Charles V and Philip II as artisan-engineer. To supply water to the royal palace in Toledo and to the city, Turriano constructed a machine that raised 16,000 liters per day through a height of 100 meters. The project has been called the greatest engineering achievement of the sixteenth century. It was valued then as much as it is praised now. At a time when the average laborer earned 25,000 *maravedi* per year, and the average professor at Valladolid, where professorial salaries surpassed those elsewhere else in Spain, received a salary of 187,000 *maravedi*, Turriano received three million maravedi to construct the machine plus a further 700,000 per year to maintain it. To be sure, this was a contract, but in the sixteenth century patronage helped determine every contract of this sort. In any event, it indicates the demand for technical services and the rewards they received.[30]

Let us, however, at a conference devoted to Isaac Newton, move the spotlight closer to home. Newton could have known Jonas Moore, who was born in England about 1617; he undoubtedly knew about Moore. The son of a yeoman, Moore launched his career in the 1630s down a tried and true path when he became clerk of the vicar-general of the diocese of Durham. The civil war soon destroyed those hopes. Significantly, Moore turned to mathematics. Recall that he was a young man without an estate who was making his own way in the world, by that time with responsibility for a family. Although his calculations are forever closed to us, he must have understood that a living could be gained with mathematics. We next catch sight of him as a teacher of mathematics in London. At this point, the drainage of the fens commenced. Patronage enabled Moore to become surveyor to the project, and through the project he made both his reputation and his fortune. He produced a great map of the fen region. Though not a hydraulic engineer, he made himself competent in the field and later found some employment in it. Moore came from a Puritan background, and he attached himself to the Protectorate. By ordinary standards he should have had trouble at the Restoration, but ordinary standards did not apply to those with special talents. Instead of being cast out, Moore began to be consulted on making rivers navigable. He surveyed the estate of the duke of Bedford. By 1663, he was a client of the duke of York, who sent him to work on the fortifications and harbor at Tangier. He surveyed the Thames for the navy. Through the partronage of York, he was appointed surveyor of the ordnance, which involved him in fortification, on which he also wrote. That is, Moore at least touched on all of the four major mathematical technologies. Because of his technical expertise, this son of a yeoman was knighted and died as rich Sir Jonas Moore, the patron of John Flamsteed.[31]

Edmond Halley moves the spotlight still closer to Newton. His story begins altogether differently, for Halley was the son of a wealthy father who sent him to Oxford as a fellow commoner. Like other men of means to whom a degree offered nothing, Halley left without waiting to gain one. At least by the time of his stay in Oxford, astronomy and mathematics had seized Halley's affections. He now conceived the project of mapping the southern skies, and his father cooperated by supplying a handsome stipend, which he continued after Halley's return from St. Helena. Married, Halley settled down to what promised to be a life free from financial concerns and devoted to science. Then disaster struck in 1684 with the death, probably the murder, of his father. Debate continues about Halley's finances, but his actions make it clear that his security had disappeared. Perhaps there was a ten-year period of partial grace. Although he took a lowly position as clerk to the Royal Society, which was frequently unable to pay his minimal salary, some evidence suggests that at this stage Halley was not yet desperate. If there was a period of grace, it came to an end by 1694, when Halley filed a suit against his stepmother in an attempt to protect his patrimony. From that time on, Halley was consciously seeking patronage and utilizing his mathematical skills to gain it.

Already in 1693 he had formed a plan with Benjamin Middleton for a voyage around the world to gather information, especially on magnetic declination, to improve navigation. The two petitioned the navy to furnish them with a ship. Middleton would undertake to man and supply the ship; Halley would furnish the technical expertise. This was clearly a patronage relationship. Nothing is known about Middleton, though it is worth noting that Middleton was the family name attached to the New River project and that in 1695 Halley surveyed the New River. Be that as it may, something delayed the expedition, and Halley was now in real need. Newton came to his aid with a position in the Chester mint and offered suggestions about an engineer's appointment in the military. Eventually the naval expedition came through, though Middleton had now dropped out of sight, and the navy financed it all. Halley had a reputation in mathematics. He was known to be working on the problem of longitude, both via magnetic declination and via lunar positions. Although he had never been to sea, except as a passenger to and from St. Helena, Samuel Pepys, who knew, recorded about 1695 that no one else anywhere combined so much "of the science and practice (both) of navigation."

In all, Halley made three voyages for the navy. He produced a pioneering map showing the lines of constant magnetic declination in the Atlantic. He charted the English Channel. Above all, he established himself in the navy's eyes. Almost immediately after the third voyage, the Earl of Nottingham, the principal secretary of state, who was responsible for naval affairs, secured Halley's appointment to inspect ports along the Adriatic coast of the Holy Roman Empire. Although he had no military experience, he was deemed qualified to advise on the fortification of a port that the Royal Navy might use with safety, and eventually even to direct the fortification of the port he chose. Halley made two trips to the Adriatic. Before his return from the second, John Wallis, Savilian Professor of Geometry at Oxford, died. In 1691, Halley had tried to gain the Savilian Chair of Astronomy; his reputation of skepticism in matters of religion had doomed his effort. In 1703–1704, however, Nottingham was his sponsor. No questions about his religious opinions were raised, and Halley remained Savilian Professor of Geometry for the rest of his life. Later another set of patrons added the position of astronomer royal to his support. Because of his mathematical skills, Halley survived the family disaster handsomely.[32]

I now turn briefly to Newton himself. Again it is a different story. Newton never knew financial need. He inherited a competent estate, to which he added the income first of a Trinity fellowship and then of the Lucasian chair. He never had to worry about his support, and the mathematical technologies did not figure prominently in his life. Nevertheless, his life also offers evidence of the demand for the technical competence that a mathematician could supply. By the late 1660s, within an interested circle, Newton had demonstrated his superb mathematical talent, and people did not forget. In January 1681, John Adams was undertaking a survey of the whole of England. The Royal Society requested that Newton furnish Adams with technical assistance, and he did write at least one letter about the project.[33] There is no evidence of his further participation, but the request itself is of interest. More than thirty years later, a different problem arose. Newton was by then the famous Sir Isaac Newton, lionized author of the *Principia*. In 1714, when a group of merchants petitioned Parliament to establish a prize for a method of determining longitude at sea, a parliamentary committee turned to Newton for advice. When Parliament did in fact set up the prize, it needed a group of authorities to judge proposals. Parliament created a Board of Longitude, on which, inevitably, Newon sat, for the rest of his life, reading proposals, most of them inane, on how to determine longitude.[34]

My theme has been the demand in European society of the sixteenth and seventeenth centuries for mathematics, and the importance of four technologies, all heavily mathematical. The technologies impinged directly on the needs of the established authorities, and through the authorities it refracted into the career patterns of mathematicians. The seventeenth century is known as the most creative period in mathematics since the age of classical Greece, perhaps partly because European society at the time made a greater demand for mathematical expertise than any previous society ever had.

NOTES

1. Charles Singer, E. J. Holmyard, A. R. Hall, and Trevor I. Williams, eds., *History of Technology*, 6 vols. (Oxford: Oxford University Press, 1954–1978), 1:522–48; 2:681–92. Carlo Maccagni, "Galileo, Catelli, Torricelli and Others: The Italian School of Hydraulics in the 16th and 17th Centuries," in Günther Garbrecht, ed., *Hydraulics and Hydraulic Research: A Historical Review* (Rotterdam: Balkema, 1987), 81–82. Constantino A. Fasso, "Birth of Hydraulics during the Renaissance Period," in Garbrecht, *Hydraulics*, 71–74. Roberto Pracchi, *Lombardia* (in the series *Le regioni d'Italia*), 2d ed. (Torino: Unione Tipografico, 1971), 289–304. Umberto Toschi, *Emilia-Romagna* (in the series *Le regioni d'Italia*), 2d ed. (Torino: Unione Tipocrafico, 1971). Luca Beltrami, *Vita di Aristotile da Bologna* (Bologna: Luigi Baltrami, 1912), 16.

2. Singer, *History of Technology*, 3:308–10. Giuseppe Barbieri, *Toscana* (in the series *Le regioni d'Italia*), 2d ed. (Torino: Unione Tipografico, 1972), 243–56. Daniela Lamberini and Luigi Lazzareschi, *Campi Bisenzio: Documenti per la storia del territorio* (Firenze: Edizioni del Palazzo, 1981).

3. Sergio Escobar, "Il controlla delle acque: Problemi tecnici e interessi economici," in Gianni Micheli, ed. *Storia d'Italia, Annali 3: Scienza e tecnica nella cultura e nella società del Renascimento a oggi* (Torino: Einaudi, 1980).

4. Alfeo Giacomelli, "Le aree chiave della bonifica bolognese," in *Problemi d'acque a Bologna in età moderna*, Atti de 2° colloquio, Bologna, 10–11 ottobre 1981 (Bologna: Istituto per la Storia di Bologna, 1983), 123–72. Centro studi "Girolamo Baruffaldi," *La pianura e le acque tra Bologna e Ferrara: Un problema secolare* (Ferrara: Casa Rurale ed Artegiana di Cento, 1983).

5. Singer, *History of Technology*, 2:619–23, 681–89; 3:94–96, 301–308. Anton Albert Beekman, *Nederland als Polderland* (Zutphen: W.J. Theime, 1932). R. H. A. Cools, *Strijd om den Grond in het lage Nederland* (Rotterdam: Nijgh & van Ditmar, 1948). H. A. M. C. Dibbits, *Nederland-Waterland: Een historisch-technisch Overzicht* (Utrecht: NVA Oosthoek, 1950). Johan van Veen, *Dredge, Drain, Reclaim: The Art of a Nation* (The Hague: Nijhoff, 1962). G. P. van de Ven, ed., *Man-Made Lowlands*, 2d ed. (The Hague: Matrijs, 1994).

6. Singer, *History of Technology*, 3:316–18. See also Fasso, "Birth of Hydraulics," 77–78, and Adolph Kleinschroth, "Simon Ruffenstuel and His Hydraulic Work around the Year 1600," in Garbrecht, *Hydraulics*, 89–92.

7. Veen, *Dredge, Drain, Reclaim*, 44, 51–52.

8. Hunter Rouse and Simon Ince, *History of Hydraulics*, new ed. (New York: Dover, 1963). Maccagni, "Galileo, Catelli, Torricelli." Cesare S. Maffioli, *Out of Galileo: The Science of Waters, 1628–1718* (Rotterdam: Erasmus, 1994).

9. John R. Hale, "The Early Development of the Bastion: An Italian Chronology c.1450–c.1534," in J. R. Hale, J. R. L. Highfield, and B. Smalley, eds., *Europe in the Late Middle Ages* (Evanston, IL: Northwestern University Press, 1965), 466–94; "The End of Florentine Liberty: The Fortezza da Basso," in Nicolai Rubenstein, ed., *Florentine Studies: Politics and Society in Renaissance Florence* (Evanston: Northwestern University Press, 1968), 501–32; "Terra Firma Fortifications in the Cinquecento," in *Florence and Venice: Comparisons and Relations*, Acts of Two Conferences at Villa I Tatti in 1976–1977, organized by Sergio Bertelli, Nicolai Rubenstein, and Craig Hugh Smyth, 2 vols. (Florence: La Nuova Italia, 1979–1980), 2:169–87; Simon Pepper and Nicholas Adams, *Firearms and Fortifications: Military Architecture and Siege Warfare in Sixteenth-Century Sienna* (Chicago: University of Chicago Press, 1986); Christopher Duffy, *Siege Warfare: The Fortress in the Early Modern World* (London: Routledge & Kegan Paul, 1979). Experts cite Enrico Rocchi, *Le Fonti storiche dell'architettura militare* (Roma: Officine Poligrafica, 1908), as the best authority on these matters; I have not yet been able to lay my hands on the book.

10. Ernst Crone et al., eds., *The Principal Works of Simon Stevin*, 5 vols. (Amsterdam: Swets & Zeitlinger, 1955–1966), 4:200.

11. Ibid., 4:86.

12. E. G. R. Taylor, *The Haven-Finding Art: A History of Navigation from Odysseus to Captain Cook* (New York: Abeland-Schuman, 1957). David W. Waters, *The Art of Navigation in England in Elizabethan and Early Stuart Times* (London: Hollis and Carter, 1958).

13. David Woodward, "Medieval Mappaemundi," in J. B. Harley and David Woodward, *The History of Cartography*, 2 vols. (continuing) (Chicago: University of Chicago Press, 1987–), 1:286–370. John Marino, "Administrative Mapping in the Italian States," in David Buisseret, ed., *Monarchs, Ministers, and Maps: The Emergence of Cartography as a Tool of Government in Early Modern Europe* (Chicago: University of Chicago Press, 1992), 5–25.

14. Tony Campbell, "Portulan Charts from the Late Thirteenth Century to 1500," in Harley and Woodward, *History of Cartography*, 1:371–463.

15. O. A. W. Dilke, "The Culmination of Greek Cartography in Ptolemy" and "Cartography in the Byzantine Empire," in ibid., 1:177–200 and 258–75.

16. Ibid., 1:464–509. Leonard Digges, *A Booke Named Tectonicon* (London, 1599).

17. See most of the separate papers in David Buisseret, *Monarchs, Ministers, and Maps.*

18. N. D. Haasbroek, *Gemma Frisius, Tycho Brahe, and Snellius and Their Triangulations* (Delft: Rijkscommissie voor Geodesie, 1968).

19. Josef W. Konvitz, *Cartography in France, 1660–1848: Science, Engineering and Statecraft* (Chicago: University of Chicago Press, 1987).

20. R. A. Skelton, *Maps: A Historical Survey of their Study and Collecting* (Chicago: University of Chicago Press, 1972), 18.

21. Hale, "Early Development of the Bastion," 466.

22. Cools, *Strijd om den Grond*, 85.

23. P. Hamon, "Finé," in *Dictionnaire de biographie Francaise* (Paris: Letouzey et Ané, 1933–) 13:col. 1370–71. Denis Hillars, "Oronce Finé et l'horologue planetaire de la bibliothéque Sainte-Genevieve," *Bibliothéque d'humanisme et renaissance*, 33 (1971): 320ff. Leo Bagrow, *A. Ortelii catalogus cartographorum*, 2 vols., Ergänzungsheften Nr. 199 & 210 zu "Petermanns Mitteilungen" (Gotha: J. Perthes, 1928–30), 1 (Nr. 199): 63–69. Lucien L. J. Gallois, *De Orontio Finaeo Gallico Geographico* (Paris: Leroux, 1890). R. P. Ross, *Studies on Oronce Finé (1494–1555)* (Ph. D. diss., Columbia University, 1971).

24. M. Fernandez de Navarrete, *Disertacion sobre la historia de la nautica y de las mathematicas* (Madrid, 1846). Francisco Sousa Viterbo, *Trabalhos nauticos dos Portugueses nos sécolos XVI e XVII* (Lisboa: Academia Real des Sciencias, 1898), 171–83. Luciano Pereiro da Silva, "Os dois doutores Pedro Nuñez," "As obras de Pedro Nuñez—Sua cronologia bibliografica," and "Pedro Nuñez espoliado por Alonso de Santa Cruz," in *Obras completas*, 3 vols. (Lisboa: Divisao de Publicaçoes e Biblioteca Agencia Geral Das Colonias, 1943–1946), 1:139–58; 3:263–72; 3:163–84. Rudolfo Guimaräes, *Sur la vie et l'oeuvre de Pedro Nuñez* (Coimbra, 1915). A. Fontoura da Costa, *Pedro Nuñez* (Lisbon, 1938). W. G. L. Randles, "Pedro Nunes and the Discovery of the Loxodromic Curve, or How, in the 16th Century, Navigating with a Globe Had Failed to Solve the Difficulties Encountered with the Plane Chart," *Revista da Universidade Coimbra* 35 (1989): 119–30.

25. *Dizionario biografico degli italiani* (Roma: Istituto Della Enciclipedia Italiana, 1960), 32:659–62. Jodoco Del Badia, *Egnazio Danti cosmografo e matematico e le sue opere in Firenze: Memoria storica* (Florence, 1881). V. Palmesi, "Ignazio Danti," *Bollettino della R. deputazione di storia patria per l'Umbria* 5 (1899): 81–125. R. Almagia, *Monumenta Italiae Cartographica* (Firenze, 1929), 41–49, 525. John Marino, "Administrative Mapping in the Italian States," 23.

26. *Dictionary of National Biography* (reprint, London: Oxford University Press, 1949–1950), 5:976–78. *Biographia Britannica*, 2d ed. (London, 1778–1793), 5:239. Anthony à Wood, *Athenae Oxonienses*, 4 vols. (London, 1813–1820), 1:414–15, 636–39. Louise Diehl Patterson, "Leonard and Thomas Digges: Biographical Notes," *Isis* 42 (1951): 120–21, and "The Date of Birth of Thomas Digges," *Isis* 43 (1952): 124–25. Francis R. Johnson, *Astronomical Thought in Renaissance England* (Baltimore, 1937); "Thomas Digges," *Times Literary Supplement*, April 5, 1934, 244; "The Influence of Thomas Digges on the Progress of Modern Astronomy in 16th Century England," *Osiris* 1 (1936): 390–410. Francis R. Johnson and S. V. Larkey, "Thomas Digges, the Copernican System, and the Idea of the Infinity of the Universe in 1576," *Huntington Library Bulletin* no. 5 (April 1934): 69–117. E. G. R. Taylor, *Mathematical Practitioners of Tudor and Stuart England* (Cambridge University Press, 1954), 175.

27. E. J. Dijksterhuis, *Simon Stevin* ('s-Gravenhage: Nijhoff, 1943); *Simon Stevin: Science in the Netherlands around 1600*, an abbreviated English version of the original Dutch work (The Hague: Nijhoff, 1970). Edmond R. Kiely, *Surveying Instruments* (New York, 1947), 224.

28. Peter Barber, "England I: Pageantry, Defense, and Government. Maps at Court to 1550," in Buisseret, *Monarchs, Ministers, and Maps*, 49–53.

29. Ibid., 53.

30. José M. Lopez Piñero, *Ciencia y tecnica en la sopciedad española de los siglos XVI y XVII* (Barcelona: Labor universitaria, 1979), 84, 240–44. Lopez Piñero et al., *Diccionario historico de la ciencia moderna en España*, 2 vols. (Barcelona: Ediciones Peninsula, 1983), 2:375–78.

31. Frances Willmoth, *Sir Jonas Moore: Practical Mathematics and Restoration Science* (Woodbridge, England: Boydell, 1993).

32. A. Armitage, *Edmond Halley* (London 1966). C. A. Ronan, *Edmond Halley, Genius in Eclipse* (Doubleday, 1969). Eugene Fairfield MacPike, ed., *Correspondence and Papers of Edmond Halley* (Oxford, 1932); *Dr. Edmond Halley (1656–1742): A Bibliographical Guide to His Life and Works* (London, 1939). Charles H. Cotter, "Captain Edmond Halley, R.N., F.R.S.," *Notes and Records of the Royal Society* 36 (1981): 61–77. Edward Bullard, "Edmond Halley (1656–1741)," *Endeavor* 15 (1956): 189–99. A. H. Cook, "Halley in Istria, 1703: Navigator and Military Engineer," *Journal of Navigation* 37 (1984): 1–23; "The Election of Edmond Halley to the Savilian Professorship of Geometry," *Journal for the History of Astronomy* 15 (1984): 34–36. Gerald Funk and Richard S. Westfall, "Newton, Halley, and the System of Patronage," in Norman J. W. Thrower, ed., *Standing on the Shoulders of Giants: A Longer View of Newton and Halley* (Berkeley and Los Angeles: University of California Press, 1990), 3–13.

33. H. W. Turnbull, J. F. Scott, A. R. Hall, and Laura Tilling, eds., *The Correspondence of Isaac Newton*, 7 vols. (Cambridge: Cambridge University Press, 1959–1977), 2:335.

34. Richard S. Westfall, *Never at Rest: A Biography of Isaac Newton* (Cambridge: Cambridge University Press, 1980), 589, 834–37.

Index